Fluid Mechanics
of the Atmosphere

This is Volume 47 in the
INTERNATIONAL GEOPHYSICS SERIES
A series of monographs and textbooks
Edited by RENATA DMOWSKA and JAMES R. HOLTON

A complete list of the books in this series appears at the end of this volume.

Fluid Mechanics
of the Atmosphere

Robert A. Brown
DEPARTMENT OF ATMOSPHERIC SCIENCES
UNIVERSITY OF WASHINGTON
SEATTLE, WASHINGTON

ACADEMIC PRESS, INC.
Harcourt Brace Jovanovich, Publishers
San Diego   New York   Boston   London
Sydney   Tokyo   Toronto

Academic Press, Inc.
San Diego, California 92101

United Kingdom Edition published by
Academic Press Limited
24–28 Oval Road, London NW1 7DX

Library of Congress Cataloging-in-Publication Data

Brown, Robert Alan, date
    Fluid mechanics of the atmosphere / Robert A. Brown.
        p.   cm. -- (International geophysics series)
    Includes bibliographical references and index.
    ISBN   0-12-137040-2   (alk. paper)
    1. Dynamic meteorology. 2. Fluid mechanics.   I. Title.
    II. Series.
    QC880.B776   1990
    551.5'15--dc20   90-775
                                                                  CIP

● For information about our audio products, write us at:
    Newbridge Book Clubs, 3000 Cindel Drive, Delran, NJ   08370

Printed in the United States of America
90 91 92 93   9 8 7 6 5 4 3 2 1

Contents

Chapter 3   Methods of Analysis

Chapter 4   Tensors and Relative Motion

Part II   The Governing Equations for Fluid Flow

Chapter 5   Conservation of Mass—Continuity

Chapter 6   Momentum Dynamics

Chapter 7   Conservation of Energy

Chapter 8   Vorticity

Chapter 9   Potential Flow

Chapter 10   Perturbation Equations

Chapter 11   Boundary Layers

Preface

This text presents the fundamental equations which govern most of the flow problems studied by atmospheric scientists. The equations are derived in a systematic way which is intended to encourage critical evaluation. This goal is a result of widespread observations that very frequently the atmospheric scientist uses "seat of the pants" approaches, often quite effectively, but without deeper analysis. This is a result of the nature of the atmospheric laboratory, where the potential set of significant physical factors is enormous, the amount of hard observational data is sparse, and the mathematical apparatus is often obtuse and evolving. The few pertinent analytical results are often found inadequate to explain actual phenomena. Under such circumstances, the tendency is to seek simple correlations from whatever observations exist.

This situation is changing. Now, data are being produced at a much faster rate than are being analyzed or absorbed. Fundamental equations are being used in numerical analyses and many "results" are being shown without any recognition of the underlying mechanisms which produce the response.

The goal of this text is twofold. First, to furnish the student with a background familiarity with the underlying physics behind the mathematics. Second, to explore some systematic methods of relating these physics to atmospheric problems.

The greatest challenge in presenting this material is overcoming the syndrome known as "tacit knowledge oversight." That is, ignoring procedures or knowledge which is so absorbed in the subconscious of applied researchers, such as this author, that they assume it is universal. Such knowledge is then implicitly assumed in the student. There is a conscious effort in this text to identify and elucidate this knowledge. The goal is to move the level at which concepts or equations are taken on faith to the most basic level of knowledge practical. This is because the geophysical scientist will quickly get in trouble if the so-called primitive equations are accepted on faith. One or another of the underlying assumptions in the derivation of the equations is often in question during applications to the immense variety of geophysical applications. If the derivation is known from fundamental conservation principles, a new corrected derivation can be made for particular situations. This principal is vital in the application of eddy-viscosity approximations. The departure point for the derivations includes certain tenets of physics such as conservation of mass and energy, Newton's laws, the continuum, scaling, and the inertial frame of reference.

The organization of this material is a critical factor in whether the student will learn it. The selection of order of presentation is complicated by the choice between basic ideas in teaching plans. Unhappy with selecting between traditional "linear" versus global conceptualizing, I have opted for a mixture. In order to make it easier to remember the many complex procedures, a hierarchical procedure is adopted when possible. Sequences of steps are grouped into units characterized by concepts. The similarities and generalization of the conservation principles applied to a fluid flow are emphasized.

Since very few students absorb this material the first time around, I have attempted to provide more than one look within the text. Hence there is Chapter 1, wherein the rudiments of the concepts are presented, and a first tour through the material is made. It should be emphasized that the student is not expected to be able to interpret these concepts after reading this chapter, only to be exposed to them. The process of interpretation requires the understanding of complex procedures and the manipulation capabilities which are presented in subsequent chapters. It is felt that early awareness of the concepts of the larger categories will help understanding by placing each topic in the overall scheme of things. This will also aid in the hierarchical organization of the large quantity of knowlege to follow. Such knowledge will allow the student to "fast forward or reverse" through the line of information, from fundamental physics to practical use. A good organization of knowledge plus an appreciation of the conceptual categories will assist the application of this knowledge.

This is difficult material, particularly for those uninitiated into the realm of fluid mechanics. Nevertheless, it is only a beginning. There are texts which probe the mathematical difficulties of existence and uniqueness of the equations. Other texts develop the equations to handle the concepts of instability theory, bifurcating solutions, strange attractors, chaos, fractals, and the myriad of other exotic frontiers of fluid dynamical mathematics. Most students will be content to view the material in this text as an overall fundamental look at the underlying physics behind the equations. They will proceed to apply these equations in approximate forms, either analytical or numerical. They should remember in their learning and applications, if it were easy, it would have all been done.

Just as the world is divided into those who understand science and those who don't, atmospheric sciences and oceanography are divided into those who understand fluid dynamics and those who don't. One can get along well without this understanding, but with it a world of awareness is discovered. The full extent of atmospheric and oceanic dynamics' vast domain is opened to study only in direct proportion to the depth of understanding of fluid dynamics. There are dues to be paid in terms of hard work. But the rewards are great.

This material was developed as a first year course for graduate students in atmospheric sciences. There is one chapter per week, the last chapter being optional. In our department, *Atmospheric Science, An Introductory Survey,* by Wallace and Hobbs, is used in the course taken simultaneously to this fundamental fluid dynamics course. *An Introduction to Atmospheric Physics,* by R. Fleagle and J. Businger, can also be used for a simultaneous course. The student then proceeds to large-scale dynamics (e.g. Holton's *An Introduction to Dynamic Meteorology*), small-scale dynamics (i.e. boundary layer meteorology), or/and numerical modeling courses. This text was designed to fulfill a need for better understanding of the basic origins of the equations used in these later applications.

I would like to thank Professors R. A. Fleagle, J. R. Holton, and J. M. Wallace for reading the material during its evolution and their suggestions. Tristin Brown contributed significantly to making the text understandable to first year students. Thanks also to the many students who participated in the development of the course and provided valuable feedback.

R. A. Brown, February, 1991

Part I | Fundamentals

We all have a basic understanding of what is a *fluid*. A fluid flows. It changes shape to conform to its container. It deforms continuously under an applied force. We know that the atmospheric air and the oceanic water are fluids. Many might know that our bodies are 97% fluid. Few realize that the earth's mantle can behave like a fluid, or that even the flow of stars in a galaxy can be described as fluid flow.

The term fluid includes both *gases* and *liquids*. Despite the very large difference in density between air and water, we will find that the flow of both fluids is described by the same basic equations. The essential difference is in their *compressibility*. There are added complications in the equation for a gas because of the gas's higher compressibility.

Fluid dynamics is evidently the study of the flow of fluids. It is similar to classical solid dynamics in that motion under the action of forces is studied. Thus we will apply the same principles—conservation of mass, momentum, and energy. But the dissimilarity is in the profound differences in the consequences and applications of these basic principles when they are applied to a fluid domain. Therefore, Part I, which includes Chapters 1–4, introduces the new terminology and techniques required in the study of the physics of fluids.

Both a liability and an asset of this body of knowledge is that it is bounded by the gray areas that lie beyond current levels of comprehension. A complete analysis of turbulent flow is beyond our comprehension. However, the studies that have dealt with the whittling away of the vast unknown territory of turbulence have produced much basic knowledge. The analysis of the inherently nonlinear equations that govern the flow of fluid has led to the development of whole branches of mathematics. These include the study of complex variables, potential flow, conformal mapping, and chaos theory. Most recently, the concept of chaos as a degree of order in a turbulence field arose from the study of atmospheric dynamics. Many of the exotic terms such as strange attractors, bifurcating cascades, and coherent structures have their observational heritage in atmospheric dynamics. However, this is tough stuff. Few if any understand it the first time around. All are satisfied with a certain level of attainment short of complete. Thus there is ample room for theoretical progress and for application of known principles to the immense laboratory of atmospheric and oceanic fluid flow. Chapter 1 provides a first look and a relatively fast overall coverage of the material. Chapter 2 introduces the important mathematical concepts.

The subject matter of this text may be especially difficult because it forms the nuts and bolts and the tools of fluid dynamics. It is often removed from the stimulation that can be provided by the study of an actual phenomenon or a physical problem.

The equations we use are about 150 years old. The applications to atmospheric dynamics began about 100 years ago. Yet, only relatively recently has there been progress in relating solutions of the equations to real atmospheric phenomena. This is partly due to the peculiar difficulties found in geophysical applications: the *data are sparse;* the basic *coordinate system is rotating* (and therefore noninertial); and the *flow is turbulent* on many scales.

As a result, atmospheric science has been essentially an empirical science for most of its existence. The principal forces that cause the flow are divined or inferred from theoretical knowledge and then related to each other with a parameter that must be determined through observations. There are a limited amount of observations available to establish the empirical relations. This state has made *any* theoretical progress in solving the governing equations quite valuable. Such results provide clues and guidelines for the *parametrization* of simple cause and effect. In fact, the relatively crude but basic theoretical methods of *scaling* and *dimensional analysis* can provide useful information even without knowledge of the governing equations. These concepts are discussed in Chapter 3.

Currently, huge quantities of data from satellite-borne sensors, coupled with computer data processing, are filling the empirical data gap. Ironically, this achievement has only increased the emphasis on the need for more theoretical fluid dynamic solutions. These are vital to organize the new data and to explain the new phenomena being observed.

Newton's law relates forces and acceleration in an inertial frame. A *rotating frame of reference* is a noninertial frame. The law must be modified to account for the acceleration of the noninertial frame. A virtual force—

the *Coriolis force*—can be added to allow us to use equations that were derived for a noninertial frame of reference. There are some analytic solutions to the equations that govern geophysical flows affected by the rotating frame of reference. These often reveal new dynamics from that found in the nonrotating system. In particular, the mathematical solutions that include *waves* and *vortices* appear more often in rotating than in nonrotating frames of reference. Observations of geophysical flows are strikingly full of waves and vortices.

The mathematical solution for a flow field that includes random waves and vortices, called *turbulent* flow, has not been found. However, new approaches toward successful parametrizations are often suggested by observations. Progress in understanding the unpredictable nature of turbulent atmospheric flow fields could lead to better comprehension of many problems in all branches of geophysical fluid dynamics. Also, the ideas that apply to the study of turbulence have a wide range of application beyond the field of fluid dynamics.

The first step in the study of turbulent fields is to establish the scales. There is always a choice of scales. One must choose the principal scales of interest in the problem, the scales of turbulence that are important, and any critical scales for the flow solutions. Therefore, scaling principles are discussed early, in Chapter 3.

The atmosphere circles the globe and is nominally 100 km deep. Frequently, specific phenomena of interest occupy only a much smaller domain of activity, such as a storm system, a cloud cluster, or a thin layer. However, the dynamics of a particular local atmospheric problem cannot always be completely separated from the evaluation of the larger-scale flow field in which it is embedded. For instance, if one is interested in the study of clouds, snow, rain, and hail, the local application of microphysics is inseparable from the dynamics of the surrounding flow field. One must understand both the cloud-scale flow dynamics and the nature of the large-scale flow that produces the pertinent cloud dynamics—and how they fit together.

For those who study air pollution, small-scale environments, air–sea interaction, or most wave-generation problems, the main dynamics takes place within a thin *boundary layer* domain. Here, even the basic assumptions made in deriving the governing equations and the proper approximations for this domain are sometimes in question. Therefore, an understanding of the ideas behind the derivation of the equations and the related limits on their range of applicability is vital to such studies.

We seek a smooth solution for the entire domain—boundary layer plus the large-scale free flow. Thus, the solutions for the boundary layer domain must be properly fitted into the solution for the larger-scale flow. This must be done such that the boundary conditions match at the seam between the two solutions. One result affects the other. Even though many problems that are of special interest to some atmospheric scientists involve only very local forces and effects, an understanding of the background flow field is nevertheless often essential. This flow field is generally determined by the fundamental equations of motion.

If the field of interest is large-scale weather and climate, these subjects are studied in large part by numerical manipulation of the basic equations for fluid flow. The fundamental nature of these equations is emphasized by their description as the "primitive" equations. However, in the prediction of regional weather, and even in the analysis of many climate problems, it has been found that the details of the small-scale effects eventually become important. In the concepts of *chaos theory*, the smallest perturbation conceivable can change the nature of the large-scale solution. Computer capacities are growing quickly, but numerical models for the large-scale domains that have grid sizes small enough to include the small-scale turbulence are not in the foreseeable future. Since the economics and time restraints on these computational solutions place a severe limit on the degree to which the local dynamics can be included, parametrization of "subgrid" scale phenomena is essential to these models. Thus the dynamics of the eddy fields, the boundary layers, and the small-scale storms must be understood well enough to achieve an effective parametrization of their influences on large-scale weather and climate.

The goal of this text is to achieve an understanding of the fundamental equations for fluid flow. A solid understanding of the assumptions made in the derivation of the equations is especially important for those who will use the results in geophysical applications. This is because the equations were classically derived for laboratory flows and ideal conditions. When used for geophysical flows the limits set by the assumptions are often reached. These applications are made to scales from centimeters to thousands of kilometers, from seconds to years. To assess the validity of the application of the equations to these untested domains, one must be familiar with the nature and origin of each mathematical term. In Part II, starting with Chapter 5, each basic equation is derived from basic principles. These include the basic equations of stability theory—the perturbation equations—and the equations for the special domain of the boundary layer.

One major difficulty is met in our study when we attempt to write the three-dimensional, viscous, rotational frame of reference version of the equations in component form. There is a great proliferation of terms as 3-D vector forces are applied to 3-D vectors that represent velocities or areas. This becomes a bookkeeping problem, and the equations become unwieldy to write. However, the concept of *tensors*, and the methodology of *index notation* help the analysis. They restore a compact and manageable cast to the equations. Therefore, we accept the notion that the final understanding of the equations will be better for the effort made early to learn some very specific rules of tensor analysis, and we introduce these techniques in Chapter 4. They will allow efficient expressions for the governing equations and greatly reduce the labor of deriving them.

Also, there is an emphasis in this text on the dynamics that produce waves, vortices, and turbulence. While these dynamics occur at all scales, the specific description of any phenomena will depend on the particular scale of the average flow. This dependence occurs because individual terms in the equations may be of increased or diminished importance depending on the different scales. For example, large-scale flows may ignore the effects of small-scale turbulence, whereas small-scale flow may ignore the effects of the rotating frame of reference. When these junctures occur, we will usually pursue the small-scale applications, leaving the large-scale dynamics to the many excellent texts on that domain. Still, the material developed here can bridge the gap between the smallest-scale and the largest-scale dynamics. To illustrate the methods and possibilities for the application of each concept, we will introduce examples at the first chance.

Chapter 1 is a survey and general discussion of the topics in this book. It introduces the concepts and attempts to fit them into an overall atmospheric fluid dynamics picture. It is meant to provide a simple first look at as many of these concepts as possible. After the details of each of the following chapters have been studied, or a section "mastered," the first chapter could be profitably read again to gain the strongest sense of the large picture.

Chapter 1 | Fundamentals of Fluid Dynamics

1.1   Atmospheric Fluid Dynamics

This section introduces the topics in fluid dynamics that are of particular interest to the atmospheric scientist. Most of the material holds for the oceanographer, although with different coordinates and nomenclature. To begin the discussion of fluid motions we will introduce several terms that should be generally familiar to the student. However, these terms will not be fully defined until later in the text.

The first terms needed to describe the flow are *laminar* (meaning orderly) and *turbulent* (meaning random and chaotic). These words describe the two basic contrasting flow regimes. They will be used throughout the text. Their precise definition will evolve as we study the flow character. Other noteworthy terms that occur often and will require much effort to define are

Inviscid: Lacking viscous forces
Internal stress: Forces per unit area on the fluid at any point due to the adjacent fluid
Vorticity: A measure of the angular velocity or spin of the fluid at a point

These terms will be carefully defined later, and only a general feeling for their meaning is needed in this chapter.

Geophysical flow and classical fluid dynamics have two basic differences in emphasis. First, geophysicists are often concerned with a rotating frame of reference, even when dealing with small-scale flows such as that which takes place in the planetary boundary layer (PBL) domain. When compared to the description in a fixed frame of reference, the rotating frame gives rise to an acceleration at every point in the field. This acceleration is the price paid to have a simpler motion with respect to the rotating coordinate system. The easy example of a merry-go-round reminds us of this fact. A person walking toward the center of rotation has a direct linear path in the coordinate fixed on the rotating platform. This is much easier to describe than the path in a frame of reference fixed in space. However, the person feels the force of acceleration toward the outer rim. If we want to use the rotating frame as a reference, we must somehow account for this acceleration.

Newton's laws dealing with particle acceleration were derived for a nonaccelerating frame of reference, called an *inertial* frame. This frame works fine for most physics problems. However, adjustments must be made when we want to use these equations in the rotating system. We will face this requirement when we make a force balance in an earth-based rotating frame of reference.

The second factor in geophysical flow is turbulence. It is present at all scales of geophysical study. The methods of dealing with this turbulence are an essential part of geophysical fluid dynamics. Classical dynamicists can often ignore this facet as it is negligible for many physics and engineering applications. In addition to the problem of a rotating frame of reference, other interesting features are pertinent to geophysical problems. Some of them are

- Solutions of great practical importance are found by ignoring many of the terms in the basic equations for fluid flow, which results in very simplified approximate equations.
- Some large-scale flows can be studied by introducing analogies to laboratory-scale results.
- The rotation of the planet has a great influence on the flow patterns of geophysical flows (especially the large-scale motion of the ocean and atmosphere). It also affects the processes of transformation from smooth to turbulent flow.
- Experimental results in the atmosphere, the ocean, and the marine interface have been sparse and inaccurate. This is rapidly changing with the introduction of satellite sensors and computer processing. The result of the abundant data and statistical and graphical methods of analysis is new points of view for many geophysical phenomena.
- The basic equations are nonlinear. Generally, this means that the governing equations for a problem can only be solved by numerical means.

The first task assigned to the first large computer was the integration of the equations of motion for the atmosphere. However, the accuracy of such numerical solutions is limited because very small-scale effects ultimately influence the large-scale motion. When calculating the flow in the very large domains required by atmospheric problems, there is a practical limit to the small-scale processes that can be calculated. For instance, keeping track of millimeter-scale turbulence elements across a kilometer-thick layer cannot be done at present. The number of grid points and calculations required would exceed the capacity of even the biggest computers available today, or soon. Global circulation models currently have typical grid sizes of 400 km and not less than 100 km. There is a challenge to successfully account for terms that depend on subgrid scale effects. These terms must be well enough understood theoretically to enable the development of ways to account for them in the model. Furthermore, the general efficiency of the numerical calculation will be greatly increased by a firm understanding of the importance of each of the terms.

All of these factors help make atmospheric physics a particularly exciting and challenging area. But before we can attempt to apply theory to practical problems we must learn certain classical fundamentals taken from the wealth of fluid dynamic information available. We start by introducing the basic concept behind the equations of motion for the flow of a fluid.

1.2   Newton's Law in a Rotating Frame of Reference

Newton's law in classical mechanics concerns forces and motion. It is the
basis of the equations used to describe the flow in the atmospheric fluid,
air. Its elementary and familiar form, with boldface type symbolizing a vector, is

$$\mathbf{F} = m\mathbf{a}$$

For specialized use, the exact form of the equations will depend on the forces
that are important to the flow problem. We will investigate these forces in
detail. However, for a first look in this chapter we will be content to simply
name them. Usually, our significant forces will consist of at least a force
due to pressure differences, $\mathbf{F}_p$, the force of gravity, $\mathbf{F}_{g'}$, and an internal
friction force, $\mathbf{F}_\tau$. These may be written into Newton's law as

$$\mathbf{F} \equiv \sum \mathbf{F} = \mathbf{F}_p + \mathbf{F}_{g'} + \mathbf{F}_\tau = m\mathbf{a} \tag{1.1}$$

Newton's laws are for a primary inertial, or astronomical frame of reference, which is a reference fixed with respect to the stars. For geophysical
flows, one might assume that the natural coordinate system would be a
spherical system centered in the earth. Yet this is a practical system only
for atmospheric problems that deal with spherical domains at a radius much
greater than that of the earth. For most atmospheric flow problems, the preferred coordinate system is one that is fixed with respect to the surface of
the earth. This is an (x, y, z) orthogonal coordinate system with the x-y plane
tangent to the earth's surface and the z direction denoting the height above
this plane as shown in Fig. 1.1.

There always exists an acceleration on particles confined to the earth's
atmosphere. If these particles were to move in a straight line, they would
soon depart the earth. When our reference is the surface of the earth, it is
always rotating with respect to a fixed axis in the center of the earth. To
keep the frame of reference at a point on the earth's surface, the frame of
reference must constantly be accelerated. Our frame of reference is thus
noninertial. Frequently in atmospheric flow dynamics, the scale of the flow
is so large that the effects of the rotation of the earth becomes an important
factor. Otherwise, the acceleration can be assumed negligible and Eq. (1.1)
is valid.

Some feeling for the rotation of the coordinate system can be obtained
by considering the surface-based coordinates at the poles versus those at the
equator in Fig. 1.1. At the poles, the coordinates are rotating once each 24
hours around the z-axis, which approximately coincides with the earth's axis
of rotation. At the equator, there is no rotation about the vertical coordinate,

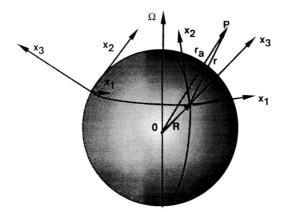

Figure 1.1 Description of point P in the rotating coordinate system $\mathbf{r}(x, y, z)$ and the fixed coordiante system $\mathbf{r}_a$. The distance $\mathbf{r}$ is greatly exaggerated for clarity.

but there is a rotation about the earth's north–south axis at a period of once per 24 hours. These rotations of the coordinates fixed on the earth's surface require accelerations of the coordinate system. From Newton's law, the accelerations can be balanced by an equivalent force. The rotation at the equator gives rise to a centrifugal force. The rotation about the polar axis produces another force, named after the French mathematician G. G. de Coriolis. There is a component of this rotation at all latitudes except the equator. Thus, the farther you travel south from the north pole, the less the influence of the Coriolis force. Eventually it reaches zero at the equator. As one proceeds south, it will increase with opposite sign until a maximum strength is again felt at the south pole. The opposite progression exists for the centrifugal force. It is zero at the poles and maximum at the equator. Both of these forces can be added to the force balance written in the rotating coordinate system. They are then called *virtual* forces.

We will use the coordinate system based on the surface of the earth. To accommodate the accelerations in this system to allow the use of Newton's inertial reference framed laws, the concept of a *virtual force* will be used. The addition of this "force," $\mathbf{F}_v$, accounts for the effects of rotation in the noninertial rotating frame of reference so that it appears as an inertial one. Thus, to write Newton's law in our noninertial earth-based coordinate system, we will have a new force balance,

$$\mathbf{F} = \mathbf{F}_p + \mathbf{F}_g + \mathbf{F}_\tau + \mathbf{F}_v = m\mathbf{a} = m\,d\mathbf{u}/dt \qquad (1.2)$$

In Eq. (1.2), $\mathbf{F}_v$ represents only the Coriolis force. The centrifugal force has been absorbed into the $\mathbf{F}_g$ term. This can be done since it is small compared to $\mathbf{F}_{g'}$, and is nearly aligned with $\mathbf{F}_{g'}$. In fact, it always points in a

direction normal to the axis of rotation. But on most of the globe, particularly nearest the equator where it is maximum, the centrifugal force is nearly
coincident with the gravitational force. Consequently the centrifugal force
is absorbed as a small addition (with a negative sign) to the gravity force,

$$\mathbf{F}_g = \mathbf{F}_{g'} + \mathbf{F}_{ce}$$

where g' indicates "pure" gravity force. This leaves $\mathbf{F}_v = \mathbf{F}_c$, the Coriolis
force.

Example 1.1

Consider the pendulum shown in Fig. 1.2. (a) Based on your intuitive feeling for acceleration, draw in the acceleration vectors at each of the positions
of the ball on the end of a string in Fig. 1.2. (b) Then derive the direction
of the acceleration by considering the incremental change in velocity at each
point. (c) Finally, consider Eq. (1.2) as the definition of acceleration.

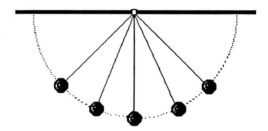

Figure 1.2   A pendulum consisting of a ball suspended on a (rigid, weightless) string.

Solution

(b) Intuitive ideas of acceleration generally are associated with velocity, although the distinction between velocity and velocity increment sometimes
gets lost. From this standpoint, the acceleration at the high points of Fig.
1.2 is sometimes thought to be zero. However, when an incremental velocity
change is obtained by considering the velocity vector at the next instant, the
correct acceleration vectors quickly result (Fig. 1.3a).

(c) When the acceleration vector is seen as simply a vector in the opposite
direction of the net force, the vector sum of the gravity force plus the force
in the string yields the acceleration direction (Fig. 1.3b).

In one case the acceleration is determined as the incremental change in

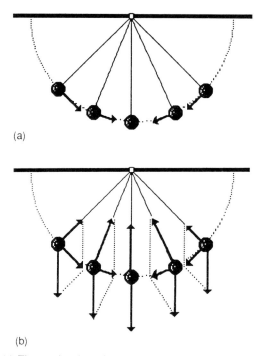

(a)

(b)

Figure 1.3   (a) The acceleration of the pendulum ball. (b) The forces and acceleration on the pendulum ball.

velocity; in the other it is the force per unit mass. There are evidently several ways of looking at the problem, and some produce easier solutions than others. Formulation in terms of fundamental definitions is often the safest and surest route to a solution. We will develop the fluid dynamic version of equation 1.2 in Chapter 6. There, the balance will often be written in terms of force per unit mass, so that the equation is a balance of accelerations.

Example 1.2

Consider a ball that was projected in the air somehow and is now flying through the air. (a) Show the important forces on it. (b) If it is whirled on a string at a constant speed such that $\mathbf{F}_{ce} \gg \mathbf{F}_{g'}$, show the forces on the ball. (c) If the coordinate system is now rotated so that the ball appears to stand still, write the balance for Newton's law.

Solution

(a) The main forces on the ball will be the force of gravity and a friction force due to air resistance. We know that the first force will assure the ball's return to earth. The second force slows objects up. We can assume that the latter acts in the opposite direction to the velocity (Fig. 1.4).

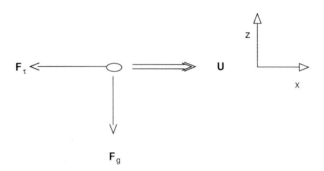

Figure 1.4   Forces (due to gravity and air friction) on a ball in free flight with velocity **U**.

The forces do not balance, and according to Newton's relation, $\Sigma \mathbf{F} = m\mathbf{a}$, there will be an acceleration in the same direction as $\Sigma \mathbf{F}$.

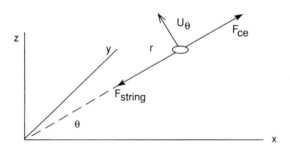

Figure 1.5   Forces (centrifugal, gravitational, and in the string) on a ball being swung on a string at velocity $\mathbf{u}_\theta$.

(b) When there is rotation we expect a new force to be present, called the centrifugal force (Fig. 1.5). This is a consequence of Newton's relation, since it states that if there is no force, there is no acceleration, and a body will continue to move at constant velocity in a straight line. When the body is moving in a circle, there must be a force providing an acceleration. For motion in the x–y plane at constant speed u_i and radius r (length of the string), there is a normal acceleration, $a_n = r(d\theta/dt)^2$. The speed is

$$u_i \equiv ds/dt = r\,d\theta/dt$$

Hence,

$$a_n = u_\theta^2/r \equiv \mathbf{F}_{ce}$$

The forces balance and the ball is in equilibrium, as shown in Fig. 1.5. We note that if the string were cut, the ball would accelerate outward due to the centrifugal (outward) force. Checking our approximation, where we have assumed horizontal motion, we note that actually, $\mathbf{F}_g + \mathbf{F}_{ce}$ must balance the force of the string, $\mathbf{F}_s$, so that the string direction must have a z-component. When $\mathbf{F}_g \ll \mathbf{F}_{ce}$, the dominant force balance is in a horizontal plane and 2-D motion in the $z = 0$ plane is a good approximation.

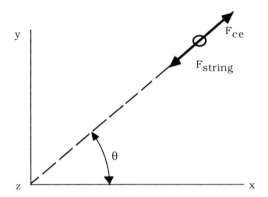

Figure 1.6   Forces on a ball swung on a string assuming the gravitational force is relatively small. The coordinate system is rotating so that the ball appears fixed.

(c) When the coordinate system is rotated about the z-axis in Fig. 1.6, to a person standing in the coordinate system the ball would appear stationary. However, the force on the string would still be required to keep the ball from flying off. This is by definition not an *inertial frame of reference* and Newton's law does not apply. However, if a virtual, or pseudo force, $\mathbf{F}_{ce}$, called the centrifugal force, is added, then the equation will be balanced. The sum of the forces can be equated to $m\mathbf{a}$.

Example 1.3

Consider a tank that is half full of liquid (Fig. 1.7). If the tank is sitting on a rotating platform, discuss and sketch the forces on a small "box" of fluid extracted from the center of the liquid. Discuss the forces and flow after the rotation is stopped.

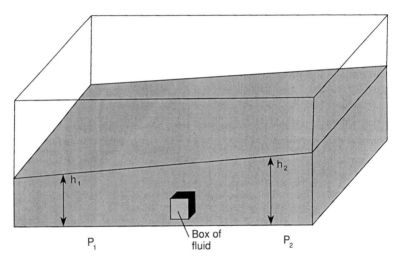

Figure 1.7   An imaginary small box of fluid in a container of fluid rotating in a large frame of reference.

Solution

A variation of pressure in the fluid along the axis of the box exists because there is more weight of water at the deeper sections. We know that the fluid is piled up at one end due to the rotation. We can assume the air pressure on the top surface is constant since the difference in the weight of air above the ends of the box would be insignificant. The pressure gradient produces a force on the small box of fluid, F_p, which is acting on the box in the direction of lower pressure, toward the center of rotation. The box is rotating, therefore there is a centrifugal force directed away from the axis of rotation. In a rotating frame of reference we must include a virtual centrifugal force F_{ce}, which acts to move the liquid outward. These two forces, F_p and F_{ce}, constitute the primary horizontal force balance on the fluid. Though gravitational force pulls the liquid downward, this is balanced by an equal and opposite force exerted by the floor. Friction will act on the box of fluid to retard the motion, but only when there is flow. In equilibrium then, $\mathbf{F}_p = -\mathbf{F}_{ce}$.

Thus, in the equilibrium state, the two horizontal forces balance, while the vertical gravitational force is balanced by the floor support, $\mathbf{F}_{floor}$ (Fig. 1.8).

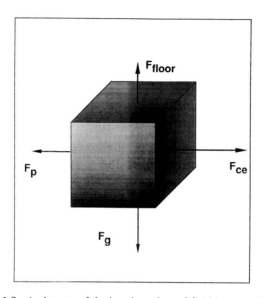

Figure 1.8 A close-up of the imaginary box of fluid in a rotating fluid.

If the rotation is stopped, there will now be a velocity difference between the rotating fluid next to the floor surface and the static floor. A friction force that is proportional to the velocity difference will propagate into the fluid, slowing the fluid down. As the fluid slows, the centrifugal force ($\frac{1}{2}U^2$) decreases, and the unbalanced pressure gradient force produces acceleration toward the low pressure. The final state is reached when the flow has reached uniform depth, U and $\mathbf{F}_P$ are zero, and there is static equilibrium.

This example illustrates the different equilibrium states for a rotating and nonrotating system. The pressure gradient force is required to balance the centrifugal force when the system is rotating.

We will see in later chapters that, for applications to global scale dynamics, we must pay particular attention to the variation of the virtual forces with latitude. This is a consequence of the change in the horizontal component of the earth's spin as it varies from one revolution per 24 hours at the poles to zero at the equator. For the time being, however, let us assume that the virtual Coriolis force is a constant. This is a good approximation for flow that does not move great distances latitudinally. The general mathematics of the Coriolis forces is given in Chapter 6.

The form of Newton's law given as Eq. (1.2) is quite general. However, this expression is deceptively simple, for each of the terms can be a complicated function of the fluid velocity. Thus, we will spend much of our time establishing simpler versions of Eq. (1.2) that are valid in special circumstances. These include the cases where

1. Horizontal components of velocity are much greater than vertical, so that $\mathbf{F}_g \approx 0$, and

$$\mathbf{F}_p + \mathbf{F}_c + \mathbf{F}_\tau = m\,d\mathbf{u}/dt$$

2. The rotating frame of reference has negligible effect; consequently, $\mathbf{F}_c \approx 0$, and

$$\mathbf{F}_p + \mathbf{F}_\tau + \mathbf{F}_g = m\,d\mathbf{u}/dt$$

3. There are no viscous effects—$\mathbf{F}_\tau \approx 0$

$$\mathbf{F}_p + \mathbf{F}_c + \mathbf{F}_g = m\,d\mathbf{u}/dt$$

4. The acceleration is negligible (steady-state) and $\Sigma \mathbf{F} = 0$, so

$$\mathbf{F}_p + \mathbf{F}_c + \mathbf{F}_g = 0$$

Combinations of any of these assumptions will lead to even simpler equations to solve.

1.3   The Laminar Flow Regime and Potential Flow

The main goal in this text is to derive and understand the mathematical equations that describe geophysical fluid flow. The equations must connect measurable dynamic and thermodynamic variables to the complete flow pattern in a specified domain. These known values form the boundary conditions. The domains in atmospheric problems are very diverse. They can include the realm of a cloud, a planetary boundary layer (PBL), a storm with a diameter from tens of kilometers to 1000 km, or any region of the globe up to an entire planetary atmosphere. The time scales must cover from microseconds, for small-scale turbulence analysis, to billions of years for studies in climatology. We need to make the task of describing these diverse flow fields more manageable. One way to do this is to define certain relatively simple flows that have relatively easy mathematics yet important practical applications.

1.3.1 Laminar Flow

When the flow is orderly and predictable, the flow pattern is often amenable to mathematical solutions. These flows are called *laminar*. In laminar flow, the individual fluid particles move along their trajectories independent of the particles in the adjacent layers. For instance, in two dimensions the flow is made up of individual layers (laminae) sliding past one another to produce velocity gradients normal to the flow. There are many types of laminar flow. They include *uniform flow,* where the flow field is everywhere parallel and constant; *shear flow,* where **u** changes with one or more coordinates; and *curving flow,* such as circular motion or periodic motion, where **u** has a constant or periodic change. Figure 1.16 is an example.

1.3.2 Inviscid Flow

An inviscid fluid is one where the internal friction force is negligible in comparison to the other forces. All fluids have some internal friction that provides a resistance to motion. When a force is applied to a fluid, it continuously deforms. The fluid nearest to the force accelerates the most, and internal friction drags along adjacent fluid. However, the acceleration cannot continue indefinitely, and eventually the drag of the liquid reaches a point where it balances the applied force. Therefore, the degree to which a fluid behaves without frictional effects depends on the speed of flow and the effectiveness of the fluid in transmitting the internal force to adjacent fluid. Faster flows, and fluids with small internal stress, behave inviscidly. The relative behavior of some common fluids is shown in Fig. 1.9.

The governing equations will be derived in this text in the most general form for geophysical flows. That is, we will include all of the terms that may occur in the most general applications. In particular, we will include

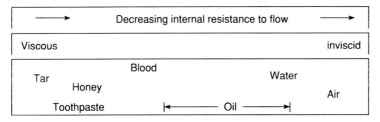

Figure 1.9 Chart showing values of internal resistance to flow of some specific fluids.

the term due to internal friction. However, we will find that the complete equations are quite complicated. They are high-order, nonlinear differential equations. They do not yield analytic solutions. Luckily, in most applications not all terms are important, and the process of solving the equations often begins by reducing the number of terms.

One of the greatest simplifications to the equations takes place when the internal stress forces arising from the fluid flow are negligible compared to the other forces in the problem. We will see that these internal stresses can be characterized by a quantity called *viscosity*. When they are small, the viscosity is small, and the flow is called *inviscid*. The governing equation will be as stated in Section 1.2:

$$\mathbf{F}_p + \mathbf{F}_c + \mathbf{F}_g = m\,d\mathbf{u}/dt \tag{1.3}$$

Example 1.4

Horizontal geophysical flows above the surface (or below in the ocean) are often well described by a simple balance between the horizontal pressure gradient force and the inertial force arising from the rotating frame of reference—the Coriolis force,

$$\mathbf{F}_p + \mathbf{F}_c = 0 \tag{1.4}$$

These simple flows are called *geostrophic*.

Discuss the approximations made in Eq. (1.2) to obtain Eq. (1.4).

Solution

Starting with a balance including all of the forces, apply the inviscid approximation,

$$\mathbf{F}_p + \mathbf{F}_g + \underset{\downarrow}{\mathbf{F}_\tau} + \mathbf{F}_c = m\mathbf{a} \qquad (Inviscid)$$

$$0$$

Now consider the flow as horizontal—two-dimensional in a plane normal to the gravitational force,

$$\mathbf{F}_p + \underset{\downarrow}{\mathbf{F}_g} + \mathbf{F}_c = m\mathbf{a} \qquad (horizontal\ \text{flow})$$

$$0$$

Finally, when the flow is steady,

$$\mathbf{F}_p + \mathbf{F}_c = m\,\mathbf{a} \qquad (steady\text{-}state\ flow)$$
$$\downarrow$$
$$0$$

This leaves the geostrophic force balance, $\mathbf{F}_p + \mathbf{F}_c = 0$

Classical inviscid hydrodynamics has been around in close to complete form for a long time (see Lamb, 1887). In fact, the inviscid equations of Euler were published in 1755. In certain cases, one can employ elegant complex variable mathematics to obtain good, usable solutions. This method uses singularities in the flow field, which represent fluid sinks, sources, and vortices to produce some very common flow fields. The flow fields can be used as a first approximation for flow around a wide variety of objects in what is called *potential flow theory*. In this theory the potential referred to is a scalar field variable which, when differentiated, yields the velocity field. Thus, when the scalar potential field is known, the velocity field is also known. We will see that the potential flow for the uniform flow around a sphere, as shown in Fig. 1.10, can be solved for analytically. The power in this simple solution comes from the fact that this relatively simple potential flow solution can be transformed into a solution for the flow around complicated shapes, such as an airfoil. The flow pictures that result from these solutions agree perfectly (within experimental accuracy) with observations such as those of Fig. 1.10 and 1.11.

1.3.3 The Potential Function

The field of one parameter can often be derived from that of another by using differentiation or integration. For instance, if you know the velocity $\mathbf{V}(x, y, z, t)$ you can calculate the acceleration, $\mathbf{a} = d\mathbf{V}/dt(x, y, z, t)$. We can say that one field serves as a potential for the other; that is, if the potential field is known, the derived field can be found. Potentials used to represent force fields are common in electrical, magnetic, and gravitational fields. In the case of a fluid flow, any parameter field that yields a force on the fluid can be a potential for the flow.

The earth's gravitational field produces an important force on each element of the atmosphere. The force is related to the separation distance—in this case to the earth's radius plus the height of the element above the surface. It is convenient to define a *gravitational potential* that is related to the

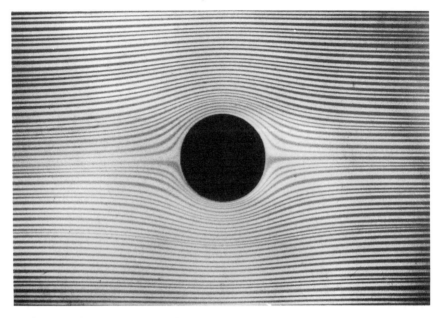

Figure 1.10   The dye pattern of the flow past a circle. Such a pattern requires a specific flow environment (in this case a creeping flow of a high-viscosity fluid in a narrow gap to approximate two-dimensional flow). (Photograph by D. H. Peregrine; from "An Album of Fluid Motion," assembled by M. Van Dyke and published in 1982 by Parabolic Press, Stanford, California.)

height. The derivative of this potential with respect to height (z) will yield the gravitational force **g**.

In the case of a fluid flow, we are dealing with the motion of a great many molecules. Their myriad interactions must be averaged and related to some mean flow parameter that is observable. The *pressure* is such a parameter. In the atmosphere, it has the virtue of being easily measured. The difference in pressure, which is force per unit area, gives rise to a net force across a small box of fluid. We will define this hypothetical box as a fluid *parcel* in Section 1.8.

An example of pressure force in action is found wherever some outside force has acted to set up a horizontal pressure gradient. This appears as a variation in depth of the fluid. The outside force might be the centrifugal force of a whirling body of fluid. Or it could be the gravitational pull of the moon on the ocean. The gravitational force on the body of fluid acts to establish a constant surface level. This state corresponds to a minimum potential energy, equilibrium state in the fluid. When the static equilibrium has been disturbed due to some external forcing, the degree of imbalance

Figure 1.11   Symmetric plane flow past an airfoil. Streamlines are shown by colored fluid introduced upstream in a water flow tunnel. [Photograph courtesy of ONERA from H. Werké (1974). "Le Tunnel Hydrodynamique au Service de la Recherche Aérospatiale," Publ. No. 156, ONERA, France.]

can be related to the horizontal pressure gradient in the fluid. In this case there exists a potential represented by the scalar pressure field such that the velocity will be directed along the negative pressure gradient, from high to low pressure. Flow will take place until there is a uniform pressure and a static liquid state. The vigor of the flow is proportional to the difference in pressure, or the pressure gradient.

The velocity of the flow can be proportional to the derivative of the pressure field in any direction. Thus, in the absence of other forces, when a scalar pressure field is differentiated with respect to time in three directions (x, y, z) it yields a vector field $\mathbf{u} = (u, v, w)$. There can be a different scalar variation in each direction. The velocity vector can be written as proportional to the gradient of P. This is written as

$$\text{grad } P \equiv \nabla P$$

Here, ∇ is an operator producing the vector gradient of the scalar field. It is called "del" and in three-dimensional Cartesian coordinates it is written

$$\nabla \equiv (\partial//\partial x, \partial/\partial y, \partial/\partial z)$$

There are other potential functions from which velocity can be determined. We will find that in the case of inviscid flow, where $\mathbf{F}_\tau = 0$, there exists a class of laminar flow solutions called Potential Flows. In these flows, the velocity can be obtained from a scalar potential function, say φ, so that

$$\mathbf{V} = \nabla\varphi$$

The basic flows in this class include

(1) Parallel Flow

(2) Linear Flows

(3) Vortex Flows

(4) Waves — periodic motion

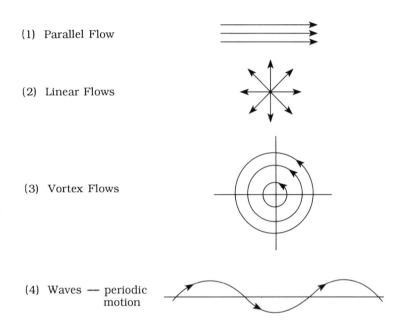

Observations have shown that these simple flows have many geophysical counterparts. We will study these simple solutions as a special case of the general equations in Chapter 9.

1.4   Waves, Vortices, and Instabilities

Waves and vortices are special flows that may fall within the laminar flow group. These flows often serve as a transition step between laminar and turbulent systems by presenting a departure from a parallel flow regime. The point of departure from laminar flow is an instability point in the initial mean flow regime. In some cases the instability leads to unchecked wave growth, breaking waves, and chaotic or random turbulence. However, in special cases the waves, or a combination of waves that form a vortex, come to equilibrium and form important laminar flow regimes.

Some laboratory examples of waves are shown in Figs. 1.12 and 1.13. Atmospheric flows are shown in Figs. 1.14–1.17. These flows are characterized by steady, periodic motion in the domain and time scales of interest. An excellent example of similar flows on vastly different scales is shown in Figs. 1.18 and 1.19.

The beautiful laminar flows in Figs. 1.12–1.17 exist only under special conditions. If the flow conditions are altered due to increased velocity or temperature, a new flow picture may result. The mathematical solutions to the equations must also change accordingly. When there are observed waves and vortices in a flow field, the equations must yield solutions that contain these phenomena.

The process of change from laminar to wavy or turbulent flow can be developed in the basic flow equations and is known as *instability analysis* or *instability theory*. Instabilities develop for several reasons. One or two of these reasons can be physically understood by relating the conditions for the instability to take place to elementary physical axioms. These are basic principles that define preferred states, such as the minimum energy or maximum entropy states. Thus it often happens that outside forces have acted to establish certain distributions of density or velocity that are unstable.

Figure 1.12 A Karman vortex street behind a circular cylinder. Streaklines are shown by electrolytic precipitation in water. (Photograph by Sadatoshi Taneda; from "An Album of Fluid Motion," assembled by M. Van Dyke and published in 1982 by Parabolic Press, Stanford, California.)

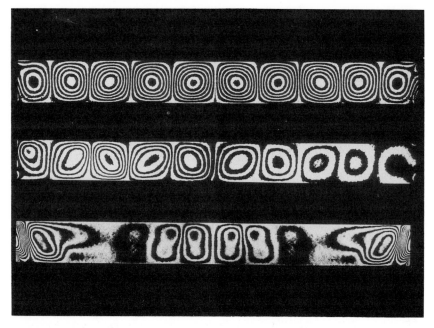

Figure 1.13 Buoyancy-driven convection rolls. These are side views of convective instability patterns in silicone oil. At the top is the classical Rayleigh–Bènard flow pattern for uniform heating leading to rolls parallel to the shorter side. In the middle, the temperature difference and hence the amplitude of motion increase from right to left. At the bottom, the box is rotating about a verical axis. [Photograph from H. Oertel Jr. and K. R. Kirchartz (1979). "Recent Developments in Theoretical and Experimental Fluid Mechanics" M. Müller, K. G. Roesner, and B. Schmidt, eds.), pp. 355–366. Springer-Verlag, Berlin.]

Flow will then take place to rearrange the energy or entropy state. The tendency of a fluid to seek a constant level in a gravitational field is one example. The rearrangement process can take place in various ways, depending on the dynamic forces involved in the system. It can involve laminar or turbulent flow. It will start as laminar, but under certain conditions, transition to turbulent flow takes place. In particular the transition can occur explosively. That is, there may be a very short time constant for the growth relative to pertinent observation times. Or the growth may take place in stages, with a very slow or infinite time constant relative to other mean flow times.

In laboratory flows, when the frame of reference is rotating, observations indicate that a slow transition regime often prevails. The change from laminar to turbulent can take place in steps, with each step existing in equilibrium under certain conditions. Then, the waves and vortices become

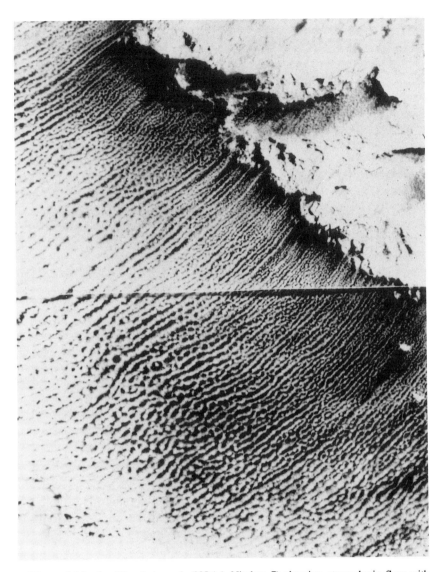

Figure 1.14 Satellite photograph (NOAA Nimbus 7) showing atmospheric flow with organized parallel "streets" of cumulus clouds sitting atop the planetary boundary layer. The flow is from over the oceanic pack ice (top) to over the sea, with cloud street separation about 2–3 km near the ice, 5–6 km at 100 km downstream.

Figure 1.15 An example of von Karman vortices shown in the cloud patterns downstream of Guadalupe Island off the coast of Baja California. This skylab photograph shows a cloudless area over the island, a cyclonic and an anticyclonic vortex immediately downstream, followed by two cyclonic vortexs. (photograph courtesy of O. M. Griffin, NRL.)

Figure 1.16 A laminated layer of clouds in the atmosphere.

Figure 1.17 A Voyager 1 photo of Jupiter showing the Great Red Spot and the turbulent region surrounding it. The smallest details seen are about 100 km across.

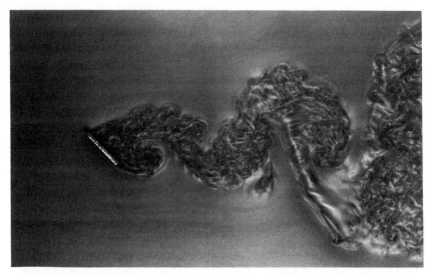

Figure 1.18   Laboratory flow of aluminum flakes suspended in water past an inclined flat plate. The plate is several centimeters long and the Reynolds number is 4300. (Photograph by B. Cantwell, reproduced with permission from the *Annual Review of Fluid Mechanics* **13**; copyright 1981 by Annual Reviews Inc.)

important to the basic mean flow description. If this occurs, there may be important stages in the transition wherein waves and vortices form a quasi steady-state mean flow solution. These states depend on the boundary conditions. Since a rotational frame of reference is "built in" for most geophysical flows, we expect waves and vortices to be seen in the atmosphere. Indeed such flows are very important in the atmosphere. They are seen in domains from the kilometer scales of tornados and PBL helical vortices to the 1000-km scales of cyclones and hurricanes.

The frequent presence of waves in the atmosphere and ocean has led to the use of generic terms such as *gravity waves* or *internal waves* to describe certain sets of often-observed waves. These generic terms say nothing about the origin of the waves. The waves are generally in steady state with negligible growth or decay. Thus, they are often considered to be a part of the background mean flow. In other words, their time scale of evolution is considered very short, and the time of decay very long, compared to the time scale of the local phenomena under investigation. Their existence in a select problem is often noted with the generic name. However, if something is known about the waves, a better name can be used. The characteristic of gravity waves is that their amplitude is opposite the force of gravity. The

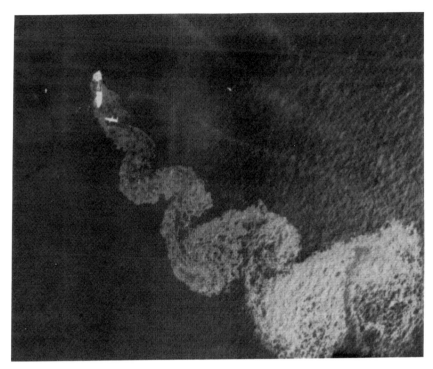

Figure 1.19   The tanker *Argo Merchant* aground on Nantucket shoals in 1976. The ship is inclined about 45° to the mean current, and the leaking oil shows a wake pattern remarkably similar to Fig. 1.18. Re $\approx 10^7$. (A NASA photograph courtesy of O. M. Griffin, Naval Research Laboratory.)

name internal waves is used simply as a distinction from surface waves, which occur at the boundary of a fluid. Ocean surface waves are a familiar example of the latter. A more descriptive term should be used if the cause of the waves can be determined: *buoyancy waves* for those caused by convective instabilities, *topographic waves* for those forced by flow over variable terrain, or *dynamic instability waves*.

The instabilities and the waves depend critically on boundary conditions. In fact, multiple solutions may exist for very small changes in boundary conditions. If the measurement accuracy is not good enough to specify the boundary condition well enough, the correct solution cannot be predicted. Some systems with closely related multiple solutions may oscillate randomly between the solutions. This is because the changing flow may alter the boundary conditions. Thus the conditions for the initial wave growth are changed by the instability perturbations themselves. This is a basic characteristic of nonlinear systems. If a numerical integration is used to

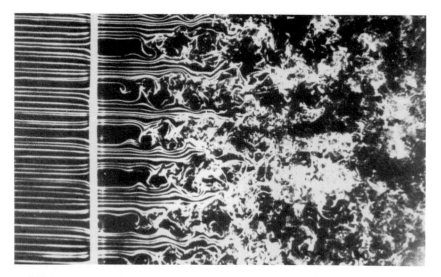

Figure 1.20 Turbulence being generated by a grid. Smoke shows laminar streamlines passing through a grid (1-inch mesh size) and becoming turbulent downstream. (Photograph by Thomas Corke and Hassan Nagib; from "An Album of Fluid Motion," assembled by M. Van Dyke and published in 1982 by Parabolic Press, Stanford, California.)

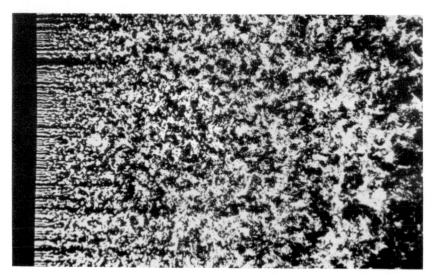

Figure 1.21 Homogeneous turbulence behind a grid (0.1-inch mesh size). At about the middle of the photograph the merging unstable waves have formed an approximation of ideal isotropic turbulence. (Photograph by Thomas Corke and Hassan Nagib; from "An Album of Fluid Motion," assembled by M. Van Dyke and published in 1982 by Parabolic Press, Stanford, California.)

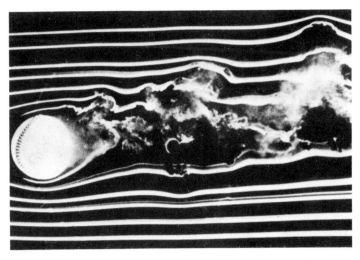

Figure 1.22 A spinning baseball (at 630 rpm) in a laminar flow at 77 ft/sec. Smoke is used to show the streamlines, separation, and asymmetric turbulent flow in the wake. (Photograph taken by F. N. M. Brown, courtesy of the University of Notre Dame.)

determine the solution, the intrinsic round-off error may be enough to cause this trend toward alternating solution regimes. This can lead to random oscillation between various attractor solutions. Such a regime, with random oscillations between multiple solutions to the same equations and boundary conditions has been called *chaotic flow*. The term *turbulence* is then reserved for the completely random oscillations with no preferred solutions.

We must often investigate waves in geophysical flows by examining their source in instability theory. One finds that the instability waves frequently produce spinning, or vortical flows. Thus, vorticity dynamics becomes an important topic for study. When the waves break and the vortex elements are random and unpredictable, the field is turbulent.

1.5 Turbulence and Transition

In this text, the stated goal is to obtain the basic mathematics for describing the flow phenomena that one is likely to encounter in the atmosphere. The derivation of the equations should be as general as possible, so that the resulting equations will encompass any new phenomena that arise. The subject of turbulent flow is particularly demanding of a good knowledge of fundamentals. Although fluid dynamics has traditionally been a cutting edge for new mathematics, the problems of turbulence have dulled many mathematical tools (and mathematicians).

Turbulence is the term for fluid flow wherein the path of an individual parcel of fluid is random, and thus unpredictable. Examples of "turbulent" fields are shown in Fig. 1.20. It is sometimes stated that geophysical flows are always turbulent. This is certainly true in that turbulence exists in a wide range of scales. However, steady and organized average flows can often be extracted from the turbulent field. These are the flow solutions we are seeking. To recognize how to find them and separate them from the turbulence, we require a good working knowledge of the turbulent field.

Turbulence is a very important factor in the flow near any boundary where the air flow comes to rest. However, it also can be found in the free atmosphere on many scales. Sources of turbulence can be found at the edge of a rapidly growing cloud, and at boundaries between adjacent, different laminar flow regimes in clear air. There is also turbulence on larger scales including the synoptic (thousands of kilometers), where the random, large-scale eddies accomplish a net poleward heat transport.

One might ask, if everything is turbulent, then how can these flows ever be predictable? The answer is that explicit turbulent motion is not predictable. On scales of flow where turbulence is the dominant motion, the flow is unpredictable. Even on scales where the forces due to turbulence are small, such as synoptic scales, predictability by numerical equations is generally limited in time and space by the build-up of forces due to the neglected turbulent effects. That is, the small turbulent fluxes through the boundary layer eventually add up to influence the large synoptic-scale motion. In addition, certain kinds of "turbulence" are well behaved in the average. Although the individual parcel motion is unpredictable, the aggregate mean flow can be predicted. Often, the average flow behavior cannot be determined explicitly. In these cases, it can either be accounted for by statistical mechanics applied to the turbulent field, or by relating the turbulent effects to mean flow parameters. In this text we are concerned only with the latter method, which will be discussed as eddy parametrization.

In the pictures shown, the point where laminar flow breaks down to turbulence is obvious. The dramatic difference seen in flow character between laminar and turbulent flow suggests that there must be corresponding differences in the mathematics. However, the random, chaotic flow cannot be described by our Newtonian mathematics. Today it is fair to say that a given flow that is apparently turbulent is describable or predictable only to the extent that laminar (orderly) characteristics can be extracted from the turbulence. Often, for certain flows such as the boundary-condition-sensitive chaotic flows, this can be done only in a statistical sense. However, the laminar character can be as simple as a steady uniform mean flow existing when the higher-frequency variations are averaged to zero over a long time. The nature of the flow in each problem faced must be resolved by scale considerations and averaging. One person's turbulence may be another's laminar flow.

Both wave and vortex flow regimes can be waystations between laminar and turbulent flows. Based on the observations, a laminar flow changes to a turbulent one under specific circumstances, as was shown in Figs. 1.20 and 1.21. These observations also show that the realization of a particular flow solution, which may be pure laminar, wavy laminar, or turbulent, depends on three characteristics of the flow. They are (1) the flow velocity, (2) the characteristic length scale, and (3) a measure of the internal ability of the fluid to "communicate" between layers via the net gravitational forces. This communication between the flow in one layer and an adjacent layer takes place on an intermolecular scale, or an eddy scale in the case of turbulent flow. The details involve the random exchange of molecules or eddies across the layers. The average effects are expressed in terms of mean flow parameters. They are characterized respectively by parameters called viscosity and eddy-viscosity. These parameters must be determined empirically.

A nondimensional combination of the three parameters representing the three effects was found to completely characterize a flow. For instance, doubling either the flow velocity or the length scale, or halving the viscosity, all produce the same flow picture. If other parameters are kept the same, the same flow pattern will result whenever the product of velocity and length scale is a constant. We will find such behavior to be of fundamental value in the flow analysis methods in Chapter 3.

It is clear from these observations that there must be a fundamental difference in the governing equations for laminar or turbulent flow regimes. The laminar solution is often described well by the inviscid flow equations, which were developed early by Euler. However, when turbulence is a factor, the equations must include new terms. They must account for the turbulent eddies in some fashion. If this can be done without destroying the basic orderly flow, we may have a modified or quasi-laminar flow. The problem of defining the flows into two categories, laminar and turbulent, is clearly not an easy one.

Example 1.5

It is of interest to see how various sources have approached the definition of turbulence. Here are some definitions of and comments about turbulence.

Webster's defines turbulence as "full of commotion or wild disorder; violently agitated; marked by wildly irregular motion." Turbulent flow is defined as the random motion of layers of a fluid, causing high resistance to movement through the fluid."

A mathematics book may give the following definition: "Turbulence—a field of random or chaotic vorticity."

Schlichting (1960)[1] wrote:

> It is not very likely that science will ever achieve a complete understanding of the mechanism of turbulence because of its extremely complicated nature. . . . the most essential feature of a turbulent flow is the fact that the pressure and velocity are not constant in time, but exhibit very irregular, high-frequency fluctuations. The velocity can only be considered constant on the average and over a longer period of time.

Von Karman made the statement:

> To my mind, there are two great unexplained mysteries in our understanding of the universe. One is the nature of a unified generalized theory to explain both gravitation and electromagnetism. The other is an understanding of the nature of turbulence. After I die, I expect God to clarify general field theory for me. I have no such hope for turbulence.

Saffman (1981)[2] wrote:

> During the past 45 years, much effort has been spent trying to determine the statistical distribution and in particular the spectra of the vorticity distribution in turbulence. However, the most exciting recent development is the growing belief, suggested by modern experimental investigations, that the vorticity fluctuations are not quite so random or disorganized or incoherent as was commonly thought. The vorticity is perhaps collected into coherent structures or organized eddies, and it is now proposed that turbulence should be modeled or described as the creation, evolution, interaction and decay of these structures. Turbulence is then thought of as the random superposition of organized, laminar, deterministic vortices, whose life history and relationships constitutes the turbulent flow.

In this text we will address turbulence right from the beginning of our development of the equations. Often, the first quarter or semester of a course in fluid mechanics is concerned with laminar flow and the neat mathematics of potential flow and stream functions. This is true in engineering and aerodynamics courses where these flows have great practical applications. They

[1] Schlichting, H. "Boundary Layer Theory," McGraw–Hill, 1960.

[2] Saffman, P. G. Vortex Interactions and Coherent Structures in Turbulence. *In* "Transition and Turbulence," R. E. Meyer, Ed. Academic Press, 1981.

also have some usage in atmospheric science. However, in atmospheric flows, turbulence is generally either present, nearby, or threatening. So instead of proceeding from the simple to the complex, we will take the approach of immediately deriving a fairly complete set of equations for fluid flow from basic principles. The laminar and potential flows will emerge as special cases for specific conditions. We will consider only the specific cases that have areas of application in the atmosphere or ocean.

One of the most interesting aspects of turbulence is its onset. This can often be predicted very precisely as a characteristic of the laminar flow, in what is called the transition problem. This subject has received increased interest as new aspects of turbulence have been discovered. One new perspective is due to the recognition of the existence of organized laminar structures buried within the mean flow, which exist independent of the smaller-scale random turbulence. These structures, which are generally vortices, can often be described as a characteristic feature of the transition process.

Transition and turbulence are dependent on the internal shears or stresses of the fluid. The shears are greatest where the flow is adjacent to boundaries. Thus, the study of boundary layers is inseparable from transition, turbulence, and viscous effects.

1.6   Boundary Layers

In many situations, when the flow is assumed to be friction-less, the relatively simple flow solutions of *potential theory* (developed in Chapter 9) predict the flow quite well. The velocity potential-flow solution for the flow around an object rather successfully predicts the flow pattern. This solution shows that the flow upstream from the object is identical to that far down stream, as though no disturbance was in the flow. It predicts zero drag on the object. However, if there were no drag on a ball or an airplane in flight, our playfields and commerce economics would be considerably changed. Balls would travel farther and faster and planes would fly with far less energy requirements. Although zero drag on an object in a flow is a nice idea in the spirit of perpetual motion, it is clearly in violation of observational evidence. Even in observations where the flow pattern appears to match the potential solution well, there is a measurable drag force on the object. This was called D'Alembert's paradox. However, it is no longer a paradox, as the conflict was removed by the *boundary layer* concept.

The boundary layer is a region with at least one dimension that is very small compared to that of the average flow field. Generally, the boundary layer is found in the regions adjacent to a solid body. In this region, the fluid experiences the layer-to-layer interaction that must ultimately bring it to a halt at a surface. If this region is sufficiently thin, then it can be ignored when the large-scale flow around an object is calculated. This may allow potential theory, which ignores viscous effects, to be used to determine the general flow picture quite accurately. However, the potential solution will provide no information on the surface frictional drag. To get the drag, a separate solution for the thin layer must be found that includes viscous effects. The phenomenon of separate solutions for separate scales is a frequent event in fluid dynamic analyses. In the atmosphere, the land or sea surface forms the pre-eminent boundary. However, boundaries can exist between any structures or regimes. These may be solid, like buildings, or nebulous, like clouds. The boundary may be simply the line between different air masses.

One example was shown in Fig. 1.11. The flow over a wing of thickness about 20–80 cm has a boundary layer that is a few centimeters thick. The potential flow solution yields a good approximation for the flow field by ignoring the boundary layer thickness compared to that of the wing. However, one can expect that the boundary layer thickness may become important near the trailing edge, where the airfoil thickness approaches zero.

The development of the boundary layer concept dates from 1904. The credit is generally given to L. Prandtl, as he presented the basic scaling ideas in a Heidelberg lecture in 1904. It is interesting that Ekman's mathematical solution for the boundary layer flow in a rotating frame of reference was also published that year. Although Ekman's solution was for the PBL flow in the ocean, it also applies directly to the atmospheric PBL.

Ekman's equations for the PBL assume steady-state, horizontal flow, so that

$$\mathbf{F}_p + \mathbf{F}_\tau + \mathbf{F}_c = 0 \tag{1.5}$$

His boundary conditions require that there is a layer next to the surface where the air velocity goes to zero. Ekman's solution indicated that the effects of the surface decay exponentially with height; that is, the layer is thin.

However, the ocean and atmospheric flows were not very well known from observations. The real progress and understanding (mathematics) of boundary layer flow came from Prandtl's laboratory experiments and theories. These were created for two-dimensional flow over a flat plate. The idea was simple: Next to a boundary there exists a thin layer wherein the fluid comes to a halt due to viscous action. In this layer the effect of the internal stress is important and Newton's law for steady-state flow must include the stress force. But the layer is so thin that the freestream flow is unaffected by friction or the flow within the boundary layer.

There are many boundary layers. There are many definitions of boundary layers. One fairly general definition of a boundary layer is: "The region wherein the forces imposed by the adjustment of the flow to the presence of the boundary are comparable with other forces acting on the fluid in the equations of motion." This definition is broad enough to include the case of boundaries contained in the freestream, where the adjustment from one temperature or velocity regime to another can take place in a thin layer. The common character of such layers is that they are thin compared to the overall region of concern in the flow problem. The heat or momentum flux contributions through the layer can be important for energy balances and forces transmitted to the surface by the freestream.

In geophysical flows boundary layers are generally the regions where the friction, turbulence, and stress are important. Thermal boundary layers are associated with temperature gradients. Velocity boundary layers are related to velocity shear. The latter is always the "boundary layer" in this text unless specified otherwise. It is the region where the boundary significantly affects the velocity (slows it up), and we call it the planetary boundary layer (PBL). Other boundary layers can be defined for layers of pollution, moisture, or any other passive quantity as the important parameter. These "boundary layers" (e.g., the "pollution layer") may or may not coincide with the boundary layers defined by the velocity characteristics. Furthermore, the velocity boundary layer itself may be subdivided.

In general, the PBL has two fairly distinct regions. One is a strong shear region near the actual boundary (the surface of the earth or ocean), called the surface layer. In this region, the flow is in planes parallel to the surface. The boundary surface is a source of turbulent eddies produced by mechanical friction. In this layer there is not much room for vertical motion $[w(0) = 0]$. Eddies must be small. They might be expected to grow in size in proportion to distance from the solid surface.

The second region is a much thicker layer, and here the effect of the earth's rotation becomes important. It is called the Ekman, or mixed layer. In this region, the eddies can be as large as the thickness of the layer. In fact, vigorous vertical mixing by large eddies is a dominant characteristic of this part of the layer. Each of the regions of the boundary layer can have different temperature distributions, with different buoyant forces. The local buoyancy influences the eddy size and distribution. Some sketches of the PBL characteristics are shown in Figs. 1.25 and 1.26.

Example 1.6

Consider the flow in the PBL, from the top, $z = H$, where geostrophic flow prevails, to the bottom surface, where $\mathbf{U} \to 0$. The Coriolis force is proportional to the wind speed, $\mathbf{F}_C = f\mathbf{U}$, where f is a constant. Sketch the force balance on particles at various heights in the PBL. The flow is horizontal and the pressure gradient force $\mathbf{F}_P$, is constant. See Fig. 1.23.

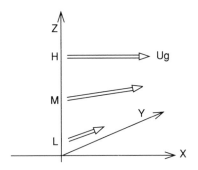

Figure 1.23   Velocity vectors at various levels in the PBL.

Solution

At geostrophic height (above the PBL) we can plot the force balance and the velocity in the x–y plane in Fig. 1.24.

Height H:

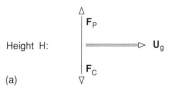

(a)

Figure 1.24   (a) The geostrophic force balance between pressure gradient and Coriolis forces. $\mathbf{U}_g$ is the geostrophic wind.

In the upper PBL, the influence of the surface is felt, slowing the wind slightly. The friction force $\mathbf{F}_\tau$ is small and in the opposite direction of the velocity. To have the vector sum of $\mathbf{F}_C$ and $\mathbf{F}_\tau$ balance, the $\mathbf{F}_C$ must turn slightly counterclockwise. Since the Coriolis force must be perpendicular to $\mathbf{U}$, 90° to the right (northern hemisphere), it too must turn:

Height M:

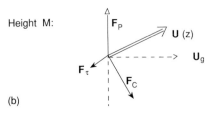

(b)

Figure 1.24   (b) Forces at the mid-level of a PBL. These now include the force of friction.

In the lower PBL, F_τ is increasing as the surface is approached. F_C is decreasing as U is getting small. Since the angle of turning of U between U_g and the surface is about 0° to 30°, F_τ is no longer opposite the U direction:

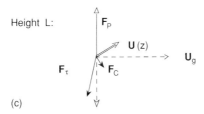

(c)

Figure 1.24 (c) Forces near the surface of a PBL. The Coriolis force is quite small and the friction force must nearly oppose the pressure gradient force. Note that the friction force is the gradient of the stress force, which may be nearly aligned with U. These are discussed in detail in Chapter 11.

Finally, friction brings the flow to a halt $[U(0) \rightarrow 0]$. Since F_τ is now balancing F_P, it is nearly perpendicular to U when last seen. However, the stress on the top and bottom sides of a slab of air are always approximately in the U direction. The stress force F_τ is the gradient of this stress, just as the pressure gradient force F_P, is the gradient of the pressure. The net result is that the viscous stress force acts to turn the flow throughout the PBL with height in the direction of F_P, or toward low pressure. We will examine these forces on a parcel of fluid in Chapter 6.

The boundary layer can be quite large—we spend our lives in a PBL except for occasional air travel and aquatic diving. (See Figs. 1.25 and 1.26) One can experience the marine thermal boundary layer in a large body of water. Due to radiational heating and limited mixing, the top few inches is often much warmer than the lower layers. In the ocean, the PBL is usually tens of meters thick. In the atmosphere it is on the order of a kilometer.

The atmosphere and the ocean frequently contain masses or layers of fluid that have become differentiated due to variable forcing. The forcing is usually thermal, such as that due to different heating rates in adjacent regions. This creates internal boundary layers. One example of this that is frequently visible due to the associated clouds is the velocity gradient regions at the edges of the jet streams. Quite often there are regions of high velocity shear between different layers of fluid in the freestream ocean or atmosphere. Fronts are another boundary between two air masses. For this reason, even if one intends to study only large-scale flows in the freestream above the complex

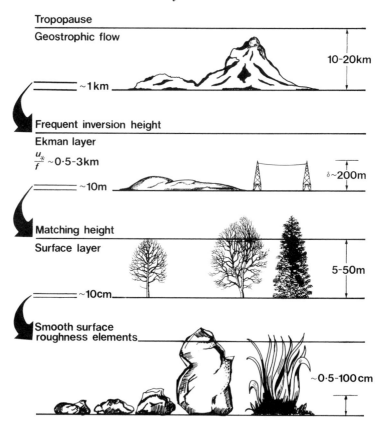

Figure 1.25 Sketch of various scale heights through the planetary boundary layer. Each higher scale presents a new regime and a new balance of forces in the governing equations. Some characteristic scale parameters are shown. They emerge from the mathematical solutions for the layers. (Courtesy of Adam Hilger Ltd.)

boundary layer region, one cannot ignore the processes involved in thin layers. There are always these thin regions where flow adjustments are taking place to accommodate strong gradients.

The principles that have been developed to study instability waves and turbulence in the boundary layer also operate in thin layers on all scales. Synoptic scale fronts and regions of cyclogenesis have thin layers, instabilities, steady-state waves, and vortices on many scales. The processes that create a cyclone, tornado, or mixing on a thin boundary layer inversion may be quite similar.

Since the boundary layer involves transition between one state and another, it is a complicated domain. This applies to both the observations and

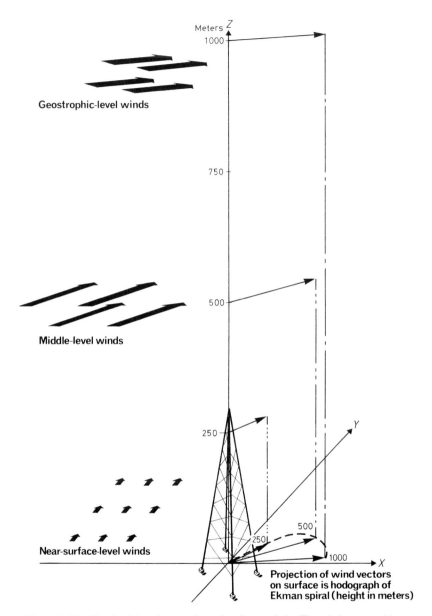

Figure 1.26 Sketch of the planetary boundary layer winds. The winds turn and increase with height. Their projection on the surface plane is a *hodograph*. Typical heights are shown in meters. (Courtesy of Adam Hilger Ltd.)

the mathematics. Still, the equations we shall derive appear to be adequate to describe most flow phenomena. Also, with the help of computers, the fluid dynamicist is becoming adequate to the task of solving these equations. For instance, the dynamics of the boundary layer can be used to explain why a ball curves in flight or why, to reduce drag, one makes wings and ship hulls as smooth as possible, but roughens golfballs.

We have noted that the PBL is characterized by turbulence and the importance of the viscous term in this region. It is therefore extraordinary that one of the first solutions for a special case of the general equations of motion was that for the turbulent boundary layer in a rotating frame of reference. However, to understand the approximations made in this solution, we must first examine the concepts that allow the application of forces to an element of fluid.

1.7   Historical Development

Problems that appear in the atmospheric sciences almost invariably involve airflow to some degree, either as a principal component, as in the weather, or passively, as in air pollution studies. To understand, explain, and predict weather and climate, the problem is to solve for the basic atmospheric flow. We may not need a complete understanding of the flow solution to predict and even explain some phenomena. Nevertheless, it is always of benefit to understand the related flow physics.

Much of meteorology was done by simply observing, systematically plotting, and discovering trends or patterns that could be used to predict future patterns. This process still goes on. However, it is fair to say that we do not fully understand a phenomena until we have placed it into a mathematical framework based on the fundamental principles of the physics of fluids. For instance, we might attack the problem of the flux of heat through a layer by simply relating the flux to the mean flow parameters. The amount of heat that flows is proportional to the mean temperature difference. But we will better understand the phenomena if we know something about the molecular motion or the turbulent eddies that are doing the fluxing. The challenge is like that of the physician who seeks to understand the cause and the cure rather than merely treating the symptoms. Furthermore, once we have derived the appropriate mathematical model, new discoveries often follow.

1.7.1   Newton's Law Applied to a Fluid

A few of the prominent individuals who worked on the early development of the fluid dynamics equations include:

Name	Dates	Topics
I. Newton	1700s	Mathematics, viscosity concept; Law of Motion for a particle
L. Euler D. Bernoulli	1750s	(Law of Motion applied to fluids) Equations for inviscid flow
L.M. Navier G.G. Stokes	1827 1845	Equations for viscous fluid flow
Boussinesq	1877	Turbulent mixing; eddy viscosity
O. Reynolds Rayleigh G.I. Taylor	1880–1890 1887–1913 1915–1970	Transition to turbulence Instability theories Geophysical applications; rotating flows

The application of Newton's law to fluids without viscosity was fairly complete by the mid-1700s due to the work of Euler and Bernoulli. However, the inviscid equations were inadequate to explain many phenomena. Two big effects had to be added. The first was the viscous forces and the second was turbulence.

Navier and Stokes developed the application of Newton's law to the flow of a viscous fluid. The acceleration, $\mathbf{a} = d\mathbf{u}/dt$, was defined for a small aggregate of fluid called the *parcel,* which we will carefully define in Section 1.8. The forces include internal body forces acting on each element of the fluid and external forces acting on the surface of the parcel.

1.7.2 *The Experiment of Osborne Reynolds (1883)*

Osborne Reynolds was the first to carefully observe and quantify the change in flow behavior as it changed from laminar to turbulent regimes. He injected dye into a laminar fluid flow in the laboratory, marked the transition to turbulence by taking note of the dye motion, and correlated the change with a combination of average flow parameters. In the experiment, shown in Fig. 1.27, the channel height h, the mean flow velocity U, and the density of the fluid could be changed (e.g., by changing fluids).

In addition to different densities, different fluids can have varied resistance to flow. This characteristic can be depicted by the constant *viscosity* μ. This quantity will be carefully defined in Section 1.11. Reynolds found transition to be a function of the nondimensional combination $\rho U h/\mu$. This important parameter is called the *Reynolds number*. The procedure called *dynamic similarity* exploits the power of dimensionless numbers to describe flow phenomena. It is discussed in Chapter 3.

In addition to recognizing the important parameters and forces involved in the transition from laminar to turbulent flow, Reynolds provided the basic

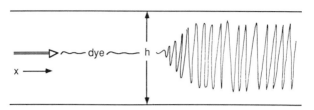

Figure 1.27   Sketch of dye marking transition from laminar to turbulent flow in a Reynolds' type experiment.

equations for the analysis of turbulence. However, before we can obtain these equations, it is necessary to carefully develop the concept of a parcel of fluid.

1.8   The Fluid Parcel

When discussing the dynamics of a discrete body—a molecule, a billiard ball, or a satellite—the initial conditions and the forces must be given. One can then determine the body motion from Newton's laws. However, when describing a fluid motion we are faced with huge numbers of molecules. In general, we are not interested in the individual molecular dynamics, so we will ignore the details of the molecular motion and speak only of average, macroscopic effects. These are the effects we can measure with a wind gauge, barometer, thermometer, or other such instrument. However, we need to define a fluid element such that these averages have a significant meaning. For this purpose we introduce a theoretical model of the fluid called the parcel. We will consider fluid properties—pressure, temperature, density, etc., associated with the parcel of fluid. The *parcel* must satisfy the following conditions:

1. Large enough to contain sufficient molecules for a well-defined average of the properties.
2. Small enough that the properties are uniform across the parcel; i.e., there are definite values of pressure, density, temperature, and velocity associated with the parcel.
3. Uniquely identifiable for short periods of time.

This concept of a parcel is basic to fluid dynamics. It will allow us to apply the fundamental conservation laws to a fluid flow. For now, it allows us to define pathlines, streaklines, and streamlines, which can be described by the following processes.

1. Identify a parcel with a spot of dye. The trajectory of this parcel (e.g., in a time-lapse picture) is a *pathline*.

2. Continuously inject dye at a point. The dye will mark a series of parcels that have occupied that point. This line is a *streakline*.

3. *Streamlines* are defined as being tangent to the direction of the flow at a given time. Thus there can be no flow perpendicular to the streamlines. The streamline is a theoretical concept and cannot be marked by dye in the general case. If the flow is unsteady, the streamline will vary from moment to moment and will not coincide with pathlines or streaklines. The concept is valuable mainly in steady flow where all three types of lines coincide.

Example 1.7

Consider the time-dependent flow of water from a leak in a water tank situated above the ground as shown in Fig. 1.28. An instantaneous picture of the flow at several times is shown. Discuss and label the streamlines, pathlines, and streaklines that could be distinguished in the flow of water from the tank.

Solution

In this case, the *streamline* is indicated by the frozen picture. This is because, at the instant the picture is taken, each particle of fluid can be imagined as having moved an increment of distance Δs in the time Δt, in the direction of $\mathbf{u} = $ limit $(\Delta s, \Delta t \rightarrow 0)$ of $\Delta s / \Delta t$. Thus each particle is displaced an infinitesimal distance along, or tangent to, the velocity vector. In this example the stream line is changing with time due to the decreasing pressure in the tank. Four such lines are shown in Fig. 1.28.

Identification of a *pathline* requires concentration on a particular particle, which we can assume was at point P at time t_0. At any instant, the velocity vector is tangent to the streamline. However, there is an acceleration component in the negative z-direction due to the force of gravity. Thus the velocity vector of a particular parcel is rotating with time. The position of the parcel must be such that the velocity change in the interval $\Delta t = t_1 - t_0$ results in a new velocity tangent to the new streamline. At later times the particle will occupy points in the sequence of streamlines as indicated in Fig. 1.28 by a heavy dashed line.

A *streakline* is obtained by imagining dye injected at a particular point for an interval of time. If dye were injected at the hole from the initial instant t_0 to t_3 and then viewed at a later time, the streak would occupy a locus

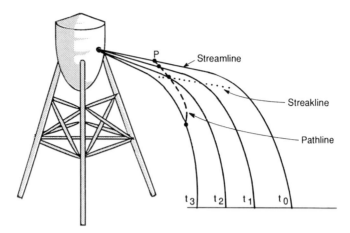

Figure 1.28   The flow "lines" with snapshots at times t_0, t_1, t_2, and t_3 of the flow from a leaking tank. A streamline, streakline, and pathline are shown.

across the instantaneous streamlines as shown in the figure by the dotted line. It traces the history of a marked mass of fluid. Each subsequent dyed particle occupies a position in a later streamline. At a given time, each is at a different height according to the vertical distance that it has fallen under the acceleration of gravity.

1.8.1   *The Parcel at a Point in a Field*

While aggregates of the elements in a particular parcel must experience nearly the same forces, usually a parcel in a different region of the fluid will be exposed to different forces and different initial conditions. This creates a very difficult problem of keeping track of each of the separate parcels and their individual histories. It therefore becomes practical to describe the fluid flow as a field of variables.

   In the field description, the dynamics and thermodynamics of each parcel is given as it occupies a specific point in space-time. Thus a field of velocity or temperature can be specified for each point in the fluid. Conveniently, this is what we usually measure—the value of a parameter at the point of our instrument. It is also what we are typically interested in—the wind, temperature, or moisture at some location.

   The large advantage of the field description is offset a bit by the fact that we are no longer dealing with the time history of a single parcel. Instead, we must consider the state of any parcel as it passes through any point. This requires us to give special consideration to the rate-of-change of the macroscopic quantities in our parcel. The change in perspective from concern with a particular parcel moving through the field to the consideration of an ensemble of parcels that occupy every point in a space–time continuum of the flow domain can be confusing. This is especially true when we sometimes find it convenient to revert to a description of the changes in a specific parcel as it passes through a particular point in the field. This switch in perspective is necessary because we must write our force balance on a particular parcel, yet we must consider how the forces are changing at the moment that the parcel occupies any point in the field. We can choose an arbitrary point and time; we then generalize the result to apply to all parcels that successively occupy all points in the field. This yields a field description of the parcel parameters. The velocity, density, pressure, and temperature will be given as a function of space and time.

For instance, one aspect of the viewpoint of the parcel as a box at a point can be seen by considering the changing shape of our fluid parcel as it moves from one region to another. It is also moving from one force balance to another. We can arbitrarily assume that the parcel is a cube at point P in Fig. 1.29. But the shape of this same aggregate of fluid might be quite distorted immediately before and after it occupies position P.

We will assume that the laws of dynamics and thermodynamics are applicable at all points in the flow field, or to any finite region or aggregate of the fluid. In the case of steady, uniform flow fields, the domain can be extended to complete volumes defined by the problem. This is called the *control volume approach*, and it requires more boundary information than is generally available in free flows. This approach averages over the entire volume using an integral version of the basic equations. The method yields approximate answers for the volume as a whole. It is convenient for use on

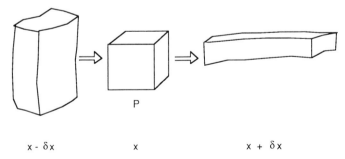

x - δx   x   x + δx

Figure 1.29   The parcel in the neighborhood of point P in a flow of variable pressure and velocity. The parcel is assumed to be a cube at point P.

confined flows. Much effort is devoted to this concept in engineering fluid dynamics. However, in geophysical fluid dynamics we are usually concerned with unconfined flow fields and we require details of the flow field inside the volume of study. This leads us to the differential approach to the conservation equations, which we will develop in this text.

1.9   Continuum and Averaging

1.9.1   Scale of the Domain

One might imagine that Newton's equation of motion could be written for each molecule, and then kinetic theory could be employed to describe its motion. The fluid motion could then be determined by averaging the motion of all the molecules. However, this procedure requires much more initial condition information than we usually have. In addition, it provides much more information than we generally need. And finally, it is a lengthy, tedious process. Fortunately, we can establish the mean flow equations for a fluid flow by applying Newton's law to our finite aggregate of molecules, the parcel. To do this, we need to define the conditions for the existence of a parcel with the characteristics described in the last section. If these conditions are not satisfied in our problem, then the extension of Newton's law for particles to an amorphous mass of fluid in a parcel may not be valid. In this case the derivations and the equations found in the following chapters may not be appropriate. It is prudent to check the conceptual requirements for the parcel before applying the equations to a problem.

In our description of the fluid state, new properties defined as averages over a specified volume are used instead of individual molecular mass and kinetic energy. Density is the sum of the mass of the molecules in a designated volume, that is, mass per unit volume. Kinetic energy is half the average molecular mass times velocity squared in this volume. This is fine for well-defined volumes, but when we want to describe continuous space and time changes (i.e., derivatives), we are interested in values for an elemental parcel that is incrementally small. For instance, in a field description, we are concerned with the density at a point. However, as the incremental volume δV, approaches 0, no averaging can be done, because there are very few molecules in δV. At some point either there is a molecule occupying the point or there isn't. Fortunately, there is usually a value of δV where enough molecules are contained to make a meaningful average, yet δV can still be considered infinitesimal compared to the field dimensions.

There is a limit on the size of the parcel on the large side too. If the flow field under consideration is such that there is a gradient in number or type of molecules, then as δV is increased, the different-sized parcels will encompass variations in density. The parcel must be small enough so that we can call δV a point with respect to any such variations in the field.

Hopefully, there is a range of δV between the two restrictions on its size, a range where the parcel density does not vary significantly. We can imagine that this process will fail at high enough altitudes where the air density is very thin. We must carefully define the conditions on the fluid such that our parcel definition is valid.

1.9.2 Continuum

The fluid continuum defines a domain of fluid where the parcel has uniform properties. On the small side, this simply requires the parcel to be much larger than molecular dimensions. However, the process of defining the derivatives of the mean flow parameters involves taking the differences of the independent variables (x, y, z, t) as they become infinitesimally small. Thus the parcel is forced to be small enough to not experience any significant mean flow variations. Figure 1.30 is a sketch of the expected variations in the density as volume increases. Wide variations occur at the small-volume end when the parcel is small enough to contain only a few molecules. At the larger end, the parcel may experience gradients in density due to large-scale variations. In this case, the density again would depend on the size of δV. We will call the region in between, where ρ does not depend on δV, the continuum. A similar sketch would apply for the average velocity of the molecules in a parcel.

The continuum hypothesis can be stated as follows: A fluid continuum exists in the range of scales wherein variations in the macroscopic fluid characteristics are small with respect to the mean variations, yet are not

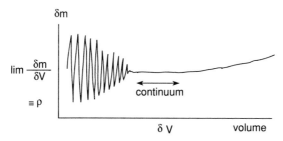

Figure 1.30 The average density (mass per unit volume) as the volume increases. Variation due to molecular spacing for small volumes, due to environmental changes for large volumes.

influenced by the microscopic variations. The macroscopic fluid characteristics are usually density, pressure, temperature, and velocity. The microscopic motions are due to molecular motion. This implies that the scale of
the parcel in the continuum is much smaller than characteristic dimensions
of the flow problem being studied, but large with respect to the scale of
microscopic variations. In other words, it excludes volumes so large that
variations in parcel properties show up due to the nonuniform behavior in
the flow domain. It bars volumes so small that characteristics are affected
by variations in the number of molecules contained in the parcel.

The parcel is an abstract physical entity. We can assume all physical
quantities are spread out uniformly over the volume of this continuum parcel, and any parcel can be a representative of a series of parcels with the
same properties. The series of parcels can then yield average states when
sampled by repeated measurements, such as those taken at

1. the same location in a steady-state field,
2. nearby locations of separation of the order of a parcel dimension, or
3. at any location in a direction in which the mean flow parameters do
not vary.

The molecules are in constant motion, entering and exiting our elemental
parcel. This motion is crucial in the processes of diffusion and heat conduction. The diffusion of momentum leads to the idea of internal friction.
If the size of the elemental volume is very much greater than the mean free
path, then sufficient numbers of molecules are contained within the volume
that the δV can be considered as in equilibrium with its surroundings. In
Chapter 4 we will rely on this equilibrium to treat the parcel as an "elastic"
entity in establishing the symmetry of the stress forces. Air at normal temperature and pressure contains $2.7 \cdot 10^{19}$ molecules per cubic centimeter,
and the mean free path is about 10^{-6} cm. Thus there is a secure range for
δV where even the smallest laboratory measurement scale is large.

From a practical viewpoint, measurements are made with instruments that
have specific size (an opening diameter, length, or time). The instrument
samples a certain volume of fluid, depending on its size and length of measurement, which generally produces an average over this region and time.
If the instrument is sampling within the continuum range, reducing the size
or time of the measurement (say, by 10–100%) will not affect the reading.
In this case, a representative average for the parcel can be obtained at each
position in the flow field and the field characteristics could be established
point by point. One can then see the mean variations which take place on
the large scales of the domain of the flow being investigated.

The mean variations will appear in the equations as derivatives. The fundamental theorem of calculus defines the derivative as

$$\underset{\delta z \to 0}{\text{limit}} \frac{\varphi(z + \delta z) - \varphi(z)}{\delta z} \equiv \frac{d\varphi}{dz}$$

where φ is some continuous function of z.

The limits δz and $\delta(\varphi) \to 0$ are important to this definition. Since we are dealing with differential equations for the flow variations, we should be on guard as to whether our derivatives exist. To approximate the limit, $\delta\varphi$ and δz must get very small in a volume that is infinitesimal with respect to any mean flow variations to have the mean flow derivatives well defined. This puts an upper bound on δV. It cannot be so large that the derivative varies due to the macroscopic changes, that is, due to variations in the flow field we are studying. On the other hand, if the macroscopic variation we are studying takes place on a scale small enough to force the parcel size to be so small that fluctuations in the basic properties occur in the parcel, then the continuum doesn't exist.

An example of when the continuum can fail to exist for a particular scale occurs when the dimensions of the flow problem approach the scale of molecular interaction. This length is the mean free path—the mean distance a molecule travels before hitting another molecule. This scale may be approached in the study of rarefied gas dynamics. Air is rarefied at high altitudes. Thus in problems dealing with the flow in the boundary layer on rockets, satellites, and other objects in thin gases, the continuum must be checked. The critical parameter in this case is the Knudsen number (Kn) which is defined as the ratio of the mean free path to the boundary layer scale. When Kn $\to$ 1, the flow is called free molecule flow and calculations are the rightful domain of kinetic theory. When Kn $\to$ 0, then we have a continuum. In the first case, the flow must be determined with the law of probability and the methods of statistics. In the latter case, we can deal with Newton's laws applied to matter in bulk. The methods should relate to one another in some domain. Indeed the first and second order expansions in the kinetic theory approach yield the same equations as does the bulk method.

Even in a rarefied gas flow, it may be possible to obtain uniform macroscopic characteristics at a "parcel-size point" by extending the length of time of the measurements. This requires the existence of conditions of uniformity on the mean flow such that the time and space averages can be interchanged. If measurements are made over a long time period, sufficient molecules might be encountered to form an average even though few were contained within the required small spatial dimension for the parcel. For geophysical flows, this is a simple and common practice in turbulence data analysis. Thus, there may exist parcel to parcel variations in flow properties at a point, but if large enough numbers of parcels are sampled, well-behaved averages can result. However, definite time or space intervals must always be attached to the average.

In the *eddy-continuum hypothesis,* we assume that the continuum is determined with respect to the small-scale turbulent eddies rather than the molecules. Since the turbulent eddies are many orders of magnitude larger than the molecules, this will place a severe restriction on the definition of a continuum.

In a closer examination of turbulence characteristics, additional problems for the definition of a continuum arise. One problem is that in a turbulent flow field the motion at any point can influence the flow at a distant point via the pressure field. Thus, there can be a variation in the nature of the turbulence in various directions other than that of the main flow. Also, decay and dissipation of the turbulence takes place at different rates for different turbulence characteristics. The description of turbulence can get quite complex. In this text we assume that these effects are small. However, in some problems they must be addressed.

Example 1.8

In applying the eddy viscosity assumption to the boundary layer regions, discuss the allowable size of the eddies in order to have a well-defined vertical shear for both (a) the surface layer and (b) the entire PBL. For the surface layer, consider a depth of $h \approx 10^4$ cm, whereas the PBL has a scale of $H \approx 10^5$–10^6 cm.

Solution

(a) To define the mean dU/dz within the surface layer, δz (the parcel dimension) might be assumed to be small enough at 10^3 cm, an order of magnitude smaller than the layer depth. However, this means that if we need 1000 turbulent eddies to determine a mean, then they must be no larger than 10^2 cm, so that 1000 will occupy a box with 10^3-cm dimensions.

If the eddies are larger, then a mean might still be defined by measuring at a point for a sufficient interval of time. If the wind velocity was 10 m/sec and maximum eddy size was $5 \cdot 10^2$ cm, 1000 eddies would be sampled in about 10 min.

(b) In the PBL, similar reasoning leads to $\delta z \leq 10^3$ cm, and the turbulent elements should be no larger than about 10 m.

Larger eddies are frequently encountered in the PBL. Since they can have very different characteristics, each averaging scheme must be tailored to the eddy spectrum.

These numbers show that the eddy-continuum hypothesis is a borderline assumption in many cases. Each flow situation must be carefully evaluated with respect to the definition of a continuum.

It is clear that careful definitions of the averaging process are necessary. This involves a large number of definitions in statistics. However in the applications to the atmosphere, we can restrict the statistical concepts to a pertinent few. This allows us to avoid some subtleties required in general statistics. However, some definitions are needed to warn the atmospheric data analyzer that the conditions for the mean to exist are often open to question.

1.9.3   Averaging

In our discussions on scaling and the continuum, we had to introduce a value called the average or mean value of the flow parameters. We will put the definition of the average on a formal basis in this section. The analysis of turbulent flow is built on the definition of the mean. First, a mean must be defined. Then turbulence can be considered as a departure from the mean. To get a mean, a specific domain (e.g., a time interval) must be chosen. Thus the mean is not unique. It is a function of the chosen domain. There are many concepts for defining a domain for the mean. However, we will not be concerned with all of the details of the averaging process in this text. Only a brief definition of commonly used concepts will be given here.

Generally, records of the wind components, u, v, and w, temperature T, and water vapor q are measured. Intrinsic averages are produced by these measurements that depend on the geometry and sensitivity of the instruments. These measurements typically can vary from 40 times per second for aircraft turbulence measurements to once per hour for some synoptic recording stations.

In atmospheric flows, we frequently find it necessary to sort out the organized waves or the random turbulence from the mean flow. This can be done only with respect to averaging times or spaces. Thus we are concerned with the value of a parameter over a specified volume and/or time interval. A simple average is the sum of the number of measurement samples divided by the number of samples. In this section we will denote space averages with $< \ >$ and time averages with an overbar.

An average over a spatial volume V of any physical parameter M can be written

$$<M> = (1/V) \iiint\limits_{V \text{ space}} M \, dV \qquad (1.6)$$

This is a special case of the general definition that allows for M to be a function of velocity $\mathbf{u}$, and thereby vary over the space of the domain. The average must depend on $\mathbf{u}$. The variation can be handled with a *probability distribution function* (PDF), $F\{\mathbf{u}(x, t)\}$. The PDF is defined by looking at the statistical average of a large number of flow samples. This is an extension of the definition of probability, as

$$\text{Probability} \equiv \frac{\text{The number of times an event occurs}}{\text{The total number of events observed}}$$

When the event is a continuous variable, like the wind speed, we define the PDF as

$$\text{PDF} \equiv \frac{\text{The number of times that } u_i \leq u \leq u_i + \Delta u}{\text{The number of observations}}$$

The average may then be written with the PDF as a weight factor in the averaging integral:

$$\langle \mathbf{M}\{\mathbf{u}(x,t)\} \rangle = \iiint\limits_{V \text{ space}} \text{PDF} \{\mathbf{u}(x,t)\}\mathbf{M}\{\mathbf{u}(x,t)\} \, dV \qquad (1.7)$$

This is a statistical or probability average, called an *ensemble*.

In special cases the average is obtained from the simpler form, shown in Eq. 1.6. These cases are obtained when the parameter being averaged is uniformly distributed. When the flow parameter is distributed uniformly over space, it is called *homogeneous*. When the distribution is uniform with respect to time, it is called *stationary*. The attributes of a statistically *stationary* flow over a *homogeneous* domain allow the simple mean value with respect to time to be used in place of the more rigorous statistical average.

If one assumes that the characteristics of homogeneity and time dependence are such that the time average is

$$\mathbf{M} = \lim_{t_1 - t_2 \to 0} \frac{1}{t_2 - t_1} \int_{t_1}^{t_2} M(x, t) \, dt \qquad (1.8)$$

then this average arises from the *ergodic* hypothesis.

Sometimes the flow variables are stationary, random functions of one of the coordinates. When the flow has such homogeneity with respect to a particular spatial coordinate, then mean values that are independent of *that coordinate* can be calculated. This can reduce three-dimensional problems to two or one dimensions.

Once the mean has been determined the fluctuations M' can be calculated about this mean:

$$M = M + M' \qquad \text{where} \qquad M' = 0$$

These fluctuations can represent either the random turbulent motion or that of an organized periodic disturbance.

Fluid dynamicists who work in the laboratory or with mechanical applications of fluid flow often automatically assume that a continuum exists. This is because within a cube of dimension 10^{-3} cm, there are over 10^{10} molecules of air; and in a cube of dimension 10^{-4} cm, there are still 10^{7} molecules. These values are certainly sufficient to establish a good average distribution and molecular composition. These numbers are huge enough to allow us to ignore the microscopic particle variations when writing equations for the mean properties. The continuum hypothesis is apt.

The mean properties are temperature, pressure, or velocity. These properties are the expressions for molecular kinetic energy, momentum exchange, and velocity averaged over a parcel volume. The equations are field equations that describe the change in the variables across the domain.

The equations are expressions of laws derived from observations. They include the conservation laws for mass, momentum, and energy. However, they are not complete. There are more unknowns than equations. We must also specify:

1. The character of the fluid, as an equation of state;
2. A relation for the internal stress forces; and
3. Boundary conditions.

Thus the assumption that a continuum exists with respect to molecular fluctuations is generally secure for the study of atmospheric dynamics. However, in geophysics, turbulence is an important factor in many flow problems. Fortunately, geophysical scales are often so large that we can treat turbulence as a random motion of small eddies buried in a mean flow. This suggests that we can avoid worrying about the individual eddy by treating it the way we did the molecule. We use the same averaging process to relate the net eddy mixing effect to mean flow properties. The transport of mean fluid properties due to the random migration of the eddies must be correlated to the mean flow. In this case, the eddy-viscosity approximation is used.

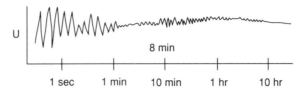

Figure 1.31   A sketch of a typical value of velocity obtained from a buoy that measures at 10 times per second when it is averaged for the interval shown.

As with molecular viscosity, it relates the diffusive action of the small turbulent eddies to the mean flow velocity gradients. A crucial factor in the success of this analogy is the ability to define a mean flow that is independent of the small-scale turbulence. That is, there must exist an eddy-continuum.

We can examine these new terms with respect to a geophysical parameter by considering an ocean buoy measuring wind velocity at 10 m above the sea surface. The record of $U(t)$ might look like Fig. 1.31.

The value of the average will evidently depend on the averaging time. In practice, eight minutes is used as a minimum time for averaging. The measurement is considered repeatable, or *steady-state* with respect to a time interval T, if $U(t) = U(t + T)$. It is *stationary* if this holds true for $T \to \infty$. In the given record, the average will remain stationary for intervals up to about one hour, where large-scale changes may appear. Thus we can use 8-min averages taken as often as possible up to the 1-hr interval. This will result in an ensemble of repeatable measurements taken under essentially the same conditions. This provides a good basis for statistical averages.

1.10   The Equation of State for a Perfect Gas

1.10.1   Introduction

The state of the fluid can be fixed by the relationship between several variables. The variables are generally pressure, temperature, and another that gives a measure of the composition of the fluid. The usual parameters for characterizing composition is the density in the atmosphere and the salinity in the ocean. Temperature stands in for thermal energy, the basic term in energy conservation. The equation of state expresses the relation between these variables for parcels that are in equilibrium.

We will accept a postulate for the relation between the properties of the fluid. This relationship will define the *state* of the fluid. A complete treatment of the state postulate must include an abstract axiomatic approach that begins with the second law of thermodynamics. This is a "conservation" concept that is independent of the first law of thermodynamics. Since these concepts do not play a major role in most atmospheric fluid dynamics problems, we will simply review some of the important points. The topic is thoroughly covered in many texts.

Unique values for all of the thermodynamic variables are specified for the fluid when its state is known. These properties may be altered by a change in state called a *process*. If the process consists of a succession of equilibrium states, then it can be reversed. In general, the effects of viscous forces and heat conduction are to make a process *irreversible*.

However, our ideal models deal with *reversible* processes only. For instance, we assume that most atmospheric dynamic processes are too rapid for heat transfer effects to take place. Therefore we can ignore the effects of heat transfer on the dynamics. The result is a reversible process involving state variables which is called *adiabatic*. This means that the state change takes place without heat transfer. The processes of evaporation and condensation are notable exceptions to this assumption. These processes are common in atmospheric flows. In these cases, the latent heat must be included.

1.10.2   Temperature, Pressure and the Perfect Gas Law

In atmospheric dynamics we are concerned most often with the macroscopic (sometimes called bulk) properties of the fluid. In this case, vast numbers of molecules contribute to the average. Common instruments inherently measure a very large-scale average property by the nature of their size. The molecular domain and concern over a continuum would seem to be left far behind. However, laminar flow based on molecular theory is also left far behind. We deal with turbulent flow and often try to model it in analogy to molecular theory. Thus it is valuable to be aware of the kinetic theory of gases when discussing the physical properties. This molecular model treats the gases as large numbers of spheres that move independently and collide. When in equilibrium, the velocity distribution has a mean value related to the internal energy. The science of thermodynamics introduces two concepts as axioms. The simplest involves the definition of temperature and the most difficult one defines *entropy* (discussed in section 1.10.4).

Temperature is defined as a macroscopic parameter that is a measure of the average kinetic energy of the molecules. It is a parameter that determines the heat transfer characteristics of the fluid. From kinetic theory, the relation can be expressed

$$\tfrac{1}{2} m v^2 = 4/\pi k T \tag{1.9}$$

where m is molecular mass (kg), v^2 is the mean square velocity of all the molecules, k is Boltzmann's constant $(1.3806 \cdot 10^{-23}$ J $K^{-1})$ and T is the absolute temperature.

The internal *pressure* in a perfect gas at rest is a normal force per unit area on an imaginary surface in the fluid such as that shown in Fig. 1.32.

The pressure force on the plane AA' results from the molecular activity. It is related to the momentum flux across AA' according to $\mathbf{F} = d(mv)/dt$. Each molecule carries momentum with it in its random migration. One can now write the normal stress force equation. It is equal to the mean normal component of momentum transferred by the sum of the molecules crossing the plane. This is

$$p = nmv^2/3 \qquad (1.10)$$

where n is the number of molecules/unit volume and nm is the mass per unit volume, or *density*, ρ. This relation is derived from arguments based on the average properties of the molecules and the symmetry of the velocity distribution.

If a solid plate is placed in the flow, molecules will not pass through the obstacle as they do through the imaginary plane in Fig. 1.32, instead they will bounce or rebound off the surface. The reflection characteristics of the molecules and the surface must be considered to obtain the momentum flux from the fluid to the surface. However, to have equilibrium there must be continuity of the pressure field. This dictates that the internal fluid pressure is also the pressure on the solid surface, given by Eq. (1.10).

Combine Eqs. (1.9) and (1.10) to eliminate velocity. This gives a basic relation between pressure and temperature,

$$p = nkT \qquad (1.11)$$

The density can be introduced into this relation by substituting for $n = \rho/m$,

$$p = \rho \, (k/m) \, T \qquad (1.12)$$

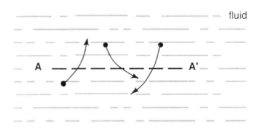

Figure 1.32 The pressure force on an imiginary plane AA' in the static fluid is due to the net molecular motion across AA'.

This fundamental equation for the basic state variables can be written in terms of the universal gas constant $R_0 = 8.314 \cdot 10^{-1} \, kmol^{-1}$, and the relative molecular mass M_r for the specific fluid,

$$p = \rho \, (R_0/M_r) T \qquad (1.13)$$

Finally, since we are generally dealing with one specific gas, air, it is most convenient to write the equation in terms of the specific gas constant for air, $R = R_0/M_r$,

$$p = \rho RT \qquad (1.14)$$

This is the *perfect gas law*. It gives an accurate relationship between pressure, density, and temperature for the gases that make up the atmosphere. The composition of air is uniform at 75.5% N_2, 23.2% O_2, 1.3% A, and 0.05% CO_2. Dry air is described well by this relation. But air can retain varying amounts of moisture, up to a saturation level that depends mainly on its temperature. For dry air $M_r = 28.96$, $R = 2.87 \cdot 10^6 \, cm^2$ $sec^{-2} \, deg^{-1}$. The state is determined if two of the properties are specified. Thus, for air at 15 degrees Centigrade and $5.3 \cdot 10^4 \, g \, cm^{-1} \, sec^{-2}$ pressure, the density must be $0.00123 \, g \, cm^{-3}$.

1.10.3   Other State Properties

Specific heat is a macroscopic parameter used for an ideal gas to relate the internal energy to the temperature. It is defined as the quantity of heat needed to raise the temperature of a unit mass of fluid one degree centigrade. For compressible gases, the amount of heat will depend on the heat transfer process. This is because some heat must go into the expansion of the fluid. Two specific heats are used for air. One is that for a constant volume process, c_v. The other is that for a constant pressure process, c_p. These correspond to the respective changes in specific internal energy e, and specific enthalpy $h = e + p/\rho$, given by

$$e_2 - e_1 = c_v(T_2 - T_1) \qquad (1.15)$$

$$h_2 - h_1 = c_p(T_2 - T_1) \qquad (1.16)$$

The specific heats can also be defined in integral form,

$$\partial e/\partial T \ \rfloor_V \equiv c_V; \qquad \partial h/\partial T \ \rfloor_p \equiv c_p$$

or integral form,

$$e_2 - e_1 = \int_{T_1}^{T_2} c_V \, dT \qquad \text{and} \qquad h_2 - h_1 = \int_{T_1}^{T_2} c_p \, dT$$

Note that,

$$h_2 - h_1 = e_2 - e_1 + (p/\rho)_2 - (p/\rho)_1$$

$$= c_v(T_2 - T_1) + R(T_2 - T_1) \tag{1.17}$$

Hence,

$$c_p = c_v + R \tag{1.18}$$

$$c_p \, dT = c_v \, dT + d(p/\rho) \tag{1.19}$$

There are tables of the gas constants R, c_v, c_p, and $k \equiv c_p/c_v$ in texts on thermodynamics of gases. These tables show that the last three vary only slightly with temperature. In atmospheric problems, the values for the standard temperature and pressure are often used. In some cases, the departures from standard values are most useful. The standards must be set for given time domains. Such standards are the grit of climatology. However the temporal variations are the most important aspect of some problems. One of the main variations we must face in the atmosphere is the change in the state parameters over vertical height ranges.

For the incompressible case, $c_p = c_v \equiv c$, ρ is constant, and $de = c \, dT$. In this case, the temperature changes due to purely thermal effects and pressure can be determined from mechanical phenomena. This results in significant simplification of the equations. In many cases air can be modeled as an incompressible fluid.

1.10.4   Entropy and Isentropic Processes

The second law of thermodynamics can be used to define a state property called *entropy*. The specific entropy may be written

$$s_2 - s_1 \equiv \int_1^2 d(q_{\text{rev}})/T \tag{1.20}$$

Here, the heat transfer q takes place in a reversible process between 1 and 2. When this process is run in reverse from state 2 to state 1 there is no net change in entropy. However, when the process is not reversible, the quantity $d(q/T)$ is positive. Then the equality in Eq. (1.20) must be replaced by $>$.

$$s_2 - s_1 = \int_1^2 dq/T > (s_2 - s_1)_{\text{rev}} \tag{1.21}$$

Real processes always contain some irreversibility, therefore entropy will always increase, even when the process is run in reverse. The amount of increase of entropy compared to that of a reversible process is a measure of the irreversibility of the process.

Entropy, for an ideal gas, is related to the other state variables by

$$s_2 - s_1 = c_p \ln(T_2/T_1) - R \ln(p_2/p_1)$$

or   (1.22)

$$s_2 - s_1 = c_v \ln(T_2/T_1) - R \ln(\rho_2/\rho_1)$$

The most common atmospheric process is adiabatic, where $dq = 0$. It is also reversible to a very good approximation. A process that is both adiabatic and reversible is called *isentropic*. From Eq. (1.22) with $s_1 = s_2$,

$$p_2/p_1 = (T_2/T_1)^{c_p/R}$$ (1.23)

We can also express the exponent in terms of the specific heat ratio, k,

$$c_p/R = k/(k - 1)   \text{(Carnot's law)}.$$

And,

$$p_2/p_1 = (T_2/T_1)^{k/(k-1)}$$

$$\rho_2/\rho_1 = (T_2/T_1)^{1/(k-1)}$$ (1.24)

$$p_2/p_1 = (\rho_2/\rho_1)^k$$

These equations are the laws associated with the *isentropic process*.

When a fluid is brought to rest isentropically, readily measured property values are obtained, called the *total*, or *stagnation* properties. This is a reversible process, with no heat transfer or work done. For a perfect fluid in horizontal flow with constant entropy, the conservation of the sum of thermal energy plus kinetic energy can be used to define stagnation values. For example we have

$$h_1 + u_1^2/2 = h_2 + u_2^2/2 \equiv h_0$$
$$c_p T_1 + u_1^2/2 = c_p T_2 + u_2^2/2 \equiv c_p T_0$$ (1.25)

or

$$T_0 = T + u^2/(2c_p)$$

Also, using (1.24) and (1.25),

$$p_0 = p[1 + u^2/(2c_p T)]^{k/(k-1)}$$ (1.26)
$$\rho_0 = \rho[1 + u^2/(2c_p T)]^{1/(k-1)}$$

The *stagnation values* are properties of the fluid, representing the maximum possible values for each parameter. Therefore, they are very convenient parameters for the nondimensionalization discussed in Chapter 3.

1.10.5   Moist Air and Virtual Temperature

In the atmosphere the gas mixture often includes molecules of water. When we include the mass of the water vapor, the state equation must include the different densities and molecular masses of air and water. However, we can define a temperature parameter that can be used to restore the basic form of the equation of state. This temperature is defined in terms of a measurable quantity that represents the amount of water vapor present, the *humidity*.

Water vapor also behaves like a perfect gas so that the equation of state may be written

$$p = \rho R^*/m_a T = \rho_a R^*/m_a T + \rho_w R^*/m_w T$$
$$= \rho RT\{1 + (m_a/m_w - 1)\,\rho_w/(\rho_a + \rho_w)\}$$
$$= \rho RT\{1 + (m_a/m_w - 1)\,q_h\} \qquad (1.27)$$

where $q_h = \rho_w/(\rho_a + \rho_w) = $ *specific humidity*, and $\rho = \rho_a + \rho_w$
If we let

$$T_V \equiv T\{1 + (m_a/m_w - 1)\,q_h\} \qquad (1.28)$$

be defined as the *virtual temperature*, we can return to the familiar equation of state,

$$p = \rho R T_V \qquad (1.29)$$

When the humidity is small, $T_V \approx T$, and in general this substitution represents a slight correction.

In atmospheric flow, we have a significant built-in vertical variation in p, ρ and T. A baseline for this variation is the adiabatic change with respect to height. Density changes over short height differences are small. With no heat addition and a constant density approximation, we can calculate the vertical adiabatic temperature change as a function of the pressure change. This temperature profile is called the *adiabatic lapse rate*. When an actual vertical temperature profile is measured, departures from the lapse rate are important indicators of the state of the atmosphere.

When the momentum equations are derived in Chapter 6, an important relation between the pressure and the height will be found. In large-scale atmospheric flow vertical velocities are typically much less than horizontal, and the flow is often assumed to be two-dimensional horizontal. This is a very good approximation for large-scale flow. Vertical velocities are very much less than the horizontal. (Of course this is not true when there is very strong convection in a weak horizontal flow.) We will find that, with this approximation, there are only two important terms left in the vertical momentum equation. These are the pressure gradient and the gravity forces, so that

$$\partial p / \partial z = -\rho g \tag{1.30}$$

This equation is exact when there is no motion and is called the *hydrostatic equation*. Integrated, it yields $p(z, \rho)$. Together with the state equation, it provides information on the thermodynamic state of the atmosphere.

1.10.6 Potential Temperature

The importance of the adiabatic temperature lapse rate as a baseline temperature has been noted. The stratification characteristics of the air are related to the difference between the actual observed temperature profile and that of the adiabatic lapse rate. Along an adiabatic temperature change, $dq = 0$. Therefore, from Eq. (1.22), the differences between observed and adiabatic temperature profiles are related to entropy change.

The *potential temperature* is designed to account for the thermodynamic effects of decreasing density with height. It is defined as the temperature that would result if the pressure were changed adiabatically to 1000 mbar. This value was chosen as it is approximately sea level atmospheric pressure. An equation for the adiabatic change in temperature with height can be obtained from a statement of the first law of thermodynamics,

$$dq = c_v \, dT + p \, d(1/\rho) = c_p \, dT - (1/\rho) \, dp \tag{1.31}$$

where q is the rate of external heat addition.

In the adiabatic case, $q = 0$, and

$$[dp/dT]_{ad} = \rho c_p$$

and from the hydrostatic Eq. (1.30),

$$[dT/dz]_{ad} = 1/(\rho c_p) \, dp/dz = -g/c_p \equiv \Gamma_{ad} \tag{1.32}$$

where Γ_{ad} is called the *adiabatic lapse rate*.

From Eq. (1.31) with $dq = 0$,

$$c_p \, dT = dp/\rho \tag{1.33}$$

This may be written

$$c_p \, dT/T = dp/(\rho T) = R \, dp/p$$

and integrated in the vertical, from $p = 1000$ mb to p and $T \equiv \Theta$ to T, to yield the potential temperature equation,

$$\Theta \equiv T[1000/p]^{R/c_p} \tag{1.34}$$

We can replace the temperature in Eq. (1.31) with the potential temperature, since

$$c_p \, d\Theta/\Theta = c_p \, dT/T - R \, dp/p$$

and

$$\{dT/dz + \Gamma_{ad}\} = T/\Theta \, d\Theta/dz \qquad (1.35)$$

This yields a compact statement of the first law for atmospheric dynamics,

$$dq = c_p T/\Theta \, d\Theta \qquad (1.36)$$

From the definition of entropy as $ds = dq/T$, and Eqs. (1.31) and (1.36) we can write

$$ds = c_p \, dT/T - R \, dp/p,$$
$$ds = c_p \, d\Theta/\Theta. \qquad (1.37)$$

Thus, lines of constant potential temperature are also lines of constant entropy.

For a thin layer the pressure will not change much and T/Θ is nearly constant. In the PBL it is nearly unity. Thus in the boundary layer, use of the potential temperature simply represents the departure of the real temperature lapse rate from the adiabatic lapse rate.

1.11   Viscosity

1.11.1   Introduction

Matter exists as either a solid or a fluid. The principal difference between the two states appears in the behavior of matter under an applied force. A solid will deform to a fixed point, and remain so deformed under the applied force until the force is removed. It then returns to its original shape. The applied force is referred to as the stress, the deformation as the strain; and they are related with experimentally determined coefficients.

The distinguishing character of a fluid is that it continues to deform as long as the force is applied, and it does not return to the original shape. In fact, a simple definition for a fluid is "a substance that deforms continuously under the action of an applied force." The fluid concept includes both liquids and gases.

When a force is applied to a parcel of fluid, the fluid elements in contact with the force move the most. Adjacent layers slide in the direction of the force at decreasing amounts in proportion to the distance from the location of the application of the force. The proportionality factor is evidently much greater in liquids than in gases. For the fluid, the applied force—the stress—

is observed to be proportional to the rate of deformation, or *rate of strain*. This quantity is simply the difference in velocity from layer to layer. When one layer is moving at a velocity different from that of an adjacent layer, there exists a strain between the layers. If the velocities are the same, then there is no strain.

When a force is applied to a fluid it accelerates and flows according to Newton's law. As the fluid flows faster, the rate of strain increases, and consequently the internal stress forces increase. The fluid can eventually come to an equilibrium where the rate of strain force balances the applied force. For example, a constant wind over the ocean surface can drag along only a fixed amount of water. We can say that there is a dynamic equilibrium where the fluid flows and creates internal stresses that resist the flow and balance the applied stress. We can expect that a solution for the flow pattern of a fluid domain requires knowledge of the stress distribution in the fluid.

1.11.2   Viscosity Fundamentals

Viscosity is an empirically determined measure of the internal forces that oppose the deformation of the fluid. The basic momentum and energy exchanges are produced in collisions between the molecules. In addition, the internal forces will include the intermolecular forces. These forces depend on the separation distance between the molecules. They will have significant value even when the molecules simply approach closely. The different *phases* of matter—solid, liquid, or gas—are related to the molecular spacing and intermolecular forces. In a solid, molecules are relatively close and experience large intermolecular forces. This is the factor that gives solids their character. In a liquid, the intermolecular forces are sufficient to hold a given volume of matter together, but inadequate to preserve shape. In a gas, molecules are far apart and intermolecular forces are too weak to hold a constant volume. The atmosphere is composed of a mixture of gases called air, and our laws must be addressed to this case. However, we will exploit the simplifications available when air behaves like a liquid. The determining factor will evidently be the degree to which air can change volume, or the measure of its *compressibility*.

If the atomic forces are considered to act only when the molecular approach is close enough to be defined as a collision, the gas is called a *perfect gas*. Empirical laws that give relations between pressure, temperature, and density, such as Boyle's and Charles' laws, were obtained for such a gas.

Corrections for the contributions from the forces involved in approaches that are near collisions can be obtained in relations such as the van der Waals equation. However, the internal resistance, called *stress,* is not completely explained with kinetic theory even when it includes such forces. Concepts such as adherence of groups of molecules seem necessary to explain the observed stress. Although some feeling for the internal workings that result in the stress can come from these considerations, a quantitative theory is not yet available. In general we could be content to consider the integrated effect of these forces as represented by the viscosity of the fluid derived from a continuum flow parametrization. However, some discussion of the kinetic theory viewpoint is instructive and it also provides valuable background for the application of eddy diffusion parametrizations to atmospheric problems.

The differences in intermolecular physics of liquids and gases are illustrated in Fig. 1.33. Air density is such that most of the time a molecule is in free flight, with a mean free path, λ, about $3 \cdot 10^{-8}$ m. Nevertheless, with about $5 \cdot 10^9$ collisions per m^2/sec, there are ample exchanges to provide excellent statistical averages of momentum exchange. The greater density of molecules in a liquid allows many more collisions and a chance for intermolecular forces to have a measurable influence.

Consider a fluid that is flowing with a velocity shear, and examine a scale where we see individual molecules on each side of a surface denoted by AA' in Fig. 1.34. The mean velocity of the fluid is the average of the myriad random motions of the individual molecules. Molecules above AA' have mean velocity u_1 and those below have mean velocity u_2 in the x-direction. Individual molecules will also have random motion in all directions. In particular, there will be molecules with a component of random motion in the z-direction.

Now consider the effect if two molecules at positions 1 and 2 exchange positions. (We consider two because on the average, as many pass upward as downward through AA'.) There is a net change of momentum on each side of AA' equal to $m_1 u_1 - m_2 u_2$. If we recall Newton's law in the form

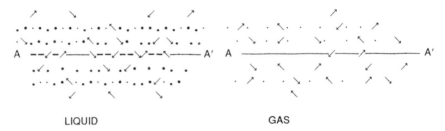

LIQUID   GAS

Figure 1.33 Sketch of a liquid and a gas molecular transport at an imaginary plane marked by AA' in the fluid. Liquid: (1) Many collisions, (2) Intermolecular forces exist across AA'. Gas: (1) Relatively few collisions, (2) No intermolecular forces exist across AA'.

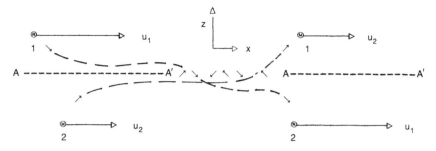

Figure 1.34   Sketch of the exchange of two molecules designated 1 and 2 across an imaginary surface AA.

$$F = \delta(mu)/\delta t, \qquad (1.38)$$

we see that an exchange of momentum across AA' is associated with a force. Thus, the rate of momentum exchange, $(m_1u_1 - m_2u_2)/\delta t$, is equal to a reaction force on AA'. When this force is averaged for all the molecules, it results in a force per unit area in the fluid. This force can be related to the internal stress τ. If the plane AA' is assumed to be one of the permeable faces of our model fluid parcel, then this force is the stress on that face of the parcel.

On some plane in the fluid, a force must be exerted to maintain the shear in the flow. The molecular momentum exchange is the process in a fluid that produces the internal stress force. This force can balance the driving force to result in flow equilibrium. If the outside force is removed, then the momentum-flux/internal-stress will make the velocity uniform. If no forces remain, a static fluid will eventually result.

The momentum exchange process requires a mean velocity shear to set up a difference in momentum plus random molecular motion in the z-direction to move the momentum. The viscosity is simply a proportionality factor that represents the effectiveness of the molecular exchange process. It empirically relates stress to mean shear, as discussed in the next section.

1.11.2.a   Viscosity from Molecular Theory

One of the successes of kinetic theory is to determine a relation for the viscosity in terms of basic molecular properties that is substantiated by observations. Statistical averages yield expressions for mean free path λ, mean molecule speed v, the average molecule, and the net flux of molecules across a plane in the fluid. Such a hypothetical plane AA', lying parallel to a mean shear flow, is shown in a close-up of two λ dimensions in Fig. 1.34. On this scale, the shear is always linear, and the parcel would define the density

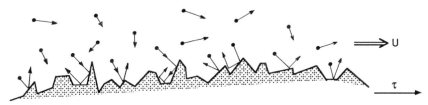

Figure 1.35 Sketch showing molecules of a fluid at a solid boundary, on the scale of a mean free path. If there is a mean velocity U, then a mean stress τ exists.

with a variation of $\approx 20\%$. Despite these approximations, kinetic theory successfully predicts:

- $\lambda = 1/[\sqrt{2}\pi d^2 n]$, where d is molecular diameter and n is molecule number density.
- Molecular flux across AA' $= nv/4$ per unit area.
- $v = [8kT/(\pi m)]^{1/2}$ where m is molecular mass.
- The average molecule crossing AA' begins at $\frac{2}{3}\lambda$ away.
- The shear stress = net x-momentum flux across AA',
  $= \delta(mu_x)/\delta t$
  $= (1/3)\ nvm\lambda\ du/dz \equiv \mu\ du/dz.$
- Hence, $\mu = [2/(3d^2)]\ [mkT/\pi^3]^{1/2}$.

Thus, viscosity depends on temperature and molecular characteristics of the fluid.

In the early days of fluid dynamics, there was considerable speculation about whether or not the fluid came to rest at a solid surface. The idea that there was at least a "slip velocity" at the surface was reinforced by the success of inviscid theory in describing the streamlines observed around various bodies. Some of these are shown in Fig. 1.10 and 1.11

Kinetic theory can be used to accommodate a prediction of slip velocity. This accommodation leads to a change in the momentum transfer across AA' and yields a momentum transfer at a solid surface, as shown in Fig. 1.35.

The momentum transfer at the boundary is assumed to be determined by the process of diffuse reflection. In this case, the average u-momentum is zero. One then obtains,

$$u_o = \tfrac{2}{3}\lambda\ du/dz$$

We can relate this velocity and λ to the continuum mean flow and the characteristic length scale. This is done using the dimensional analysis techniques explained in Chapter 3 to obtain,

$$\frac{u_o}{U} = \frac{\lambda}{L}\left[\frac{2}{3}d\left(\frac{u/U}{dz/L}\right)\right] \approx \frac{\lambda}{L} \tag{1.39}$$

This ratio of scale lengths is known as the Knudsen number. When it is small, the slip velocity is zero. It is about 10^{-7} in the atmospheric surface layers. Therefore, slip velocities become important only in very high altitudes, where the λ becomes very large as the density drops.

Since molecular velocities increase with temperature, we might expect the rate of momentum exchange to also increase with temperature. This would result in a corresponding increase in viscosity. Such a relation exists for a gas. However, it is not observed in a liquid. This is odd because the molecules are much closer together in a liquid and the momentum exchange is much greater! However, the more dense molecular packing in this case makes the intermolecular forces more important than the molecular momentum transfer. For a liquid, the increased agitation of the individual molecules apparently leads to a decreased propensity to form groups of molecules, and in turn this leads to a decreased resistance to deformation. The net result is that the viscosity goes down with temperature. The explanation of this result is based more on empirical observations than on any rigorous theoretical understanding. Thus, although we have gained some insight into the underlying mechanism of internal stress, the theory is incomplete and a parametrization is clearly needed. We will deal with the parametrization of the internal stress in detail as we derive the force balance on a parcel in Chapter 4.

1.11.3   Parametrization

In general, parametrization is simply a method of establishing a correspondence between the average effects of a process that appears in the mean flow and some other measurable mean-flow parameter. For instance, the procedure of relating the internal fluid stress to the mean-flow shear by means of a coefficient, called viscosity in this case, is typical of the parametrization technique. When a physical process takes place on scales that are outside the scale of resolution for the problem concerned, only the averaged effects of the process are used. This average is obtained by relating it to one of the flow parameters on the observable scale.

In a numerical problem, the physical process in question often takes place on a scale that is much smaller than the smallest grid size included in the basic problem. The grid size is a crucial factor in numerical calculations, since cost and time for runs go up as grid size goes down. The details of a subgrid scale process will involve basic conservation principles applicable to the smaller scale. These details may or may not be known. In the case of the internal stress caused by some unknown combination of the molecular collisions and the intermolecular momentum exchanges, the basic process is not yet understood. Fortunately, we can relate the net internal force between layers as proportional to the mean velocity gradient, viscosity as the coefficient of proportionality. This relation is an example of *parametrization*, a general technique often resorted to in science and engineering. All "laws" can probably be traced back to parametrization. The power of the law depends on the sound statistical basis of the parametrization. Its status as law is also enhanced by the degree of lack of understanding of the underlying process.

Parametrization is a respected and widely used technique in geophysical fluid dynamics. It can be employed to establish relations between a cause and an effect when both have been measured in an experiment. There is no requirement to have knowledge of the underlying physical processes that form the basis of the relationship. However, any ideas about the physics certainly would aid in finding the correct form of the parametrization. The experiment needs to measure one parameter simultaneously with the other. When a plot is made of one versus the other, a line or curve or more sophisticated statistical relation can be found to fit the data. This will involve constants, which are the parameters relating the two variables. The success of the relation will depend on the constancy of the parameters. If the parameters are the same for a wide range of the primary dependent variables, then a valuable formula is the result. We will discuss a systematic approach to parametrization in Chapter 3. For now, the definition of viscosity can be seen as a parameter that relates forces and velocity gradients in a simple experiment of continuum flow.

1.11.4   Viscosity as an Empirical Constant from Experiment (Couette Flow)

One can obtain measurements relating an applied force to the rate of strain, or flow shear, by considering fluid confined between two plates and applying a force to move one plate. (This can be done practically by confining a thin layer of fluid between two concentric cylinders of large radii, one of which is rotated.) The equilibrium flow pattern is shown in Fig. 1.36. There exists a linear variation of the flow velocity from zero at the fixed wall to the velocity of the moving plate. From another aspect, if U is a function of z only, one can consider the definition for an imaginary thin layer of the fluid with thickness dz and velocity change du. The results of the experiment show that the force required on the top plate is proportional to U and inversely proportional to the depth of the fluid, h.

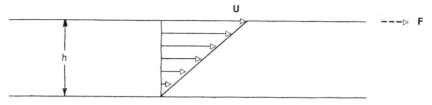

Figure 1.36 The linear velocity shear in a 2-D parallel flow between plates separated by *h*. The top plate is moved along by force **F** at velocity **U**.

When *h* is small enough, the observed linear-velocity profile indicates that the internal frictional force per unit area is constant across the layer. The shear across the layer is $U/h = du/dz$. Thus the experiment suggests that the stress τ (which is the measured force per unit area of the plate) is proportional to du/dz. The proportionality factor μ is defined as viscosity. This is Newton's law of friction:

$$\tau = \mu \, du/dz \qquad (1.40)$$

Since the viscosity occurs in the equations of motion in combination with the density as μ/ρ, we define this quantity as the *kinematic viscosity, v.*

Fluids that obey this stress-strain relationship are called *Newtonian*. Air, water, and simple fluids are Newtonian fluids. Many fluids do not follow this relationship: blood, catsup, toothpaste, some paints and plastics are examples. Viscosity may be a function of the strain rate in non-Newtonian fluids. In Fig. 1.37, observations are plotted of several natural strain rates to shear stress relations.

In atmospheric applications, when we attempt to parametrize turbulent eddy effects in analogy to molecular interactions, we define an eddy-viscosity coefficient. This is a bold assumption, which runs into a lot of criticism. One aspect of the increased complexity faced by the eddy-viscosity hypothesis can be seen in the type of fluid categorization. The eddy-viscosity

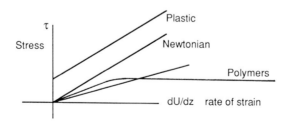

Figure 1.37 Plot of stress versus rate-of-strain relations for various categories of fluid.

may be considered to be a measure of non-Newtonian fluid behavior. This is because, unlike the uniform molecular behavior associated with Brownian motion in the fluid, eddies may vary in size and distribution. Since the variation may be a function of the velocity shear, the linear stress to rate-of-strain relation may not apply. With this caution kept in mind, the rest of our development of the equations will deal with Newtonian fluids.

1.11.5 Eddy Diffusion

Because geophysical flows, particularly PBL flows, are generally turbulent, we will discuss the eddy parametrization method in parallel with the molecular averaging process whenever it is appropriate. This is because the eddy flux is often modeled in analogy to the molecular flux. However, the turbulent eddies of aggregates of fluid parcels are clearly quite different from molecules.

The molecular development results in relations for internal stress that are strictly applicable only to laminar flows. However, the viscous forces associated with the molecular interaction are invariably negligible on atmospheric scales. The important fluxes are carried by the turbulent eddies. One result of this fact is that when turbulence is negligible, so is internal stress. The inviscid flow equations have been used with great success on atmospheric scales. However, the turbulence in many atmospheric flows is such that the characteristic scale of the turbulence can be of the same order as the dimensions of the atmospheric domain of the problem. Thus the turbulence is not easily ignored. Nor is it easily parametrized.

In many flows, turbulent eddies provide flux mechanisms that are large enough to modify larger scale atmospheric flow dynamics. The effective stress and fluxes that are a result of turbulence are especially crucial to boundary layer flows. They must be included in large-scale numerical models as the models become more accurate. The success of the weather predictions will depend at some point on the representation of the fluxes and dissipative processes related to turbulence.

In 1877 Boussinesq introduced a mixing coefficient, or eddy viscosity, K, in an analogy with the laminar flow relation between the stress τ and the velocity shear so that

$$\tau = \rho K \, du/dz \qquad \text{(the turbulent eddy based stress)} \qquad (1.41)$$

This assumes that the transport of properties (heat, momentum, etc.) is done by turbulent elements with scales much smaller than that of the basic mean flow.

We will find that the scale of the turbulence is very important. In fact,

turbulent eddies are often too large to make such a simple diffusive parametrization as Eq. (1.41). Ultimately on some scale all transport is advective.
A parcel carries along its momentum, heat, or pollution. However, advection by random motion can only be represented statistically. If the eddies
are small, then the local shear and temperature gradient may be used to
model the net transport as eddy diffusion. If the eddies are very large, then
their individual advective characteristics must be accounted for. These may
become organized, in which case they may be included in the mean laminar
flow. Or they may be random, in which case complex statistics related to
the turbulence spectrum must be employed. Again, the crucial factor in the
parametrization is the ratio of the scale of turbulent eddies to the scale of
the domain. One person's advection is another's diffusion.

In addition, several implicit assumptions are made on the nature of the
turbulent elements.

1. The turbulent elements have cylindrical or tubular characteristics. They
are vortex elements of variable size and strength that have characteristic
density, temperature and velocity.

2. The turbulent eddies transport the fluid properties in random motion
in analogy to the molecular transport. They exchange properties (momentum, heat, etc.) by rapid mixing.

K is expected to be much larger than μ/ρ to account for the greatly increased flux capabilities of turbulent flow. K is analogous to the kinematic
viscosity ν. In a similar fashion, the Fourier law for heat conduction in
laminar flow,

$$Q_{\text{lam}} = -k_h \, dT/dz \qquad (1.42)$$

has an analogous law and eddy coefficient,

$$Q = -c_p g K_h \, dT/dz \qquad (1.43)$$

A sketch of small-scale turbulence within a shear layer is shown in Fig.
1.38. If the scale is that of the PBL, the height is about 1 km. A reasonable
continuum scale is about 10 m $\leq \lambda \leq$ 100 m. A parcel with 10-m sides
will contain enough 1-m scale eddies to allow the calculation of a reasonable average, yet permit the mean flow variation to be defined on the
10–100-m scale.

However, in the PBL example, we are pushing the boundaries of the
continuum hypothesis. Eddies found in the PBL are frequently larger than
one meter and can be as large as the height of the PBL. To obtain the
average effect of such large eddies, one must resort to special averaging

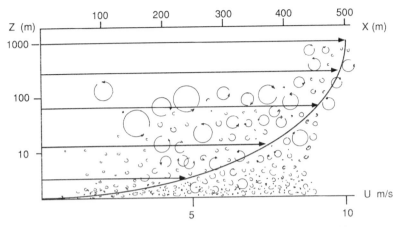

Figure 1.38   Sketch of a boundary layer containing turbulent eddies. In this case, assume that the eddies are generated by the surface roughness.

techniques. If there is horizontal homogeneity to the large eddies, then a very long sampling time may gather enough eddies to average. Another way to obtain an average over many large eddies is with an airplane flight through the PBL. The flight path must be at least 20 km long.

Boussinesq recognized that these eddy coefficients were a crude approximation and cautioned that the turbulence must be uniform and of much smaller scale than the basic (mean) flow being described. Additional problems in the application to the atmosphere arise because

1. Turbulence varies in size and intensity, hence K varies.
2. Turbulence is proportional to the mean velocity shear, hence K is a property of the flow dynamics.
3. Turbulent eddies sometimes are of a size the same order as the mean flow. Thus, the parcel cannot be large enough to contain enough eddies to provide a good average.
4. Turbulent eddies are not solid entities and the momentum exchange process must involve mixing of parcels instead of solid-body kinematics. The mixing will require some finite time interval.

Eddy diffusivity modeling has met with much success, despite its severe limitations. Various criteria have been developed that relate the diffusivity coefficient to the distance the eddy travels before adjusting to its surrounding dynamic and thermodynamic conditions. Other theories relate the diffusivity to the associated characteristic times and velocities of both the mean and the turbulent flow components. In practice, these nuances may be beyond our knowledge or our ability to obtain measurements of atmospheric turbulence. Thus, we're often forced to resort to the *ad hoc* assumption of an eddy molecular analogy for the small-scale turbulence. Proof of the validity of this assumption is then obtained only from an *a posteriori* observational agreement with the theory.

1.12 Summary

This chapter is an introduction to some new concepts, a review, and a reference section for later chapters. The topics will emerge again in the development of the equations, often as purely mathematical concepts. It is hoped that this initial exposure will make the reader familiar with the terms, if not yet comfortable with them. Generally, this takes many cycles through the material. The following chart summarizes the topics of this chapter. Elements within the chart are delineated as motivations, concepts, and definitions by various styles of type as shown in the title.

MOTIVATIONS, Concepts, and *Definitions*

OBSERVATIONS THAT CHARACTERIZE THE FLUID
DYNAMICS OF THE ATMOSPHERE

FLUID FLOW
    Newton's Law Applied to a Fluid
        The Parcel
        Continuum
        Averaging
    Pressure Force
    Internal Friction Force, *Viscosity*

WAVES AND VORTICES
    Rotating Frame of Reference
        Coriolis Force

LAMINAR AND TURBULENT FLOWS
    Potential Flow
    Transition and Turbulence
    Boundary Layers

THERMODYNAMICS
    The Equations of State
        Perfect Gas
        Temperature, Pressure,
        Entropy, Enthalpy,
        Virtual Potential
        Temperature

TURBULENCE AND EDDIES
    Eddy Viscosity

1.13 Glossary

Adiabatic   A thermodynamic process without heat exchange.

Adiabatic lapse rate   Temperature variation experienced by a parcel in
the atmosphere adiabatically moving vertically.

Body force   Force acting through a distance on every element of a parcel.

Boundary layer   Region near a boundary where viscous forces affect
velocity profile.

Continuum Domain of validity of fluid parcel concept, large enough to provide an average of small-scale variations, small enough to allow definition of mean flow derivatives.

Coriolis force Virtual force added to earth-based frame of reference to simulate inertial frame.

Eddy continuum A continuum defined with respect to turbulent eddy scales.

Eddy viscosity Parameter relating turbulent eddy momentum flux to mean shear.

Ergodic A homogeneous and time-independent flow.

Fluid Matter which continuously deforms when a force is applied; includes gases and liquids.

Fluid parcel Imaginary volume cube of fluid at an instant of time.

Geostrophic flow Flow due to balance between horizontal pressure gradient and Coriolis force.

Homogeneous Uniformity in the flow velocities and thermodynamic parameters.

Inviscid Not influenced by viscosity.

Isentropic process One without heat addition or loss.

Knudsen number λ/H, ratio of molecular mean free path (or eddy domain equivalent) to boundary layer height.

Laminar Flow with each layer independent of adjacent layers.

Newtonian fluid A fluid which obeys the stress-strain relation, $\tau = \mu \, du/dz$.

Parametrization The process of relating the change in one variable to that of another. This is usually done by making simultaneous measurements and using empirically determined parameters.

Pathline Trajectory of a particular parcel.

Perfect gas law $p = \rho RT$

Potential flow Flow wherein velocity is derivable from a prescribed function. Specific mathematical conditions are in Chapter 9.

Potential temperature Temperature resulting if pressure is changed adiabatically to 1000 mb.

Reynolds number $\rho U h/\mu$, nondimensional parameter characterizing the flow regime.

Shear Velocity gradient.

Specific heat Parameter relating internal energy to temperature.

Specific internal energy (etc.) e, internal energy per unit mass.

Stagnation values Properties of the fluid obtained at zero flow velocities. Also called total values.

Stationary Not varying with time; steady state for all time.

Steady-state flow The flow is not changing with time.

80   1 **Fundamentals of Fluid Dynamics**

Streakline   Line denoting parcels that have passed through a particular point.

Streamline   Lines everywhere tangent to the velocity vector.

Surface layer   Flow region with strong shear immediately adjacent to a boundary.

Transition   Process of flow changing from laminar to turbulent.

Turbulence   Random, unpredictable fluid flow.

Uniform flow   A flow with velocities everywhere parallel and constant.

Virtual force   Fictitious force added to noninertial frame of reference to simulate inertial frame.

Virtual temperature   Temperature that accounts for water vapor in air.

Viscosity   Parameter characterizing internal stress forces.

Vortex   Flow around a point with basically circular symmetry and tangential velocity.

1.14   Symbols

$\mathbf{a}$	Acceleration, a vector
$c_{p(\text{or } v)}$	Specific heat at constant pressure (volume)
e	Specific internal energy
$\mathbf{F}$	Force, a vector
g	Gravity
h	Specific enthalpy ($= e + p/\rho$)
K	Eddy viscosity coefficient
k	Ratio of specific heats ($= c_p/c_v$)
μ	Kinematic viscosity, g/m sec
m	Mass
ν	Dynamic viscosity, m^2/sec
ρ	Density, g/cm^3
p	Pressure
q_h	Specific humidity
R	Specific gas constant for air
R_0	Universal gas constant
τ	Viscous stress
T_v	Virtual temperature (adjusted for water vapor)
$\mathbf{u}$	Velocity, a vector

Problems

1. (a) What is the ratio of the density of water to that of air, ρ_w/ρ_a, at 10°C and absolute pressure of 103 kN/m^2? (b) What is the ratio of dynamic viscosities of air and water at standard pressure and temperature (20°C)? Likewise, the kinematic viscosities μ_w/μ_a.

2. The design of a keel of a boat requires fluid dynamics knowledge of the air and the water. Discuss

    (a) The main purpose of the keel of a typical sailboat
    (b) The same for a wind surfer
    (c) Another purpose of the keel in part (b)

3. A baseball is a smooth sphere with seams and threads forming a pattern over the surface. A knuckleball is a type of baseball pitch that is thrown such that there is no, or very little, spin. It exploits the fact that a turbulent boundary layer has less drag than a laminar one. The ball then moves in an unpredictable manner. Explain why, in 25 words or less. Estimate the best speed to throw a knuckleball on a calm evening, temperature 60°F.

Use ν @ 60°F $= 1.6 \cdot 10^{-4}$ ft^2/sec. Experimental results show transition from laminar to turbulent flow on a sphere occurs at Re $= 1.8 \cdot 10^5$; ball diameter $= 3$ inches.

4. If you are to design an experiment to evaluate a mountain-valley wind, which has a strong diurnal cycle, discuss the points you must consider to obtain a good ensemble of measurements.

5. Assume that you have a laminar flow of air in a wind tunnel. Are you (more, less, or neither) likely to have transition to turbulence if (a) you increase the height of the tunnel, keeping windspeed constant; (b) you inject 20% helium; (c) you double the speed of the wind; (d) you heat the air, with the pressures kept constant.

6. The buoyancy of a parcel of fluid simply relates to the mass per unit volume of the parcel relative to that of the surrounding fluid. A glass is full of water and ice cubes so that the cubes extend above the rim of the glass. When the ice melts, does the water spill? So, when the greenhouse effect melts the pack ice over the Arctic ocean, will this help the oceans flood the coastal cities?

7. A person is in a raft floating in a pool and throws overboard a heavy anchor. Does the water level in the pool rise, lower, or stay the same ?

8. The vertical pressure variation in a static ideal gas may be written

$$dp/dz + \gamma = 0, \qquad \text{where } \gamma = \rho g; \quad p(0) = p_0$$

Combine this with the ideal gas law to obtain an expression for the pressure variation in an isothermal layer of the atmosphere.

9. From a kinetic theory (molecules) approach, what is the (a) fluid stress, (b) pressure, (c) temperature?

10. What is fluid viscosity? Describe how it can be measured.

11. The cylindrical weight of 5 lb falls at a constant velocity inside a 6-in. diameter cylinder. The diameter of the weight is 5.995 in.. The oil film has viscosity of 7 x 10^{-5} lb sec/ft^2. What is the velocity of the weight?

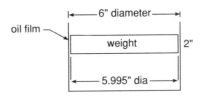

12. Viscosity of a fluid is determined in an apparatus such as the viscometer shown. Where Ω is the rotation rate (rev/sec), $t \ll R$.

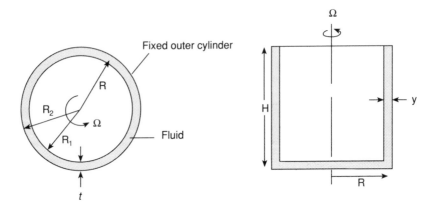

The fluid is confined between concentric cylinders with dimensions shown. The outer cylinder is fixed and the inner is rotated at a constant speed Ω, requiring a torque T. Since $t/R \ll 1$ the tangential velocity can be assumed to vary linearly across the gap. For a Newtonian fluid with viscosity μ, develop a formula for $\mu(T, \Omega,$ and viscometer geometry).

13. To use the eddy-viscosity parametrization, one must be certain that the eddies fit the parcel requirements for the derivation of the equations. Give three requirements on the eddies.

14. We often use the principle of a force being equal to a momentum

exchange. Consider an hourglass on a scale. Discuss the scale reading from the moment the sand starts to fall until the moment it has all fallen. Explain or prove your contentions. Be cautious of the endpoints.

15. The boundary layer flow in a thin layer of fluid on an inclined plane can be represented by:

$$u(z) = U[2z/H - (z/H)^2]$$

where U is constant and H is the layer thickness. Find the shear stress at the surface $H/2$ and at the free surface $z = H$.

16. The formula in problem 15 can be used to approximate a thin layer of water flowing under an airflow that drives the surface of the water along at U_A. H is the depth of influence. What is the shear stress in the water? Comment on it compared to the air stress.

17. Consider a body in a fluid flow and look at an arbitrary segment of its surface.

Discuss three forces that act on this surface.

Chapter 2 | Flow Parameters

In describing a fluid flow, we look at all variables from a field perspective. The field description presents the best flow picture, simplifies the derivations of the equations, and unfortunately adds more complexity to the derivatives (there's no free lunch).

The basic flow variable measured and used in atmospheric flow analysis is the velocity. When the velocity is given at each point and at any time in a given domain through a field description, the derivatives are known from direct or numerical differentiation. Then, with the accelerations known, the forces can be obtained from Newton's momentum law. Thus, expressions for the conservation laws can be written that apply at each point and result in a set of differential equations that completely describe the flow. The most difficult part of this process is expressing the acceleration in the Eulerian frame of reference. In this chapter we will address the velocity derivative and some of the parameters and operations involving it.

Once we have the velocity field, we find that certain combinations of the derivatives with respect to the field variables form valuable parameters. These are the *divergence* and the *vorticity*, each of which can be used effectively in describing flow phenomena. This chapter discusses some of these characteristics of the velocity field.

The effect of the earth's rotation on a frame of reference fixed on the surface is seen to cause an effective acceleration when compared to an inertial frame. When this acceleration is added to the velocity derivative as a virtual force per unit mass, we find that the equations can be written as though the rotating frame was an inertial one.

Finally, we discuss two theorems that relate values on the surface of a domain to the derivatives over the volume of the domain. We will find these to be particularly helpful in the later derivations of the equations which express the various conservation laws.

2.1 Local Time and Spatial Changes (Differentiation at a Point)

A classical physics problem is to describe the time history of a particle as it moves under the action of various forces. Newton's laws were developed to solve this problem. The solution is generally found in a *Lagrangian* frame of reference, where a particular body is identified at a particular time and its subsequent location and velocity are described as a function of time as $U(t)$. This works well for billiard balls and rockets. It might apply to fluids if we describe the individual molecular motion and interaction. However, this is impractical for most fluid problems as molecular scales are extraordinarily small compared to typical atmospheric scales. Thus we will apply the conservation principles to a conglomerate of particles that make up a much larger scale *parcel*.

One atmospheric example where we might follow a particular parcel of air in a Lagrangian manner is when it emerges from a smoke stack or other source of pollution and we are interested in the pollution path line. Another example is found in the analysis of time-dependent flow of a fluid in a specified control volume using a finite differencing numerical model, where tracking of a particular parcel is often feasible and informative.

However, most often in atmospheric dynamics we are likely to be interested in the time history of a geophysical parameter, such as the wind, at a point. Often we are interested in plotting the wind at all points in a given region, or field. This specification of the vector wind field is equivalent to determining the streamlines, since they are defined as tangent everywhere to the velocity vectors. We are less likely to be interested in identifying a particular parcel of air and following its subsequent journey. Even in the case of the point source of pollution, we are more likely to be interested in the time history of the pollution at various points in the field of flow than in the complex history of a particular parcel.

Generally our problem is to specify a flow field. That is, we write the velocity $\mathbf{U}(\mathbf{x}, t)$ to denote the velocity at any point $\mathbf{x}$ at time t. This is the *Eulerian* specification. In this perspective, there are no specific parcels of interest; rather, a field of parameters characterize the flow at each point. In the derivation of the equations for this field, we define a continuum of fluid from which we extract a representative parcel at a given point in space and time. We must consider this hypothetical particle-like parcel because we wish to apply Newton's laws for particle motion to the fluid parcel. These laws are written in a *Lagrangian* perspective wherein the change in velocity of a particular parcel is a function only of time, $\mathbf{U}(t)$. We will find that the transformation to a Eulerian frame of reference is easily accomplished using the definition of a *total derivative*. Note that the distinguishing feature between Lagrangian and Eulerian specifications is that the Eulerian expression will contain location, $\mathbf{x}$, and time as independent variables. In the Lagrangian expression, $\mathbf{x}$ is a dependent variable expressed as a function of time.

A liability of the Eulerian description is that the acceleration of the fluid element is no longer simply the rate of velocity change divided by the increment of time in the limit, as shown here:

$$\frac{du}{dt} = \lim_{\delta t \to 0} \frac{u(t + \delta t) - u(t)}{\delta t}$$

The change in definition of acceleration occurs because we are not following the particular parcel throughout the field, but rather describing the rate of change of velocity of the parcel at the instant it occupies any point in the field. Since any dependent parameter may be varying in space independently of its variation with time, the rate of change at a point will depend on the value of the parameter immediately before and after it occupies a particular point. The acceleration will be made up of the local time change plus an advective component due to the velocity gradient,

$$\frac{Du}{Dt} \lim_{\delta t, \delta x \to 0} \frac{u(x, t + \delta t) - u(x, t)}{\delta t} + \frac{u(x + \delta x, t) - u(x, t)}{\delta x}$$

Here, the capital D/Dt is used to indicate a total derivative.

Example 2.1

Data taken in large-scale modeling is often gathered by ships traversing in a selected domain. A typical experiment might involve a study of the North Atlantic and a particular ship that begins a run at longitude 10°W and latitude 40°N in the fall. (See Fig. 2.1.)

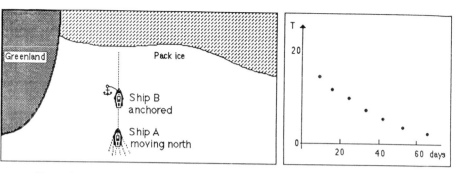

Figure 2.1   A sketch of a ship (A) cruising north past a fixed ship (B) with the daily temperature readings of ship A.

As the ship proceeds due north, one of the parameters measured is the air temperature at 1800 Z each day for a 6-week period. The cruise area and the data record are shown. Comment on the data trend. Discuss the temperature record measured at a ship anchored at P.

Solution

The temperature taken on the moving ship is steadily decreasing. Two possible causes are evident, associated with the fact that both time and the spatial coordinates are changing. The concepts of the Eulerian versus the Lagrangian specification are at issue here. Going northward toward the pack ice is moving into colder regions. In addition, the duration of the cruise is sufficient to experience the cooling effect of approaching winter. The trend measured on the moving ship is a temperature change due to its motion from warm to cold regions plus the global temperature change. We could write

$$\frac{dT}{dt}\bigg] = \frac{\partial T}{\partial t}\bigg] + \frac{\partial T}{\partial t}\bigg]$$

ship A   =   global   +   motion

or

$$\frac{dT}{dt}\bigg] = \frac{\partial T}{\partial t}\bigg] + \frac{\partial T}{\partial y}\,V\bigg] \quad \text{where } V = \frac{dy}{dt}$$

ship A   =   global   +   motion

In contrast, the anchored ship will register only the global temporal temperature change. This will be the local temperature change measured at P.

$$\left.\frac{dT}{dt}\right]_{\text{ship B}} = \left.\frac{\partial T}{\partial t}\right]_{\text{global}}$$

An observer at P could determine the time history of the temperature on the moving ship only if the $T(z)$ field and the ship speed V were known.

We should note that the temperature rate of change measured on the moving ship will be different than that measured on the anchored ship even when the ships are side by side, essentially at the same point. This is caused by an increment Δt before or after the time of coincidence; the moving ship temperatures differ slightly from those at the anchored ship due to the Δy displacement.

Consider the flow out of a high-pressure tank into a narrowing channel, as shown in Fig. 2.2.

Here the flow is rapidly varying in the x-direction and the overall picture is changing with time because the pressure in the tank is steadily decreasing. The fluid accelerates under the pressure differential force, which is constantly dropping. Any parcel at P experiences an acceleration in flow velocity due to the higher pressure immediately upstream and lower pressure immediately downstream of this point. This force due to the pressure gradient decreases as the pressure decreases (unless it is maintained by outside means). Thus this component of acceleration is decreasing with time. However, a particular parcel is actually at P only instantaneously because it is traveling with the velocity at P. It is consequently experiencing a change in velocity due to the spatial variation in $\mathbf{u}(\mathbf{x}, t)$. This part of the velocity change

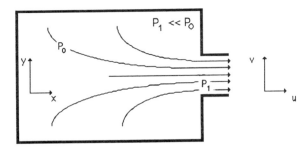

Figure 2.2   The flow out of a tank at relatively high pressure into the environment. Each parcel follows a different path.

will be called the advective component of the acceleration, since it is due to the advection of the parcel through the varying velocity field. Note that this component of the change in velocity will be present even if the entire flow field is steady. An example is the case where a constant pressure is maintained in the tank (e.g., by a pumping system). In other words, if the flow is steady state, each successive parcel that occupies any point P has the same velocity, and the overall field is unchanging. If the pressure is allowed to drop, then the overall velocity field will also drop with time, eventually to zero.

A sketch of a parcel that we identify as a cube when it is at P at time $t = 0$ is shown in Fig. 2.3 at small time increments before and after $t = 0$. Although the entire velocity field might be changing with time, it is easiest to address the advective part of the changing velocity if we consider a steady-state flow. In this case, the *flow field* is not changing with *time*. That is, the velocity, which may be different at every point in the field, is constant with respect to time at every point in the field. In a steady-state field, we could also view Fig. 2.3 as showing an instantaneous picture of a sequence of parcels that pass through point P, each attaining a cubical shape at P. Each parcel will have the same distorted shape immediately before and after being at point P because each will experience identical velocities, accelerations and forces as it moves in the steady-state field. However, a parcel *moving through the field* will change velocity constantly to adjust to the velocity at each new point. It is experiencing acceleration due to its advection through space. If we were moving with the parcel (in a Lagrangian sense), then the velocity would be changing with time, and time only. The spatial location would also be changing with time. However, in the Eulerian perspective, the parcel is an imaginary infinitesimal cube confined to the arbitrary point P at time t. In other words, in the limit $\Delta t \to 0$ if $\Delta u / \Delta t = 0$, but the limit $\Delta x \to 0$ of $\Delta u / \Delta x$ is not zero at the point P.

These two aspects of the derivative at P in an Eulerian sense can be related to the scalar change in temperature in Example 2.1 by considering the

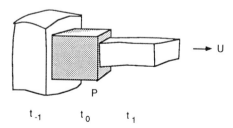

t_{-1} t_0 t_1

Figure 2.3   The parcel at a point of convergence in a flow field. It is assumed to be a cube at point P.

moving ship replaced with a parcel of air and a steady-state global temperature (e.g., at solstice). If we are describing the changing temperature of the parcel of air at the instant it is at P, the rate of change will depend on the parcel velocity through P and the temperature gradient. If there is a global change in temperature with time, it must simply be added to the advective change.

We intend to write the equations for the fluid velocity and other macroscopic parameters in an Eulerian coordinate system, determining the fluid state at each point in the flow domain. However, confusion can enter when the conservation laws are first written for a particular parcel as it moves through a point $P(x, y, z)$. Strictly speaking, this is a LaGrangian approach, since the parcel is given and we seek the change with time. However, we will consider this as the *instantaneous* force balance for this parcel as it occupies an arbitrary point P of the field. As we take the limits δt, $\delta x \rightarrow 0$, the distinction is lost between this particular parcel and the succession of parcels that occupy point P over a finite increment of time. Although we have formulated the equation in a Lagrangian sense, it is transformed to Eulerian coordinates. As long as the equations describe the field variables as a function of *position and time,* we have an Eulerian description.

If, as in Fig. 2.3, the flow field is contracting, then the forces on a parcel in the vicinity of $P(x, y)$ are acting to compress the elemental parcel. If the fluid is incompressible, when the forces push the fluid parcel inward on the sides in one direction, it can bulge outward on the sides in another direction, like a cube made of silly putty or jello. When the fluid is compressible, the density in the field is increasing in the x-direction, and an aggregate of a constant number of molecules will require less space as the parcel moves along. In both cases the shape of the parcel is changing. However, when constructing our hypothetical parcel in the Eulerian frame of reference we can assume that it is a cube at P and the distortion in the distance δx or time δt is small, vanishing in the limits. Hence in the coming chapters we will be writing force balances across a finite cubical parcel, and then determining derivatives in the limit of an infinitesimal parcel at a point.

The forces in the field may be changing with time, so that a variable acceleration exists, and the parcel velocity at any point is changing with time. In addition, the velocity of the parcel is most likely changing with distance, perhaps as the parcel moves from a high-pressure, low-speed region to a low-pressure, high-speed region. We begin to get an idea of the complexity of this change when we realize that the velocity is a vector and is often three-dimensional. Flow may be changing direction and magnitude. Thus, it is easier to begin our formulation of the derivative in Eulerian space with respect to a scalar parameter.

The temperature $T(x, y, z, t)$ of a parcel can vary with respect to each of the independent variables. We separate the total temperature change experienced by the parcel into a time-dependent local change and a change due to a spatial variation of temperature in the domain through which the parcel is moving. For the special case of heat or temperature change, the part of the derivative due to the parcel motion is sometimes referred to as the convective part. Since convection is used to describe an important class of atmospheric flow due to lighter air rising, the term *advective* is preferred here.

We can express the dual aspect of the change with simple calculus. Consider the temperature T, which varies in space and time.

$$T = T(\mathbf{x}, t) \tag{2.1}$$

Initially, let the spatial change be in the x-direction only. From the chain rule of calculus, the total change in T can be written

$$dT(x, t) = \partial T/\partial t \; dt + \partial T/\partial x \; dx$$

or

$$dT/dt = \partial T/\partial t + \partial T/\partial x \; dx/dt$$

Substituting, $u = dx/dt$ and $DT/Dt \equiv dT/dt$ (indicating *total, material, or substantial, derivative*)

$$DT/Dt = \partial T/\partial t + u \; \partial T/\partial x \tag{2.2}$$

Thus the *total temperature change* experienced by the parcels that pass through the point P is the sum of a local time variation in the temperature field plus the temperature change due to the velocity field transporting the parcels through the variable temperature field.

In the same way, let us consider the *vertical density distribution* in the atmosphere. The overall mean density is observed to decrease with height, $\rho(z)$. If there exists a vertical flow in our field of interest, the air will continually adjust to the surrounding pressure and the density will decrease in accordance with the vertical density gradient. Thus, although the density at a given height is constant (even with the upward velocity flow), a particular rising parcel of fluid is experiencing a continuous decrease in density in proportion to the vertical velocity.

Now it is also possible that the overall density is changing, say due to uneven solar heating effects. Mathematically, we can write, $\rho(z, t)$ and express the change in ρ

$$d\rho = \partial\rho/\partial t \; dt + \partial\rho/\partial z \; dz \tag{2.3}$$

If we are concerned with the change in mass for specific particles in a small region at height z, then we must consider the mass changes due to the time change plus that due to higher-density fluid entering from below and lower-density fluid leaving from above. Consider a parcel that momentarily occupies the height z. The density change with respect to time equals the local time rate of change of the field plus the change due to the parcel movement through the variable density field.

$$D\rho/Dt = \partial\rho/\partial t + \partial\rho/\partial z \, dz/dt = \partial\rho/\partial t + \partial\rho/\partial z \, w \qquad (2.4)$$

Finally, for another derivation of the total derivative, consider a parcel moving along with position $s(t) = [x(t), y(t), z(t)]$, as shown in Fig. 2.4.

Now consider a fluid property $f(x, y, z, t)$ and its variation from the chain rule for differentiation, where $(u, v, w) = (dx/dt, dy/dt, dz/dt)$:

$$Df/Dt = \partial f/\partial t + \partial f/\partial x \, dx/dt + \partial f/\partial y \, dy/dt + \partial f/\partial z \, dz/dt$$
$$= \partial f/\partial t + u \, \partial f/\partial x + v \, \partial f/\partial y + w \, \partial f/\partial z \qquad (2.5)$$

known as Euler's relation.

This is the *total* derivative. It is also called the *Eulerian* derivative (because it is used in the Eulerian description and the right side expresses the change in Eulerian coordinates), the *Lagrangian* derivative (because it gives the change of a parcel from the viewpoint of a particular parcel, the left side expressing the change with time), the *substantial* derivative, or the *material* derivative (it expresses the rate of change of a substance or material property). We will call it the total derivative. When there is no change at a point with respect to time (the Eulerian time derivative, $\partial f/\partial t = 0$), there can still be an advective part to the change, $u \, \partial f/\partial x$, $v \, \partial f/\partial y$, $w \, \partial f/\partial z$. The advective part depends upon the velocity of the parcel at the point (u, v, w), and the gradients of f in the field.

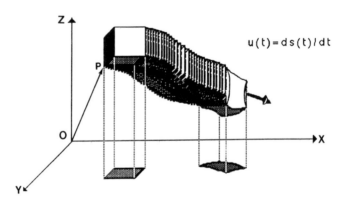

Figure 2.4 Parcels in the neighborhood of a point P on a streamline (or a given parcel at different increments of time after it leaves P). The parcel is assumed to be a cube at the initial point P located at $s(t)$.

2.2 The Advective Change and Index Notation

For a parcel in three-dimensional space, a small differential of the dependent variable $\mathbf{u}(\mathbf{x}, t)$ may be written:

$$\delta\mathbf{u} = \mathbf{u}(\mathbf{x} + \delta\mathbf{x}, t + \delta t) - \mathbf{u}(\mathbf{x}, t)$$

where $\delta\mathbf{x} = \mathbf{u}\,\delta t$. This is a compact expression for the change in velocity vector with respect to changes in space and time. However, in practice we are often treating the individual components of the velocity. We will find that when the above expression for the change in vector velocity is expanded to component form in the Eulerian frame of reference, there is a large increase in the number of terms. For instructive purposes, we will do this for the velocity in Cartesian coordinates.

$$\delta x = u\,\delta t, \quad \delta y = v\,\delta t, \quad \delta z = w\,\delta t;$$

or

$$\delta x_i = u_i\,\delta t, \quad i = 1,2,3$$

The index i can take any of three values corresponding to the three-dimensionality of space.

With the goal of writing this vector change with respect to the individual component changes as indicated by the indices, we can rearrange the total differential $\delta\mathbf{u}$. To save space, all terms are not always included here, just a sufficient number to indicate the form of the rest, which are represented by. . . .

$$\delta\mathbf{u} = \frac{\mathbf{u}(t + \delta t) - \mathbf{u}(t)}{\delta t}\,\delta t + \frac{\mathbf{u}(x_1 + \delta x_1) - \mathbf{u}(x_1)}{\delta x_1}\,\delta x_1$$

$$+ \frac{\mathbf{u}(x_2 + \delta x_2) - \mathbf{u}(x_2)}{\delta x_2}\,\delta x_2 + \frac{\mathbf{u}(x_3 + \delta x_3) - \mathbf{u}(\delta x_3)}{\delta x_3}\,\delta x_3$$

$$= \frac{\delta(u_1\mathbf{i} + u_2\mathbf{j} + u_3\mathbf{k})}{\delta t}\,\delta t + \left[\frac{u_1(x_1 + \delta x_1) - u_1(x_1)}{\delta x_1}\frac{\delta x_1}{\delta t}\,\delta t \right.$$

$$\left. + \frac{u_1(x_2 + \delta x_2) - u_1(x_2)}{\delta x_2}\frac{\delta x_2}{\delta t}\,\delta t + \frac{u_1(x_3 + \delta x_3) - u_1(x_3)\delta x_3}{\delta x_3}\frac{}{\delta t}\,\delta t \right]\mathbf{i}$$

$$+ \left[\frac{u_2(x_1 + \delta x_1) - u_2(x_2)}{\delta x_1}\frac{\delta x_1}{\delta t}\,\delta t + \frac{u_2(x_2 + \delta x_2) - u_2(x_2)}{\delta x_2}\frac{\delta x_2}{\delta t}\,\delta t \right.$$

$$\left. + \frac{u_2(x_3 + \delta x_1) - u_2(x_3)}{\delta x_3}\frac{\delta x_3}{\delta t}\,\delta t \right]\mathbf{j} + [\ldots]\,\mathbf{k}$$

$$= \frac{\delta \mathbf{u}}{\delta t} \delta t + \left[\frac{\delta u_1}{\delta x_1} \frac{\delta x_1}{\delta t} \delta t + \frac{\delta u_1}{\delta x_2} \frac{\delta x_2}{\delta t} \delta t + \frac{\delta u_1}{\delta x_3} \frac{\delta x_3}{\delta t} \delta t \right] \mathbf{i} + \left[\frac{\delta u_2}{\delta x_1} \frac{\delta x_1}{\delta t} \delta t + \dots \right]$$

where the bold $\boldsymbol{\delta}$ denotes a "total differential" similar to the use of capital D.

Anticipating that in the limit $\delta x_i / \delta t = u_i$, substitute for the velocity components $u_i = \delta x_i / \delta t$ and divide by δt.

$$\frac{\delta \mathbf{u}}{\delta t} = \left[\frac{\delta u_1}{\delta t} + \left(u_1 \frac{\delta u_1}{\delta x_1} + u_2 \frac{\delta u_1}{\delta x_2} + u_3 \frac{\delta u_1}{\delta x_3} \right) \delta t \right] \mathbf{i}$$

$$+ \left[\frac{\delta u_2}{\delta t} + \left(u_1 \frac{\delta u_2}{\delta x_1} + u_2 \frac{\delta u_2}{\delta x_2} + \dots \right) \delta t \right] \mathbf{j} + \dots$$

It is evident that this procedure is cumbersome due to the very large number of terms required. We can use the *summation convention*, where repeated indices in a single term imply a sum over the range of the index. In this case, $j = 1, 2, 3$. This consolidates the terms, so that we can write

$$\delta u_i / \delta t = \{ \delta u_i / \delta t + u_j \, \delta u_i / \delta x_j \}$$

where

$$u_j \, \delta u_i / \delta x_j \equiv u_1 \, \delta u_i / \delta x_1 + u_2 \, \delta u_i / \delta x_2 + u_3 \, \delta u_i / \delta x_3$$

When we take the limit,

$$\lim_{\substack{\delta u, \delta t, \\ \delta x \to 0}} \left[\frac{\delta u_i}{\delta t} = \frac{\delta u_i}{\delta t} + u_j \frac{\delta u_i}{\delta x_j} \right]$$

$$= \partial u_i / \partial t + u_j \, \partial u_i / \partial x_j = D u_i / D t$$

where $D u_i / D t$ is the parcel i-component of acceleration.

This can also be written in symbolic notation. (The details of the symbolic versus indicial notation and the nine-component gradient of a vector $\delta u_i / \delta x_j$ will not be discussed until Chapter 4.)

$$D\mathbf{u}/Dt = \partial \mathbf{u}/\partial t + (\mathbf{u} \cdot \nabla) \, \mathbf{u}$$

Total derivative =   local +   advective   (2.6)

The economy of space and time afforded by index notation is obvious.

Note that the direction of motion of a parcel at a point can vary depending on the mix of advective and local accelerations. The lines parallel to the vector **U** everywhere are streamlines. The local acceleration can impart a different path to the particle at each point, making pathlines distinct from streamlines. The two lines will coincide when flow is steady.

Figure 2.5   Examples of two-dimensional converging flow (a) due to constant-speed vector convergence and (b) due to constant-direction speed changes.

2.3   Divergence/Convergence

Few velocity fields have constant magnitudes and directions. Most often, the velocity changes in magnitude and/or direction from point to point. We use the terminology *divergent* when velocities tend to separate parcels at a point, *convergent* when they are coming together. The parcels of fluid in Fig. 2.5a are evidently converging, a term used in automobile traffic control to describe the situation of an onramp (a "source" of cars) feeding into a freeway flow of traffic. Those familiar with freeway traffic might recognize Fig. 2.5b as a convergent situation as well, wherein traffic is uniformly slowing down (and becoming more dense).

We obtain the mathematical representation of the divergence when we let del operate on a vector, $\mathbf{F} = F_1\mathbf{i} + F_2\mathbf{j} + F_3\mathbf{k}$, in the dot product, to get

$$\nabla \cdot \mathbf{F} = (\mathbf{i}\, \partial/\partial x + \mathbf{j}\, \partial/\partial y + \mathbf{k}\, \partial/\partial z) \cdot (F_1\mathbf{i} + F_2\mathbf{j} + F_3\mathbf{k})$$

$$= \partial F_1/\partial x + \partial F_2/\partial y + \partial F_3/\partial z$$

This may also be written

$$\partial F_1/\partial x_1 + \partial F_2/\partial x_2 + \partial F_3/\partial x_3 \equiv \partial F_i/\partial x_i$$

which is the divergence of the vector $\mathbf{F}$. It is a scalar.

When the vector is the velocity vector $\mathbf{u}$, then $\nabla \cdot \mathbf{u}$ is the divergence of the fluid at a point. If the fluid is incompressible, then the divergence would be zero unless there was a source of fluid at that point. If a source existed, $\nabla \cdot \mathbf{u}$ would be the amount of fluid diverging from the point. (Our traffic example above is essentially a compressible flow—up to the point of impact.)

Convergence is the negative of divergence. If there is a sink for the fluid at a point, the convergence equals the flow converging into the sink.

Example 2.2

Verify that the divergence of the product of a scalar times a vector may be written (for Cartesian coordinates)

$$\nabla \cdot (\Phi\mathbf{u}) = \Phi\nabla \cdot \mathbf{u} + \mathbf{u} \cdot \nabla\Phi$$

Solution

The concept to be used is the chain rule of differential calculus. We are familiar with the rule for simple variables, so we can use it on the components of the vector **u**. The Cartesian unit vectors are constants. By direct expansion,

$$\nabla \cdot (\Phi \mathbf{u}) = \nabla \cdot [\Phi(u_1 \mathbf{i} + u_2 \mathbf{j} + u_3 \mathbf{k})]$$

$$= \partial(\Phi u_1)/\partial x + \partial(\Phi u_2)/\partial y + \partial(\Phi u_3)/\partial z$$

$$= \Phi \, \partial u_1/\partial x + u_1 \, \partial \Phi/\partial x + \Phi \, \partial u_2/\partial y$$

$$+ u_2 \, \partial \Phi/\partial y + \Phi \, \partial u_3/\partial z + u_3 \, \partial \Phi/\partial z$$

$$= \Phi \nabla \cdot \mathbf{u} + \mathbf{u} \cdot \nabla \Phi$$

We should note that if the unit vectors changed directions, as they might do in some coordinate systems, then we would have additional terms.

Since the dot product of two vectors is a scalar, the divergence field is a scalar field. In general, when there is positive divergence at a point there must exist a source of fluid at that point. Otherwise, we would quickly run out of fluid, creating a void or a break in the continuum at the point in question. In our three-dimensional space, matter would have to be created at that point. However, in a two-dimensional incompressible flow, we are simply approximating the three-dimensional flow using the fact that horizontal velocities are much greater than vertical flow. Thus, there may be a *small* vertical flow that gives the effect of a source or sink to the horizontal flow without disturbing the approximation. There may also be a *strong* vertical flow at a *point,* which then is a singular point in the otherwise horizontal flow field. If we can isolate this point, the rest of the field may still be described with two-dimensional dynamics. One of the great advantages of using divergence fields in geophysics occurs when the flow is basically two-dimensional (horizontal) and incompressible. Note that with such a flow existing near the surface there can be nonzero divergence only if fluid enters or leaves the flow at the point of divergence. This can be done from the third dimension via vertical flow. An example of this is the horizontal flow in a pan with a spigot or a drain at some point in the flow. This procedure of approximating flow problems will be explored in Chapter 9. An observational example in the atmosphere is shown later in Fig. 2.13.

In the case of negative divergence (i.e., flow that is converging) in the horizontal atmospheric flow next to the surface, there must be a positive vertical flow component as the fluid is squeezed upward. If the vertical flow is relatively small, the two-dimensional, incompressible approximations may still be valid. However, the small vertical flow may have important physical effects, implying rainy or clear weather, or an up- or down-welling in the ocean. The physical interpretation for this particular combination of derivatives at a point will be further explored in connection with the conservation of mass in Chapter 5.

2.4 Vorticity

Just about everything spins—from galaxies down to particles that do not even have mass. Spin is an important characteristic of the motion. As we learn the complexities of rotational motion, we might (plaintively) ask, why is there so much spin? However, a better question might be, why not? Since motion is caused by reaction to forces, straight-line motion of a finite body requires great symmetry in the application of the force, which must pass precisely through the center of mass. Although this symmetry may often exist for body forces, it is unusual for most applied forces to be directed through the center of mass. For instance, as molecules collide, any hit off center imparts spin to both bodies. For a sliced tennis or golf ball, or a parcel of air within the boundary layer, the difference in surface stress across the body imparts spin. Any time that the net forces do not pass exactly through the center of gravity, angular rotation will be imparted to the body.

Common experience reminds us that small perturbations in the forces applied to propel ourselves in walking through the woods, or swimming in a lake, have a cumulative effect. Without guidelines to serve as constant feedback, our path will soon be a circle. We can consider linear motion as a special simple case where the angular motion can be neglected.

On the other hand, spin has some wonderful properties that help us to understand many phenomena. The fact that total angular momentum is conserved is fundamental to many analyses in physics. From the ice skater, the diver and the gymnast, we know one application of the conservation of angular momentum to the rotational motion of the human body. When a performer pulls their legs or arms into their body, the mean distance to the center of mass is shortened, and since angular momentum is conserved, spin is increased.

In the atmosphere or ocean, if a specific body of fluid that is rotating is compacted, the rate of rotation will similarly increase. In the case of a fluid made up of parcels, as each parcel moves closer to the spin axis, its speed must increase if its angular momentum is to remain constant. However, the conservation of angular momentum has singular results as the distance from the rotating parcel to the center of mass goes to zero. The speed of the parcel cannot increase infinitely, and new forces must enter the balance to slow the parcel down. This phenomenon is commonly known to exist in the physical form of the eye of the hurricane or the calm in the center of any cyclone, and it is expected to exist in the center of a tornado. (See Fig. 2.6.)

Spin enters our problems in two ways. One spin effect results in the addition of a virtual force to the basic equations. This force must be introduced

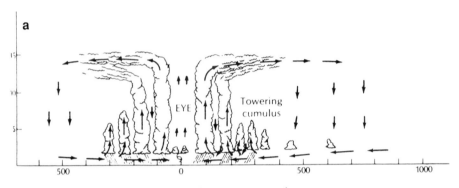

Distance from hurricane center (km)

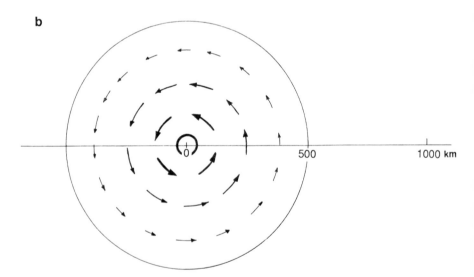

Figure 2.6   (a) Cross-section of a hurricane. Horizontal extent is 1000–2000 km; vertical extent is 10–15 km. Vertical wind vectors are 1–5 m/sec. (b) Two-dimensional windfield at the surface of a hurricane. Wind vectors are 10–90 m/sec.

to allow for the noninertial character of the rotating frame of reference used in most atmospheric dynamics problems. This frame of reference is fixed on the spinning earth and is used because of its overwhelming convenience. However, Newton's law of motion is for an inertial frame, and the virtual force is needed to account for the noninertial effects. For example, one effect of the earth's rotation is to dramatically alter the observed flows from those obtained in laboratory experiments that are generally done in a nonrotating frame of reference. The other spin effect we consider in our problems is the spin of the parcel itself, and this parameter will be a field variable.

There are other reasons why the study of geophysical flow in a rotating frame of reference differs significantly from the fluid dynamics of laboratory scale flows. One of the most startling effects of rotation is in the transition from laminar to turbulent flow. In the laboratory, transition is most often explosively sudden, as in the Reynolds' experiment. However, when rotation is applied to the laboratory system, the transition frequently proceeds through stages slowly, so that a particular stage may dominate the observed flow. The resulting flow pattern is dependent on the amount of rotation imposed. We might expect some of these transition stages to appear in the rotating frame of geophysical flow, as indeed is the case.

Spin is a general term conveniently used to relate the angular velocity created by something turning around an axis to the velocity field existing around that same axis. For instance, the spin of the earth around its axis once per day defines the angular velocity that a geostationary satellite must have to remain above a certain point of the rotating earth. The balance between the centrifugal force and the gravitational force will define the height at which this velocity must be achieved. In the coordinate system fixed on the earth's surface, all elements of the atmosphere have some spin associated with the earth's rotation in addition to any intrinsic spin that the elements may have.

Vorticity is a concept associated with the spin of the parcel at a local point. It is neatly defined by the mathematical description of the velocity field. It is the vector, or cross product, of del and the velocity vector.

$$\zeta \equiv \nabla \times \mathbf{u} \qquad (2.7a)$$

or in matrix form,

$$\mathbf{z} = \begin{pmatrix} \mathbf{i} & \mathbf{j} & \mathbf{k} \\ \partial/\partial x & \partial/\partial y & \partial/\partial z \\ u & v & w \end{pmatrix} \qquad (2.7b)$$

$$= (\partial w/\partial y - \partial v/\partial z)\mathbf{i} + (\partial u/\partial z - \partial w/\partial x)\mathbf{j} + (\partial v/\partial x - \partial u/\partial y)\mathbf{k}$$

Example 2.3

What is the vorticity for flows given by

$$\text{(a) } \mathbf{u} = (3x, 1, 0); \quad \text{(b) } \mathbf{u} = (Cy, 0, 0)?$$

Solution

We begin by plotting a few points of the fields.

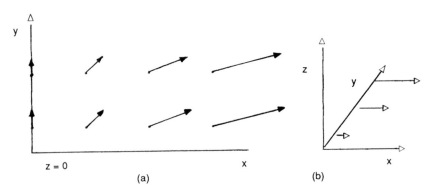

(a)   (b)

(a) The vorticity:

$$\zeta = \nabla \times \mathbf{u} = \partial v/\partial x - \partial u/\partial y = 0 - 0 = 0$$

Thus, although the velocity field is turning, the net rotation of a parcel is zero. Note that there is divergence.

$$D = \nabla \cdot \mathbf{u} = \partial u/\partial x + \partial v/\partial y + \partial w/\partial z = 3$$

(b) In this case,

$$\zeta = \nabla \times \mathbf{u} = (0, 0, -C)$$

The shear in the v-direction produces a vorticity directed in the negative z-direction.

In two-dimensional horizontal flows, $w = 0$ and $\partial/\partial z = 0$, so that only the single component of ζ is nonzero.

$$\zeta = (\partial v/\partial x - \partial u/\partial y)\mathbf{k}$$

In this case the vorticity is simply a scalar.

$$\zeta = \partial v / \partial x - \partial u / \partial y$$

Many important geophysical flows are basically horizontal, and the vorticity is a scalar field. This simplification makes it much easier to understand and apply vorticity behavior.

Another application of rotation of the fluid as represented by the vorticity will be found in the examination of the details of turbulence. The individual eddies of fluid inevitably have spin, and their behavior is determined by it. One mathematical definition of turbulence is "a chaotic field of vorticity." However, when certain large eddies in the turbulent field show persistence of a well-defined structure, they may be removed from the "turbulent" field and called "coherent" structures. These nonlinear solutions of the governing equations often produce vortices. There are conservation characteristics associated with vorticity, as will be discussed in Chapter 8. The random occurrence of such organized structures in the solution of the nonlinear equations has been studied under the general term "chaos," to distinguish these quasi-steady phenomena from the truly random character of turbulence.

Vorticity can be considered as a characteristic of particular velocity fields. It is a specific arrangement of velocity derivatives that form a scalar flow characteristic at each point in the fluid flow. We will find that it is associated with the rotation of a parcel and not with the rotation of the flow field. The term is also applied to a finite region, called a *control volume,* which is formed by a uniform aggregate of parcels with the same spin. Thus we can speak of the "vorticity" of an air mass. Only when the vorticity is zero at all points in a domain is the flow field called *irrotational.*

Vorticity characterizes the tendency of the parcel element to rotate about its center. Thus, if a parcel-size wheel (like a water wheel) with paddles or vanes is placed in the flow, it will rotate with the parcel in proportion to

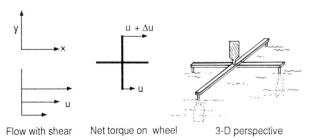

Flow with shear   Net torque on wheel   3-D perspective

Figure 2.7 If it is very small with respect to the flow domain dimensions, a paddle wheel may approximate the spin of a parcel. Flow in two-dimensional shear gives rise to differential velocity across the axis of the paddle wheel, causing rotation.

the vorticity of the parcel. The little wheel is an effective "vorticity meter" that travels with the parcel in a Lagrangian coordinate system.

One can see from Fig. 2.7 that if the "vorticity meter" is placed in a parallel flow with velocity shear, it will move with the flow and rotate due to the different velocities on opposite vanes. It is not so evident (but true) that it is possible to have a curving flow with a velocity distribution such that the wheel will move along the curved path, but will not rotate. This indicates that the flow is irrotational (Example 2.4).

Example 2.4

Discuss the vorticity for the two two-dimensional flows shown in Fig. 2.8 and describe the motion of a vorticity meter placed in the flow. In the first, the channel is narrow, so that the boundary layer effects slowing the flow are significant. In the second, the boundary layers are negligibly thin. Assume for now that at a point P in the center of the curve the velocity can be approximated with

$$u = u_0/2 - C_1 x - C_2 y$$

$$v = -C_2 x - C_1 y$$

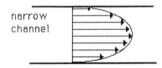

narrow
channel

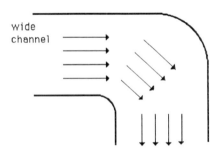

wide
channel

Figure 2.8 Examples of parallel flow in a straight and a curved channel. The straight channel is narrow so that side effects are felt. The curved channel is wide so that side effects are negligible.

Solution

We are concerned with the vorticity in specified velocity gradient fields. There are different contributions from the two terms in the 2-D vorticity expression. In the narrow channel flow, shear is present everywhere. Only $\partial/\partial y$ is nonzero and the vorticity is

$$\zeta = \nabla \times \mathbf{u} = \begin{pmatrix} \mathbf{i} & \mathbf{j} & \mathbf{k} \\ 0 & \partial/\partial y & 0 \\ u & 0 & 0 \end{pmatrix} = -\partial u/\partial y \, \mathbf{k}$$

Therefore, a vorticity meter will experience torque on the vanes extending in the y-direction. (See Fig. 2.9.) It will rotate counterclockwise above the centerline and clockwise below the centerline. Because of symmetry, $\partial u/\partial y = 0$ at the centerline. Thus if the meter is placed precisely at the centerline, there will be no torque across the wheel. Hence there is no vorticity.

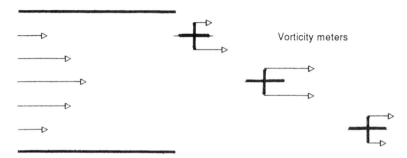

Figure 2.9 Velocities on the vane of a wheel (a vorticity meter) in the parabolic flow in a relatively narrow channel.

In the curving channel flow, both $\partial u/\partial y$ and $\partial u/\partial x$ are not zero.

$$\zeta = \nabla \times \mathbf{u} = \begin{pmatrix} \mathbf{i} & \mathbf{j} & \mathbf{k} \\ \partial/\partial x & \partial/\partial y & 0 \\ u & v & 0 \end{pmatrix} = (\partial v/\partial x - \partial u/\partial y)\mathbf{k}$$

Thus vorticity can be zero if the term in brackets is zero. If we consider a point P in the middle of the curve,

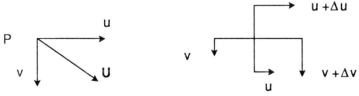

u decreases and v increases as x increases, y decreases.

$$u_0/2 - C_1x - C_2y$$
$$u = C_1x + C_2y,$$
$$v = C_2x + C_1y, \quad \text{with} \quad C_1 \quad \text{and} \quad C_2 > 0$$

Hence,

$$\partial v/\partial x - \partial u/\partial y = 0$$

We note that since vorticity involves a combination of velocity gradients, it can be zero in unexpected conditions.

We will return to our discussion of vorticity only after developing the basic equations. The concept of vorticity is intrinsically related to the mathematical description of the flow.

2.5   The Vortex

The term *vortex* is used to describe a flow field pattern. It is as distinct from the term vorticity as the term cyclone is from velocity. Cyclone is a term describing an atmospheric phenomenon that has a specific velocity field associated with it. Vortex is a term describing a vorticity concept that has a specific vorticity field associated with it. The concept of a vortex involves fluid rotating around a central spin axis. The vorticity is concentrated at the spin axis and specified at every other point in the flow field (generally as a constant, often zero). The vortex will have a specific velocity field associated with it too. In fact, we will see that a cyclone can be approximated as a certain type of simple vortex. Since there is rotation about a point, the equations written in cylindrical coordinates will take advantage of symmetries in the problem. For instance, the field of velocity around a point is frequently independent of the angle, θ. This is approximately the case in a hurricane, where it depends primarily on the radius, r.

Under special conditions, the equations of motion and vorticity reduce to very simple forms. For instance, in the case of inviscid, irrotational, steady-state motion, one can write a simple flow equation when the velocity depends only on the distance from the center of rotation, r, in cylindrical coordinates (u_r, u_θ, u_z). Observations (as first noted by Leonardo da Vinci) show that

$$Vr = \text{Constant} = C \tag{2.8}$$

or,

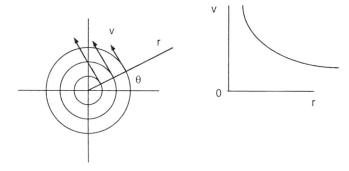

Figure 2.10 The free vortex. Flow is tangential and inversely proportional to r.

$$u_\theta = C/r$$

The flow field is illustrated in Fig. 2.10. The sign of u_θ is defined positive for counterclockwise or negative for clockwise flow—according to the right hand rule.

The flow field defined by Eq. (2.8) is called the *free vortex*. The expression for the velocity obviously has difficulties at $r = 0$, approaching infinity. However, the vorticity, which is made up of a combination of the spatial derivatives of **u**, need not be infinite at $r = 0$. In fact we will find that the vorticity of a parcel in a free vortex is zero at all points except at the center point, where it has a finite value (Example 2.5). The free vortex provides a good approximation for the velocity around a bathtub drain, the swirling flow produced by a canoe paddle, a tornado, or a hurricane.

Example 2.5

Show that the vorticity is zero everywhere in a free vortex except perhaps at the origin.

Solution

Since a vortex geometry is essentially a cylindrically rotating column of fluid, we expect the expression of vorticity will be simplest when written in cylindrical coordinates. We then need only write out the concept of vorticity from the definition [Eq. (2.7)] as the cross-product of del and velocity.

$$\nabla \times \mathbf{u} = \left[\frac{1}{r}\frac{\partial u_z}{\partial \theta} - \frac{\partial u_\theta}{\partial z}, \ \frac{\partial u_r}{\partial z} - \frac{\partial u_z}{\partial r}, \ \frac{1}{r}\left(\frac{\partial r u_\theta}{\partial r} - \frac{\partial u_r}{\partial \theta} \right) \right]$$

For the free vortex, we have

$$u_\theta = C/r, \quad u_r = u_z = 0$$

Substituting these values into the terms for the cross product in polar coordinates for $r > 0$ yields

$$\nabla \times \mathbf{u} = [0, 0, 1/r\{\partial(rC/r)/\partial r\}]$$

or

$$\zeta = 1/r\, \partial C/\partial r = 0/r \equiv 0$$

At $r = 0$, the vorticity is indeterminate, at $0/0$. The origin has an infinite velocity and must be excluded from any realistic flow. In the real vortex, the velocity drops to zero in a small but finite core region where viscosity is important and vorticity is nonzero.

For solid-body rotation, the flow dynamics are given by

$$V/r = \text{Constant} = d\theta/dt \quad \text{and} \quad V = d\theta/dt\, r \tag{2.9}$$

where $d\theta/dt$ is the angular rotation rate. This flow is shown in Fig. 2.11.

This flow is called a *forced vortex*. This case occurs in the steady flow of a rotating cylinder or dishpan filled with fluid. The frictional force from the rotating bottom is eventually transmitted throughout the depth of the fluid so that it rotates as though it was a solid body sitting on a turntable.

Similarly, one component of the earth's spin will produce an effective rotation of the earth's surface, which will in turn force a vortex flow from the surface of the earth into the atmosphere. The rotation rate is one revo-

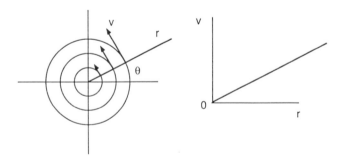

Figure 2.11 The forced vortex. The flow is tangential and linearly proportional to r.

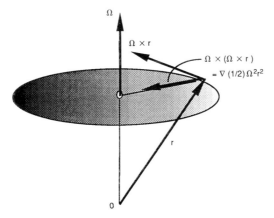

Figure 2.12   The centripetal force/unit mass. The centripetal force is directed inward toward the axis of rotation. The gravitational force would be directed along **r** to the center of mass of the earth.

lution every 24 hours at the poles and zero at the equator. This is the kernel of the rotation of cyclones, hurricanes, and tornados. Since the component of the earth's spin changes sign at the equator, these phenomena will rotate counterclockwise in the northern and clockwise in the southern hemisphere.

Example 2.6

Calculate the vorticity field for the velocity field given by

$$\mathbf{u} = (0,\, r\Omega,\, 0)$$

where $\Omega = d\theta/dt = \text{constant}$.

Solution

The vorticity in cylindrical coordinates is

$$\nabla \times \mathbf{u} = \left[\frac{1}{r}\frac{\partial u_z}{\partial \theta} - \frac{\partial u_\theta}{\partial z},\ \frac{\partial u_r}{\partial z} - \frac{\partial u_z}{\partial r},\ \frac{1}{r}\left(\frac{\partial r u_\theta}{\partial r} - \frac{\partial u_r}{\partial \theta}\right) \right]$$

For the given velocity field, $\partial/\partial r = \partial/\partial z = 0$, and there is a contribution from only the third term (the z-component),

$$(1/r)\{\partial r(r\Omega)/\partial r\} = (1/r)\{\partial r^2\Omega/\partial r\} = 2r\Omega/r = 2\Omega$$

Thus there is a constant vorticity, 2Ω, in the r, θ plane. The vorticity vector has only this one component in the z-direction.

We see that the term "vortex" can be associated with quite different velocity distributions. It is also clear that we need an explicit expression for the rotational forcing by the earth's spin in our coordinate system.

2.6 The Coriolis Term

In our expression of Newton's law, $\mathbf{F} = m\mathbf{a}$, as it applies to a fluid parcel, we are balancing the forces and the acceleration terms per unit mass. This law was derived for an inertial frame of reference. However, we have noted that this law can also be used in the earth-based frame of reference, which is rotating and noninertial, provided that a virtual force is added. This is needed to balance the acceleration terms that are effects of the rotation. Since any acceleration can be expressed in terms of its corresponding force per unit mass, this virtual force has the effect of simulating the noninertial acceleration, leaving the remaining terms to balance as though in an inertial frame. In geophysical flows the virtual forces are centrifugal and Coriolis forces. A good detailed discussion of virtual forces, and vectors in general, is found in Synge and Griffith (1959).[1]

Initially, we consider the acceleration with respect to a nonrotating coordinate system, fixed in the earth as was shown in Fig. 1.1. We must calculate the velocity as the derivative of the position vector $\mathbf{r}$ and the acceleration with respect to the velocity $\mathbf{V}$. The derivative of a vector with respect to the "absolute" coordinates is related to the derivative in the rotating coordinate system by the operator:

$$(d/dt)_A = (d/dt + \Omega \times)$$

where $\Omega = (0, \Omega_E \cos \phi, \Omega_E \sin \phi) \equiv \tfrac{1}{2}(0, f', f)$ is the vector along the axis of rotation; Ω_E is the rotation rate of the earth; and ϕ is the latitude.

Thus,

$$\mathbf{V}_A = (d\mathbf{r}/dt)_A = d\mathbf{r}/dt + \Omega \times \mathbf{r} = \mathbf{V} + \mathbf{V}_E \qquad (2.10)$$

where $\mathbf{V}$ is the relative velocity, $\mathbf{V}_E$ is the velocity due to the earth's rotation, and $\mathbf{V}_A$ is the absolute velocity with respect to the fixed coordinates.

---

[1] Synge, J., and Griffith, B. A. "Principles of Mechanics," McGraw-Hill, 1959.

Similarly,

$$(dV/dt)_A = (d/dt + \Omega \times)(V + \Omega \times r)$$

$$= dV/dt + \Omega \times (\Omega \times r) + \Omega \times V + \Omega \times dr/dt$$

$$(dV/dt)_A = dV/dt_r + \Omega \times (\Omega \times r) + 2\Omega \times V \quad (2.11)$$

"absolute" acceleration	=	relative acceleration	+	centripetal acceleration	+	Coriolis force

We see that the absolute acceleration has three parts, a relative acceleration at r plus two terms that depend on the rotation vector Ω. Rotation is defined by the right-hand rule—positive is clockwise when looking in the direction of Ω. Referring to the vector r in the rotating frame of reference in Fig. 1.1, note that the Cartesian coordinate system centered on the earth's surface can have any orientation with respect to Ω. Coordinate z can be parallel at the poles to perpendicular at the equator.

Since Eq. (2.11) is the acceleration in an inertial frame of reference, the motion of a parcel could be obtained by equating this acceleration per unit mass to the forces on the parcel using Newton's law.

$$dV/dt |_A = \sum F$$

However, since we want to use the noninertial rotating frame of reference, we will have to rearrange Eq. (2.11) to write

$$dV/dt |_r = \sum F - \Omega \times (\Omega \times r) - 2\Omega \times V$$

where $-\Omega \times (\Omega \times r)$ is the virtual centrifugal force, and $-2\Omega \times u$ is called the *Coriolis acceleration,* or force per unit mass.

The centrifugal force term is aligned with the gravitational vector at the equator. Although this vector has one component along the gravitational vector and another component tangential to the earth's surface, the magnitude of both components decreases from a relatively small value at the equator to zero at the poles. The virtual centrifugal force term (per unit mass) is $-\Omega \times (\Omega \times r)$. It is a vector directed away from the rotation axis, as it represents the tendency of a parcel to fly outward due to the rotation. It can be expressed as the gradient of a scalar potential, $\frac{1}{2}\Omega^2 r^2$, shown in Fig. 2.12.

Gravity can be written as the gradient of a gravitational geopotential, $\varphi' = -g'z$, noting that $-g' = \partial\phi'/\partial z$. Since the centrifugal acceleration term is small compared with gravitational acceleration g', it can be combined with the latter into an effective gravitational acceleration, $-g = -g' - \Omega \times (\Omega \times r)$ in the equations of motion. Henceforth, the contribution of centrifugal acceleration to the gravitational acceleration will be assumed when the force of gravity is given. On earth $g \approx 9.8$ m/sec^2. A corresponding total geopotential can also be defined,

$$\Phi = \varphi' - \tfrac{1}{2}\Omega^2 r^2$$

so that $\nabla\Phi = g$ is the effective gravity.

2.6.1 Relative Vorticity

In atmospheric science the rotating frame of reference contributes a component of planetary vorticity to the total vorticity, so that

$$\zeta_t = \zeta_r + \zeta_p \tag{2.12}$$

where total vorticity is made up of relative plus planetary vorticity. The component of planetary vorticity perpendicular to the surface (in the z-direction) is called the *Coriolis parameter,* defined by

$$\zeta_P = f \equiv 2(\text{Earth rotation rate}) \cdot \sin(\text{latitude})$$

or

$$f = 2\Omega \sin \phi \tag{2.13}$$

We can expect this term to occur as a virtual force in the earth based equations of motion, where it is customary to drop the subscript "r" on the relative vorticity,

$$\zeta_t = \zeta + f \tag{2.14}$$

2.6.2 An Example of a Vorticity Calculation

An example of atmospheric vorticity calculation is shown in Fig. 2.13. The wind field for a region of the north Pacific is observed from satellite microwave data in (a). This field is then used to calculate the vorticity (b) in the vicinity of a cyclonic storm with a frontal band between different air masses. There are strong regions of convergence associated with the low-pressure region and along the frontal band. We will learn to associate this with areas of upward flow after studying conservation of mass in Chapter 5. The vorticity field is quite variable and provides basic information on the character of the flow, as discussed in Chapter 8.

2.7 Integral Theorems

We will find two basic theorems of integral calculus of great use in the derivation of the equations of motion. These theorems relate surface fluxes to changes over the interior of the volume enclosed by the surface. The

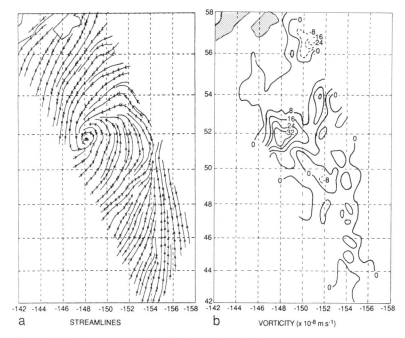

Figure 2.13 An example of windfields used to calculate streamline and vorticity fields. The wind fields are obtained from satellite data over the ocean surface. The vorticities are calculated for each grid point based on neighboring points.

connection between what is going on at the surface and the change integrated over the volume of the parcel will be frequently needed in the derivations. For instance, when we speak of the conservation of mass, momentum, or energy for a parcel we must relate the overall change over the parcel volume to the fluxes of each quantity through the surfaces. The total derivative has one part that is a change over the volume and one part that depends on the advective flow through the sides of the parcel. Also, our forces are a combination of body forces acting on the volume of a parcel and surface forces acting on each face of the parcel. The result is a mixture of volume and surface integrals in the formulation of the momentum expression. As we manipulate the equations to differential form, we will find these relations repeatedly useful.

2.7.1 Leibnitz's Theorem

As heat flows into or out of a specified domain of matter, the measure of the heat content of that domain—the temperature—will change. Likewise, the net mass, momentum, energy, CO_2, O_3, H_2S, or any pollutant contained in a domain of fluid will change if there is a net flow of the conservative quantity out through the boundaries. Our domain will initially be the parcel, but later we can expand it to be a room, a cloud, an air mass, or even a planetary atmosphere.

Thus, we are frequently interested in the time rate of change of a conservative function **f**, integrated over a volume that may be changing due to its motion with velocity **u**. Leibnitz has given us a theorem that can be used to cover these cases even in the general case where the boundary is expanding or contracting with time (**f** may be any rank tensor):

$$\frac{D}{Dt} \iiint \mathbf{f}\, dV = \iiint \frac{\partial \mathbf{f}}{\partial t}\, dV + \iint \mathbf{f}(\mathbf{u}\cdot\mathbf{n})\, dA \qquad (2.15)$$

This equation states that if the boundary of the volume is moving, then the time rate of change of any function integrated over the volume must also consider the surface integral of the flux through the surface. If the volume is not moving, the area integral is zero and the time derivative may be taken inside or outside the integral. This theorem is most familiar, and illustrative, in its one-dimensional form,

$$\frac{D}{Dt}\int_{x=a(t)}^{x=b(t)} \mathbf{f}(x,\, t)\, dx = \int_{a}^{b} \frac{\partial \mathbf{f}}{\partial t}\, dx + \frac{db}{dt}\mathbf{f}(x=b,\, t) - \frac{da}{dt}\mathbf{f}(x=a,\, t)$$

$$\underset{1}{} \qquad\qquad\qquad \underset{2}{} \qquad\qquad \underset{3}{}$$

The areas corresponding to each term on the right side can be shown in Fig. 2.14. There is a net change in the area between the times t and $t + \delta t$ that is made up of three parts: (1) due to the change in **f** with time, (2) due

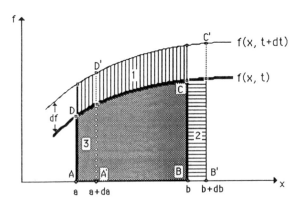

Figure 2.14 The parts of area under $\mathbf{f}(x,\, t)$ represented in Leibnitz's theorem. The function is shown at times t and $t + dt$. The total change in **f** includes areas 1, 2, and 3 as points A, B, C and D move to A', B', C' and D'.

to the change in the right-side boundary, and (3) due to the change in the left boundary.

The net value of **f** within the volume can also be changed if there are sources or sinks of **f** within V. One can write

$$\frac{D}{Dt} \iiint \mathbf{f} \, dV = \sum \mathbf{Q} \qquad (2.16)$$

where **Q** represents sinks or sources of **f**.

The local time change of **f** in the parcel plus the net **f** advected through the surface area, plus contributions from any sources or sinks in the parcel must add up to zero. Or,

$$\iiint \frac{\partial \mathbf{f}}{\partial t} \, dV + \iint \mathbf{f}(\mathbf{u} \cdot \mathbf{n}) \, dA = \sum \mathbf{Q} \qquad (2.17)$$

In the next chapters we will substitute mass, momentum, and energy for **f**. In fact, we could obtain the conservation laws immediately by direct substitution of these quantities. However, in these cases **f** and **Q** are made up of many parts, and the underlying assumptions for the fluid flow equations are best understood if the derivations are examined from a physical point of view.

2.7.2 Divergence Theorem

The *divergence theorem* (or Gauss's theorem) relates the volume integral to the surface integral in the same way that the fundamental theorem of integral calculus relates the line integral to its endpoints. It can be written

$$\underbrace{\iiint \text{div } \mathbf{f} \, dV}_{\text{Vol}} = \underbrace{\iint \mathbf{f} \cdot \mathbf{n} \, dA}_{\text{Area}} \qquad (2.18)$$

where **f** is any rank tensor, (see Chapter 4) and **n** is a unit vector outward from the volume at the surface area increment dA. This equation states that the surface integral of the component of **f** directed normal to the surface equals the volume integral of div **f**.

When **f** is a vector, for example **u**, as shown in Fig. 2.15 the divergence of **u** integrated over the volume equals the integral of the **u** component directed normally outward.

When **f** is replace with **u** in Eq. (2.18), we obtain the integral relation for velocity,

$$\underbrace{\iiint \text{div } \mathbf{u} \, dV}_{V} = \underbrace{\iint \mathbf{u} \cdot \mathbf{n} \, dA}_{A} \qquad (2.19)$$

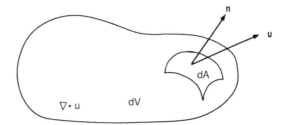

Figure 2.15 Elements of a volume and its surface that are related in the Divergence Theorem.

We will use this relation many times in the derivation of the conservation equations. It relates the advective effect of the flow in and out of the parcel to the net change of the divergence within the volume.

Example 2.7

Consider the steady nondivergent flow of an incompressible fluid through a constant area duct at a section where there is a spigot feeding 0.2-m^3/sec fluid into the flow (Fig. 2.16). The flow velocity in is $u_1 = 4$ m/sec. What is the flow velocity out? Use the Leibnitz theorem with $\mathbf{f} = \rho$ (the constant fluid density).

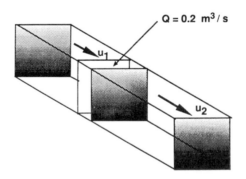

Figure 2.16 Fluid flow through a channel with inflow Q.

Solution

In the three terms of the equation for the change in ρ, there is no local change, the only component of u normal to the control volume is that through the two faces along the channel, and there is a positive flow into the volume. Thus, writing Leibnitz's theorem out in component form,

$$\iiint \frac{\partial \mathbf{f}}{\partial t} - dV + \iint \mathbf{f}(\mathbf{u} \cdot \mathbf{n}) \, dA = \sum Q$$

reduces to

$$0 + \rho(-u_1 A + u_2 A) = 0.2 \text{ m}^3/\text{sec} \cdot \rho$$

or

$$(-4 \text{ m/sec} + u_2) \, 5 \text{ m}^2 = 0.2 \text{ m}^3/\text{sec}$$

and

$$u_2 = 4.04 \text{ m/sec}$$

This is a small increase in velocity to accommodate the increased amount of mass to be moved through the same area of channel due to the influx Q.

2.8 Summary

This chapter has dealt with the concepts that make fluid dynamics different from, and some would say much more difficult than, classical dynamics. Each of the topics: total derivative, divergence, vorticity and the vortex, and the Coriolis force will be treated in greater detail in later chapters. The integral theorems will be employed in the derivation of the equations.

The concepts and their close connections can be summarized as in the following chart.

FRAME OF REFERENCE

Lagrangian

    A Specific Parcel $f(t)$

    Total Derivativve $\dfrac{df}{dt} = \dfrac{\partial f}{\partial t}$

Eulerian

    A Field Description $f(\mathbf{u}, t)$

    Euler's Relation $\dfrac{Df}{Dt} = \dfrac{\partial f}{\partial t} + \underbrace{u\,\dfrac{\partial f}{\partial x} + v\,\dfrac{\partial f}{\partial y} + w\,\dfrac{\partial f}{\partial z}}_{\text{advective change}}$

    Index Notation $u_j\,\dfrac{\partial f_i}{\partial x_j}$

DIVERGENCE $\nabla \cdot \mathbf{u} \equiv \dfrac{\partial u_i}{\partial x_i}$

VORTICITY $\zeta \equiv \nabla \times \mathbf{u}$

The Vortex

    Free

    Forced

CORIOLIS ACCELERATION $-2\Omega \times \mathbf{u}$

INTEGRAL THEOREMS

Leibnitz Theorem

Divergence Theorem

Problems

1. Write out in components i, j and $k = 1, 2, 3$.

    (a) $\partial u_i/\partial t + u_j\,\partial u_i/\partial x_j = F_i$

    (b) $\epsilon_{pqr}\,\partial s_r/\partial x_q$

2. Is the field given by

$$u = 6(xy + y^2), \quad v = 3(x^2 - y^2), \quad w = 0,$$

    (a) divergent?   (b) irrotational?

3. Consider the flow described by $\mathbf{U} = 10x\mathbf{i} - 10y\mathbf{j}$. Is this flow irrotational? What is the divergence?

4. Given:

$$u = 3x + 2yz + u_0, \quad v = 4xy + 3t + v_0, \quad w = 0$$

where u_0, v_0 are constants. Use $(\mathbf{i}, \mathbf{j}, \mathbf{k})$.

   (a) Is this a Eulerian or a Lagrangian description?
   (b) What is the "local" acceleration?
   (c) What is the advective acceleration?
   (d) What is the Eulerian derivative?

5. An airplane flies along a warm front northward at a speed of 360 km/ hr. The temperature at a ship anchored in the vicinity shows an increase of 12°C/day. A satellite measures a horizontal temperature gradient in the weather system of -0.06°C/km northward. What is the temperature gradient measured in the airplane?

6. A velocity field $\mathbf{u} = u\mathbf{i} + v\mathbf{j} + w\mathbf{k}$ is given as

$$u = x + 2y + 3z + 4t^2$$

$$v = xyz + t$$

$$w = (x + y)z^2 + 2t$$

Calculate (a) the local acceleration, (b) the advective acceleration, (c) the total acceleration at the point $(1, 1, 1, 2)$.

7. The temperature of a thermometer that drifts down a river at 10 km/ day shows an increase of 0.2°/day. A thermometer anchored at a spot in the river shows a decrease of 0.6°/day. What is the temperature gradient along the river?

8. Consider a nozzle design that has inside diameters of 9 cm at entry, and 3 cm at exit, and linearly varying cross-sectional area in between over a length of 36 cm. Flow is approximately one-dimensional incompressible with constant flow rate of 0.02 m³/sec. What is the advective acceleration at the midpoint along the nozzle?

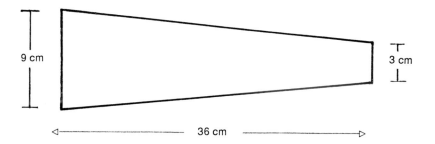

9 cm   3 cm

36 cm

9. The velocity down the center of a narrowing valley can be approximated by

$$U = 0.2t/[1 - 0.5x/L]^2$$

At $L = 5$ km and $t = 30$ sec, what is the local acceleration half-way down the valley? What is the advective acceleration. Assume the flow is approximately one-dimensional. A reasonable U is 10 m/s.

10. What are the consequences of the fact that our usual frame of reference on earth (latitude, longitude, and height) rotates once every 24 hours? How is it handled?

Chapter 3 | Methods of Analysis

The principles of scaling and dimensional analysis often fall into the category of knowledge called "tacit," meaning they are assumed to be known by all. These simple principles, which may be known by most scientists, are often taken for granted or accepted as intuitive. However, in more complex applications, systematic methods are necessary to keep everything straight and to attack problems where intuition has failed. Such problems arise frequently when dealing with a nonlinear system, as we must do in atmospheric science.

We will outline the process of "intuitive" reasoning used in the disciplines of scaling, dimensional analysis, and dynamic similarity. Each of these concepts is in universal use in fluid dynamics. They are of paramount importance in atmospheric dynamics because of the immense range of scales addressed. A flow regime can range from centimeter-scale roughness used in boundary layer studies to the 25,000-km zonal averaged domain. The time scale can vary from the 40 measurements per second in high-frequency turbulence measurements to millions of years in climate studies. Within these huge ranges of the dependent variables, there are frequently certain scales where some of the linear terms in the governing equations are much greater than the nonlinear ones. In this case, an analytic solution to the approximate linear equations may well be found. It is imperative that the range of validity of the approximation be known and adhered to in applications. That this is frequently not the case is testimony that the principles of scaling are worth a close look.

3.1 Scaling

When a flow problem is first addressed, it is generally stated in terms of a parameter, called the dependent variable, v. This variable is a function of one or more other parameters called the independent variables, $x_1, x_2, \ldots, x_n$. The geometry and boundary conditions are then specified. The task is to find a relation, $v = \mathbf{f}(x_1, x_2, \ldots, x_n)$. When the dependent variable is fluid velocity and the independent variables are spatial coordinates and time, the functional relation is that of a flow problem. The first scaling method we discuss, dimensional analysis, yields such a functional relation.

Initially, the first investigators tried the simplest of relations, a linear correspondence between two variables, v and x; $v = Cx$. Leonardo da Vinci used this formula to relate one variable to another in most of the physical phenomena he observed. He is known to have discovered the law for the velocity in a forced vortex, where velocity is directly proportional to the radial distance from the center of rotation, $u = Cr$. (He also suggested other physical applications where it didn't apply, but these have been forgotten.) Later, other investigators used more complicated functions such as log x and sin x in trying to devise a functional relation that explained various phenomena. With the advent of computers, polynomial fitting became popular, $v = C_0 + C_1x + C_2x^2 + C_3x^3 + \ldots$. In each case, the constants have to be evaluated with experimental observations.

When relating two variables with a constant coefficient, the constant may be required to have specific dimensions. For instance, if $u = Cr$, to have velocity in meters/second and r in meters, C must have dimension 1/second. If another observer then determined that the velocity also depended on rotation rate, Ω [1/sec], the relation could be rewritten, $u = C'\Omega r$, where $C = C'\Omega$ and C' is dimensionless. In a simple relationship such as this, where we can say that u is *parametrized* with respect to r, there is no requirement that the coefficients be dimensionless. However, if there had been some requirement that the constant be dimensionless initially, then this would have suggested that another variable with dimensions in inverse seconds must be involved in the relation. It may be evident to some at this point that "if all of the important parameters on which u depended were included in the functional relation, then any constant must be dimensionless." In other words, when an equation includes a constant with "balancing dimensions" it is most likely an equation with one or more missing parameters. (Of course it is possible that the dependence on the missing parameters is so minute that the constant is effectively a "universal" constant.) The motivation to have dimensionless constants, together with the requirement that the relation be dimensionally homogeneous, are the basis of a systematic process of dimensional analysis. We will discuss such a process in this chapter.

Another category of scaling analysis occurs when the basic equations for a general class of problems (e.g., fluid flow) are available. Then, every important variable is present in the equations plus the boundary conditions. Yet, this may be more than is necessary for a particular problem. The solution for some particular circumstances might depend on only a few of the many variables expressed in the equations. Then, a simpler version of the basic equations may be applicable. For instance, the governing equations of fluid flow represent the acceleration due to a sum of many forces focused on the fluid parcel. Often, not all of the forces are significant, and the acceleration can be accurately determined by considering only a few dominant forces. Intuition is a venerable method in deciding which terms are important in the balance of forces. However, a systematic method can be derived by looking at the characteristic scales of the problem. Then, although experience and intuition will still be a factor, the method can serve to organize the investigation based on dimensional analysis. This process is called dynamic similarity and is the subject of Section 3.3.

3.1.1 Parameters for Nondimensionalization

In a particular problem, each variable will generally extend over a specific range of values with limits imposed by the problem. The geometry, domain of interest, and other boundary conditions will set the scale limits on the variables in the problem. Often a variable is divided by a typical value that it assumes in a problem, a process sometimes called "normalizing." If this value is its maximum value, then the dimensionless ratio simply varies between zero and unity. If the value of the parameter used to form a ratio is an average value, then the resulting nondimensional variable will typically vary around unity. In many problems this value is never far from unity, and we can use this quality to compare different terms by simply comparing their nondimensional ratios.

Generally, a significant value for each variable can be found in the statement of the problem. In fact, if no representative value can be found for any one of the independent variables, this is a strong suggestion that the solution does not depend explicitly on the magnitude of this variable. When a representative value of a variable is evident, this value can be chosen as the number with which to divide and thus nondimensionalize the variable in the problem.

Often, the important length scale of a problem is evident, such as the diameter of a sphere, the thickness of a wing, the diameter of a jet, or the depth of a fluid. However, characteristic scales may not be obvious. Some may appear only as part of the solution to a problem. Others might be unexpected combinations of two or more dimensional parameters. These complications may limit, but do not eliminate, the usefulness of dimensional analysis. Even though much dimensional analysis is done only after the analytic solution to a problem is completed, it still provides an efficient process for organization and presentation of results.

In variable flow fields, there may be more than one choice of scale. For instance, d'Alembert, Euler, and their contemporaries solved equations for the flow field around a body, such as those shown in Figs. 1.10, 1.11, and 1.22. But they were puzzled over the prediction by their solution of no drag on the body. They had ignored the viscous term in the equations, as well as the boundary condition that the velocity must go to zero at the surface. Consequently, there was no surface stress, drag, or viscous force in the equations. The solution to the resulting inviscid Euler equations produced a flow that depended on the dimensions of the body only. This worked well in predicting the large-scale flow around a body, but it missed the effects taking place in a thin layer right next to the body. The very small scale of the boundary layer emerges only through consideration of the properties of the fluid.

When the viscous term and the boundary condition $U(z = 0) = 0$ are included, a new vertical scale for the thin boundary layer next to the body surface emerges. The new scale suggests that viscous forces become important in the force balance. This means there exists a separate solution to the new equations that is needed for the layer immediately adjacent to the body.

Example 3.1

You are interested in the flow, $U(x, y)$ around an infinite cylinder in two-dimensional uniform flow as shown in the sketch (Fig. 3.1).

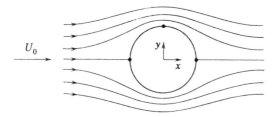

Figure 3.1 Two-dimensional flow around a cylinder; diameter d, freestream velocity U_0.

(a) Write a functional relation for inviscid flow by scaling each of the variables with U_0 and d.
(b) What are the complications when viscosity $v[L^2/t]$ (with dimensions of length-squared over time) is added to the possible dimensions for scaling the distances?

Solution

(a) U will depend on U_0, d, and the location x, y. If we write

$$U/U_0 = f(x/d, y/d),$$

we would still have the original problem, $U' = f(x', y')$, only now the terms are nondimensional as defined above. If we can solve for $U(x, y)$ (analytically or with an experiment), then the solution would be good for all U_0 and d.

(b) Now we have $U = f(U_0, d, v, x, y)$. We could write

$$U/U_0 = f[x/d, y/d, v/(U_0 d)].$$

This suggests that we could again find a solution of the form $U' = f[x', y', v/(U_0 d)]$. Now the solution will also depend on the value of $v/(U_0 d)$. This dimensionless parameter frequently appears in fluid dynamics. We have already identified it as the Reynolds number. We will see why it is an important parameter in this chapter.

There is an additional problem to scaling the distances. A scale length could be chosen based on the problem parameters as v/U_0. For typical values of $U = 10$ m/sec and $d = 1$ m, v/U_0 is very small $\approx 10^{-4}$ cm. Thus, a nondimensional $x/(v/U_0)$ would vary from 0 to 1 in a very short distance of order 10^{-4} cm. This is a suggestion that the scale determined by viscosity is very thin. The suggestion will be put on firm foundation with the concepts in this chapter, providing the basis for deriving the boundary layer concept.

We see that even without a knowledge of scaling principles, simple arguments based on scales bring about provocative questions.

For an atmospheric problem, the domain being studied can determine whether the flow is dominantly laminar or turbulent. For instance, as we move up the spatial scale in a description of the wind at a point, we may encounter a sequence of orderly (laminar) and chaotic (turbulent) flow regimes:

1. The wind buffeting you is turbulent on the meters scale.
2. There is a fairly steady average wind on the kilometer scale.
3. This average wind may be part of a large storm/cyclone, or even hurricane, with large gustiness on the tens of kilometers scale.
4. There is a large-scale average flow in the cyclone vortex on the 100–1000-km scale.
5. The storms themselves, and the high- and low-pressure systems circling the globe, move about randomly on the scale of thousands of kilometers. In the region of mid-latitude westerlies, these may pass a given point on an average of once every few days. However, the randomness of the motion of an individual storm system is evident in the moderate success of the three-day weather forecast.
6. And finally, there is a global-scale steady flow (the trades and the westerlies) and a constant equator-to-pole flux of heat on the 10,000-km scale.

There is a similar hierarchy of variability in the wind as we move up a scale in time. The magnitude of the wind will be alternately random or predictable as time intervals change from microseconds to millenniums.

As a specific example of the wide variation in atmospheric scales we can look at those scales that arise when our domain is one of the most complex, the PBL.

Time scales: Small-scale turbulence is an important factor in the PBL. Measurements of this turbulence are made at time intervals of 0.1 sec or less. But the wind given in a weather report is a few minutes average. And a climatologist may be interested only in a daily, monthly, or yearly average.

Length scales: These vary from millimeters to kilometers. The description of surface roughness includes a wide range of scales. The variety in the surface roughness on the earth includes the smooth pack ice, the variable ocean surface, forests, cities, and mountains. Also, within the PBL, waves and eddies range in scale from millimeters to kilometers.

Frequently in atmospheric flows a scale arises when waves or eddies are present. These phenomena can be characterized by their wavelength λ and period T. An important parameter is the ratio of λ to the spatial scale of the mean flow (or T to mean times). An extreme example of a large-scale eddy is found in the red spot on Jupiter (shown in Fig. 1.17), which is several thousands of kilometers across and has lasted hundreds of years. The Voyager pictures of Jupiter reveal many other smaller (1000 km) vortices embedded in the large-scale mean flow bands. The latter circle the planet and are visible through telescopes from earth. The telescope views filtered out the turbulence and revealed only the average effect of the bands. As huge as the Jovian vortices are, their scale is considerably smaller than that of the planet's atmospheric mean flow. On certain scales they might be considered as turbulent "spots" in the mean flow. For instance, one can imagine the existence of some entity on Jupiter (with climatological interests) to which a time span of 1000 years is a characteristic time, and the red spot is a brief storm. On the other hand, to a human-scale (100 years, 100s km) creature living within the domain of the red spot, the local flow of the red spot seems permanent. Thus, local flow within a turbulent eddy might be uniform and enduring with respect to relatively small scales, while the turbulent eddy might be just a small perturbation on a much larger scale mean flow.

When addressing a large quantity of data—an increasing occurrence in the age of satellite sensors—the first task is to establish an average, or mean value. However, an averaging process must be taken over a specific space or time scale. To discern a fluctuating variable with given wavelength or period λ or T, the chosen averaging scale should be less than 20% of the interval. This will produce an adequate number of points to resolve the cycle. On the other hand, if we wish to determine a mean, so that the average of the fluctuation is zero, then the averaging scale must be greater than ten times λ or T.

When addressing the problem of gathering statistics on turbulent eddies, Taylor's *frozen-wave hypothesis* provides a useful concept. The turbulent eddies are assumed to yield the same statistical mean parameters regardless of whether they are obtained at a point over a period of time, or over a distance in a very short period of time. Thus, the turbulent eddy is considered "frozen" in the mean flow, traveling with the mean flow velocity. In this case, $\lambda = TV$, where V is the mean flow velocity. Although this assumption is not always valid, it is a good approximation in many cases.

Since turbulence is omnipresent in atmospheric phenomena, it is important to be able to separate out any aspects of the data that shows signs of some regularity. This can be done by defining some simple flow characteristics:

1. If $u(t) = u(t + T)$, then $u(t)$ is *repeatable*.
2. If $u(t)$ is repeatable for $T \rightarrow \infty$, then it is *stationary*.
3. A collection of repeatable measurements comprises an *ensemble* of points.

An example of this type of flow might be the study of sea-breeze phenomena. Measurements are made at the same time every day, with $u(t) = u(t + 24 \text{ hr})$. About a hundred points are needed to get a good average. If several 10-minute averages were being measured at nearly the same time each day, one could collect an ensemble of points by measuring on successive days. There must be uniform large-scale conditions—no storms arriving during the study. This would mimic stationary conditions.

Many different problems are studied in the atmosphere with widely different scales. There is a matching amount of different definitions of the meaning for "small" and "large" scales. Again, one person's micro- or meso-scale may be another's macro-scale.

Example 3.2

Consider a typical PBL that is 2 km deep. An airplane flies through the layer taking velocity measurements at a rate of 40 per second. With the plane airspeed at 100 m/sec, the space interval is once per 2.5 m. The PBL is full of turbulent eddies with scales from centimeters to kilometers. Random motions about a mean should be plus as much as minus over a long record, with a resulting average of zero. Typically, several thousand points are required to obtain a good null average. An average over the entire record would yield the mean flow in the layer.

Consider the data record shown in Fig. 3.2: Suppose there are well-organized eddies with a known wavelength λ in the layer. How can they be identified?

Figure 3.2 Sketch of a data record containing organized eddies with wavelength λ. Segments of the record separated by λ are shown.

Solution

The averaging time must be much less than λ/V, where λ is the eddy wavelength and V is the aircraft speed. For $\lambda = 4$ km, $T \ll 40$ sec, say 5 sec, which yields 200 points.

This is not enough time to accumulate sufficient points to ensure that the smaller-scale random turbulence average will be zero. That would require at least 2000 points or 50 sec of data. One way to collect enough points is to average over 5-sec intervals each separated by a time interval equal to the period of the organized wave (in this case, 40 sec)—as emphasized in the sketch. If 10 waves are included, the equivalent of 2000 points are obtained. This is known as *compositing* because many waves are collected and averaged into one.

A basic problem in separating an organized wave motion from the turbulent fluctuations is that the organized eddy scale is often unknown *a priori*. A search must then be conducted over various lag times. The difficulties illustrate the careful attention that must be paid to the choice of averaging times. The choice depends on the characteristic time scales and periodicities of the phenomena being investigated.

=====

3.1.2 Spectral Analysis

To sort out the many different scales of phenomena that may appear in a flow, we need methods that look at the spectrum of the scales appearing in a fluctuating variable. For instance, in a typical atmospheric flow, dynamic, convective, and hydrostatic instabilities can occur on all scales. The result is that small perturbations to the mean flow find fertile conditions to rapidly grow under certain flow environments. In some cases, these waves may come to an equilibrium with the mean flow and remain as part of the mean flow solution. In other cases, the waves may break, generating turbulence in a cascade of energy to smaller and smaller waves. Finally, the flow regime may be so sensitive to the boundary conditions that it randomly oscillates between different solutions. In any case, we need to identify the different regimes. This requires data taken across the energy spectrum. High frequency data must be taken over long periods or distances to also yield low-frequency data.

A fast-response wind measurement will record a very large range of the fluctuations, random and organized. The analysis problem is to separate out the different phenomena, which we will do by analyzing the spectra of the velocity field.

When measurements are made of the amplitudes of a parameter (e.g., velocity, temperature), there usually is a wide range (a spectrum) of variation. For any given time or space interval, there will be a readily determined mean value. The variation about this mean may contain an easily identifiable periodic variation associated with an organized wave motion in the flow. Or there may be several different wave motions superimposed upon the mean. Separation of the different wavelengths by inspection may become quite difficult. When the spectrum of wavelengths is continuous and random, there is no characteristic periodicity and the flow is turbulent.

The description of the spectrum of motion can get semantically complicated in the case where an organized perturbation exists in a quasi-steady state. The perturbation is steady with respect to an interval that is significant, but that is also much smaller than the scale of the problem. Also, it may be steady only for random intervals. The predisposition toward this instability in the mean flow can be determined, however the conditions on the mean flow for its occurrence cannot be *exactly* specified. The flow that contains these *attractor* solutions has been called *chaotic*. This is distinct from a turbulent flow, where *nothing* about the flow perturbations is predictable. In this text we will be concerned only with the mean flow, turbulence, and an example of organized secondary flow.

To separate the flow variations from the mean in the equations of motion, the parameter is written as a mean plus a *perturbation* component, $\phi = \langle \phi \rangle + \phi'$. Here, the perturbation includes all deviations from the mean and it may be regular, as for a wave, or random as with turbulence.

We would like to be able to present the data in such a way that the important scales that make up the waves and turbulent eddies become evident. Since turbulence is random, a statistical approach is needed. Excellent texts on turbulent statistics include Lumley (1970) and Monin and Yaglom (1975). Here we will simply present the results and a brief definition of the important terminology.

Turbulent flow has a high, random, variability in **u**. To describe the statistical nature of turbulent flow, we are interested in

1. The relative frequency of occurrence; for example, the percent time that $|u|$ is between 6 and 7 m/sec.

2. What wavelengths (or frequencies) occur; for example, how much energy is in wavelengths between 1 and 2 cm, or between 0.01 and 0.02 $\sec^{-1}$.

To sort out the statistical characteristics of the perturbations we need some basic definitions of probability to be extended to the flow parameters. The probability of an event e_i occurring is defined in the following way:

If there are n_i occurrences of event e_i in N trials, the ratio, $n_i/N \rightarrow P_i$, a constant, as $N \rightarrow \infty$. The *probability* of event e_i is P_i. The *expected value*, E, of e_i can be calculated using the probability of e_i in,

$$E\{e_i\} = \sum_{i=1}^{N} e_i P_i \tag{3.1}$$

In the general case, we can calculate the expected value of any function of e_i as

$$E\{f(e_i)\} = \sum_{i=1}^{N} f(e_i) P_i \tag{3.2}$$

Example 3.3

Calculate the expected value of a coin flip, if we assign the value $g = 0$ to heads and $g = 1$ to tails.

Solution

There are two possibilities, e_1 = heads, e_2 = tails. The probability of either heads or tails in an infinite number of trials is $P = \frac{1}{2}$. Thus,

$$E\{g(e_i)\} = \sum_{i=1}^{N} g(e_i) P_i = \frac{1}{2} \cdot 0 + \frac{1}{2} \cdot 1 = \frac{1}{2}$$

The expected value is an average that in this case would never be obtained.

Example 3.4

What is the expected value of a die throw?

Solution

There are six sides, with values 1 to 6. Thus $g = 1, 2, 3, 4, 5,$ and 6. Each g has a probability of $1/6$ in an infinite number of trials. Hence,

$$E\{g(e_i)\} = \sum_{i=1}^{N} g(e_i) P_i$$

$$= 1 \cdot 1/6 + 2 \cdot 1/6 + 3 \cdot 1/6 + 4 \cdot 1/6 + 5 \cdot 1/6 + 6 \cdot 1/6$$

$$= (1 + 2 + 3 + 4 + 5 + 6) \cdot 1/6 = 21/6 = 3.5$$

This is not the value one expects in any single throw of the dice. It is the expected value per throw one can expect in a very large number of throws.

When considering a coin flip, there are two possible results. In the case of a die, there are six possible results. And in the case of a turbulent velocity field, there are an infinite number of possible velocities. Thus, when we consider a continuous variable such as the flow velocity, there are no discrete events to make up the e_i. Instead, we must arbitrarily select an interval of the continuous function to make up the event,

$$e_i \equiv \text{occurrence of } u_i - \Delta u/2 \le u_i \le u_i + \Delta u/2$$

In this case, n_i/N will be a smooth function of Δu, and the probability is now a function of **u**,

$$p(u_i) = \underset{N \to \infty}{\text{limit}} \underset{\Delta u \to 0}{\text{limit}} \frac{n_i/N}{\Delta u}$$

This is called the *probability density function.*

One of the ways we examine the nature of the departure of u from the mean, $\langle u \rangle$, where $u = \langle u \rangle + u'$, is by analyzing the *spectra* of u.

3.1.2.a   The Autocorrelation Function

We can determine much about the nature of a variable by considering the product of the variable times itself at a later time. For instance, we can expect a peak in this value at any time that corresponds to a periodic maximum for the variable. For example, we expect the product of the noontime temperature times itself at a lag interval of 24 hours to be quite high compared to the value at an interval of 12 hours (in the lower latitudes). We would expect a minimum in this product to occur when the initial point is chosen to be the minimum temperature times the similar value obtained at a 24-hour lag time. The correlation of a parameter times itself at a later time can depend on the initial time and the lag time.

First, we look at the velocity autocorrelation, although any dependent variable can be treated in a similar fashion. Let the ensemble of points be stationary so that $\langle u \rangle$ is constant. We define the *autocorrelation function* in terms of the expected value of the product of u' at time t, with u' at a later time, $t + \Delta t$.

$$R(\Delta t) = E\{u'(t)\, u'(t + \Delta t)\} \tag{3.3}$$

This will depend only on the *lag*, Δt. When $\Delta t = 0$, then $R(0)$ is simply twice the kinetic energy of the velocity component.

The Fourier transform will express the autocorrelation function (and the turbulent kinetic energy) in terms of the frequency ω:

$$\Phi(\omega) = 1/\pi \int_{-\infty}^{\infty} R(\Delta t)\, e^{i\omega \Delta t}\, d(\Delta t) \tag{3.4}$$

$$R(\Delta t) = \frac{1}{2} \int_{-\infty}^{\infty} \Phi(\omega)\, e^{-i\omega \Delta t}\, d\omega \tag{3.5}$$

$$R(0) = \frac{1}{2} \int_{-\infty}^{\infty} \Phi(\omega) \, d\omega \qquad (3.6)$$

Here, $\omega = 2\pi$ (period of sinusoidal oscillation) $= 2\pi f$, where f is the frequency in cycles/sec or hertz.

When data is available in space, such as that taken from an airplane, x and Δx can be substituted for t and Δt in these equations, provided that the data has spatial homogeneity. This corresponds to *stationarity*.

One idealized sketch of such a plot is shown in Fig. 3.3. This temperature spectra suggests that the thermal energy has many peaks on vastly different scales.

Actual data sets have finite length, and the longer waves, or low frequencies, are often not well defined due to limited data at the longest period or distance. For instance, in the temperature record of Fig. 3.3, there is little global data for even the 1000-year cycle, although paleoclimate methods (tree rings, isotope formation) can be used to infer global temperatures in prehistoric times. The peaks that appear in the temperature record include the daily warming of the diurnal cycle, the 3–5 day cycle of mid-latitude storms in certain regions, and an annual cycle. The small increase of a degree or two over the past 100 years may be due to the greenhouse effect. This is suggested by the correlation with the increase in atmospheric CO_2 over the same period. If sufficient data were available, we would naturally expect energy peaks at several longer frequencies associated with interglacial periods and ice ages. In fact, there is an implied peak in the temperature spectra at the glacial-interglacial frequency. This peak has been found to correspond to a peak at the same frequency in the spectra of earth's orbital perturbations. Such analyses are the grit of parametrization and prediction.

When the departure from the mean value of a variable is in the form of a wave, it has a characteristic period, frequency, and magnitude. With given boundary conditions, the wave motion is predictable. Using the wave characteristics, it is possible to identify the perturbation component in the data

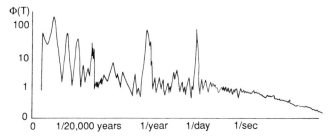

Figure 3.3 Idealized sketch of spectra of global temperature average versus time.

due to the waves. As an example, consider the flow velocity. If we let u' be the wave component and u'' be the turbulent component, then the velocity may be written as the sum of the mean plus the periodic perturbation plus the random turbulence. We can write $u = \langle u \rangle + u' + u''$. The principle that organized predictable variation may emerge from the basically turbulent field is well established. Therefore, we can explore data for organization in all scales larger than that characterizing the turbulence. We can even separate the turbulence from the organized eddies that have the same scale, using the fact that one is random and the other is organized. The autocorrelation of the organized wave will peak at its wavelength, whereas the random turbulent motion will have zero autocorrelation. The inherently turbulent PBL is a likely domain for this practice.

3.1.3   An Example of Organization within Turbulence

Observations of the spectral energy content of the PBL generally show a peak in the turbulent energy content at wavelengths from 1 to 50 meters, with a steady decay in energy at higher frequencies. Sometimes there is a secondary peak at low frequencies corresponding to a one-half to two-hour time interval. This is caused by large-scale organized waves with wavelengths from one to tens of kilometers. These longer wavelengths may be particularly difficult to distinguish in observations since they are not always advected past a fixed measuring station within the mean flow. For instance, a standing large-scale wave would not be noticed at a point, since the successive measurements taken at the point would always be at the same place in the wave. To figure out the scales requires multiple point measurements or a horizontal traverse such as that provided by an aircraft flight. Information on the intermediate scales of flow in the PBL is a relatively recent accomplishment, and available measurements still must be regarded as incomplete.

It is clear that in most flow problems organized waves with wavelengths of the order of the PBL height (1 km) and an order of magnitude larger are typically present. They may take the form of organized cells or helical vortices with horizontal axes, plumes with or without organization, or simply waves. When they are organized, they are called rolls, large eddies, coherent structures, or cells, depending on the origin and the characteristics of organization. In other words, there are currently several designations for similar phenomena, depending on whether they are found analytically, numerically, or observationally. The ambiguity in identifying the nature of a wave can be resolved only by studying the physical mechanisms of the instabilities. Then it is possible to provide a physically descriptive name (e.g., buoyancy wave) instead of a generic one (e.g., gravity wave).

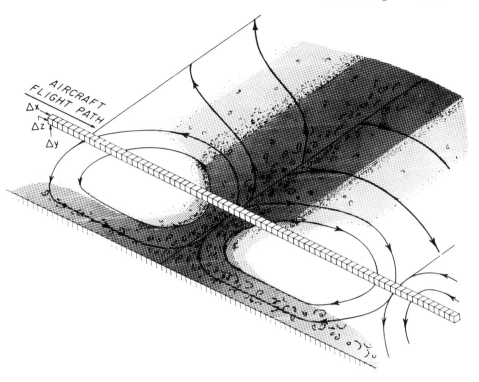

Figure 3.4 A sketch of the PBL flow including the embedded large-scale organized eddies. The concentration in convergence regions of passive constituents is shown. The ensemble of parcels measured by an airplane are indicated. The mean flow is parallel to the roll axes.

Figure 3.4 is a sketch of the helical roll phenomena in the PBL and the path of an aircraft flight designed to measure the characteristics of the roll waves, which will appear in flights perpendicular to the roll axis. Such rolls appear in the PBL due to dynamic instability of the flow, or due to convective instability. Figure 3.5 shows the spectra that would be produced by a roll-containing PBL.

Our understanding of the small-scale patterns that occur in well-studied average atmospheric flows is constantly being changed, as more and better data become available. To establish the mean flow requires only a few points and long time records, while the smaller-scale dynamics require more dense and frequent measurements. Even when organized large eddies are present in a flow the spectra derived from an observation set of that flow may not reveal a low-wavenumber energy peak corresponding to the eddies. This is often because the data set is too sparse—there are not a sufficient number

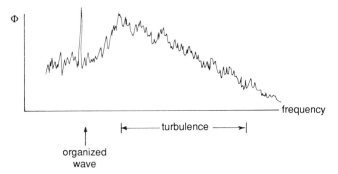

Figure 3.5 Energy spectra example for a PBL. There is a peak at a long wave and a steady decay in energy in shorter and shorter waves. There is a characteristic trailing off of the high-frequency turbulence spectral energy.

of the low-frequency events present to emerge in the spectra. For instance, in the PBL there is often a very limited number of points available to identify the waves with wavelengths greater than a kilometer. To define these waves, a long time or distance must be measured. In addition, data sets have routinely been averaged in a manner that eliminated organized long waves. This is called "despiking" (throwing away marginally supported deviant points) and "detrending" (smoothing the data). The assumption was made that the low-frequency peaks were spurious. This was assumed because these organized long waves are not always recognizable as coherent. Consequently they were often considered to be anomalous, irregular data points. When more care is given and the observer has a better understanding of what an organized long wave looks like, the consistent appearance of these waves in the data is distinguishable. Also, the incentive to find these long waves came only recently with the realization that the organization of the large eddies may be important in defining the main flow. It is expected that in the huge spectrum of atmospheric motion, other organized waveforms exist awaiting adequate data sets to identify them.

To know what to look for, we must develop some understanding of the origins, and thereby the expected scales, of the organized eddies that occur in the mean flows. New sources of data are becoming available, such as satellite-measured winds at 50-km intervals over the ocean. From these data, new wave systems that occur in known mean flows are being discovered simply by spectrally analyzing the data. The problem then is to identify the source of the waves. We will investigate the equations to be used for this in Chapter 10.

3.2 Dimensional Analysis

The process of nondimensionalization involves dividing each variable that occurs in a problem by its characteristic value. Correct scaling greatly helps data presentation. It reveals generalizing methods for the analysis of the equations in a process to be discussed later as "similarity." Although the characteristic scales may be lengths set by the boundary conditions for the problem, they may also be a combination of other parameters. For instance, a length scale may be formed from a combination of velocity and time scales, $U_0 t$. When the scales get complicated there is some loss in the physical connection of the variables (e.g., when time is nondimensionalized as tU^2 / ν). If the choice of scale is incorrectly done in the process of nondimensionalizing the equations, subsequent approximations can lead to the wrong equations. This occurs when the wrong terms of the equation are neglected because they are expected to be relatively small based on incorrect scaling. These pitfalls are best avoided by establishing a systematic procedure for the nondimensionalization.

3.2.1 Some General Rules

Dimensional Analysis (DA) has some of the same attributes as turbulence analysis. In addition to appearing to be a mysterious and incomplete topic, one must begin the analysis by an intuitive process of selecting a proper range of scales in which to define the problem. While DA is actually a complete and organized discipline (see e.g., Sedov, 1951[1]), one or more steps in the process still requires this jump of intuition. Also, a large part of the reason why DA is surrounded by the mystic is that the discipline is seldom presented rigorously. Unfortunately, this must be the case in this text as well, due to the extensive and complex material. However, we will look at the basic rules and several applications that are most pertinent for atmospheric applications.

A basic problem addressed in science in general is the orderly arrangement of various items or data. The goal is to extract identities or similarities in groups so that they can be labeled or categorized. If two quantities are to be equal, then they must have the same label. For instance, apples don't equal oranges. But if categorizing is by fruit, they may be treated as equals. And if the basic label is atoms, then they may have equalities in composition but not in structure.

By definition, if elements are to be related in an equation, there exist

[1] Sedov, L. I. "Similarity and Dimensional Methods in Mechanics," Academic Press, 1959.

restrictions on the labels to be used. DA states this requirement in a principle: *Equations must be dimensionally homogeneous.* This simply says that if the quantities are to be balanced or set equal, all terms must have the same dimension. This procedure is routinely used by scientists who work with equations, and should be by students who memorize equations for exams. For instance, when considering the famous equation for the distance moved under constant acceleration, suppose you remembered

$$x = x_0 + V_0 t + \tfrac{1}{2}at$$

but you were not quite comfortable with that last term. Using a version of scaling, or dimensional analysis, you reason as follows: since x is a distance, or length, each term on the right must be a length. x_0 certainly is, $V_0 t$ is L/t times t, leaving length, but $\tfrac{1}{2} at$ is $\tfrac{1}{2}$ (dimensionless) times an $[L/t^2]$ times $t[t]$, leaving $[L/t]$. This inconsistency in dimensions evidently would be remedied if t were squared, and you arrive at the correct formula,

$$x = x_0 + V_0 t + \tfrac{1}{2}at^2$$

If it is possible to deduce all of the significant parameters that affect the behavior of a particular variable, a relation between all of the parameters in a general equation can often be derived. This can be done with no knowledge of any basic laws of physics, making it a very powerful tool.

Example 3.5

Derive a relation for the fall velocity of a body in a vacuum under the influence of gravity, $g = 980$ cm/sec^2, as a function of how far it has fallen. There are no other significant parameters or dimensional constants in the relation.

Solution

Since the body is falling in a vacuum, there are no fluid molecules, no stress or drag, and the force is constant. Since the body is constantly accelerating under the influence of gravity, the speed will depend on how far it has fallen, Δz. If it was before Galileo's time, one might also assume that it depended on mass, m,

$$V = f(g, \Delta z, m?)$$

We want dimensional consistency:

$$V[\text{cm/sec}] = f(g, \Delta z, m) \ [\text{cm/sec}]$$

$$= f(g[\text{cm/sec}^2], \Delta z \ [\text{cm}], m \ [g])$$

Since we have specified that there are no other pertinent parameters, m cannot occur in the relation or we would have a mass (grams) dimension with no way to cancel it. There is only one combination to get cm/sec dimensions from g and Δz,

$$V = C\{g \ \Delta z\}^{1/2} \ [\text{cm}^2/\text{sec}^2]^{1/2}$$

An experiment must be performed to determine the value of the constant, which would yield

$$V = \{2g \ \Delta z\}^{1/2}$$

This example illustrates the application of step-by-step logic that often characterizes dimensional analysis. It also illustrates the benefits: only a single V at a value of Δz and g is needed to get C, whereas a series of experiments might have been done without knowledge of the functional relationship.

Care must be taken in choosing the parameters for nondimensionalization. For simple physical problems, the following rules work:

1. List all quantities (dimensional parameters) likely to influence the variables in question. Consider parameters such as time, length, gravity, pressure, velocity, density, viscosity, ...

2. List the dimensions of these quantities in a chosen set of basic units. These are measurable quantities. They have arbitrary scales and a system of units (e.g., length L, time t, temperature T, mass M).

3. Establish the restrictions that are imposed on a functional relation among the parameters by the requirement that dimensional homogeneity must be maintained.

The last step needs to be organized, and the π theorem does this.

3.2.2   The π Theorem

The process of nondimensionalization is formalized in the π theorem: A nondimensional (ND) form of the dependent variable can always be expressed as a function of the other ND independent parameters. Furthermore, *the minimum number of unrelated ND parameters that can be found is equal to the number of original dimensional parameters minus the number of fundamental units of the dimensions.* For instance, consider a velocity, with dimensions of meters per second, which is a known function of only two parameters, one with dimension meters and the other with dimension time. There are three parameters, minus the number of dimensional units—meters and seconds, that equal one ND parameter. The three variables can be combined into a single ND variable. The π theorem states that we can reduce a problem involving n parameters, which have m basic dimensions, to a problem involving only $n - m$ independent ND parameters. This reflects the fact that the more basic dimensions that are involved, the fewer independent ND combinations that can be formed. Although this is true, the definition of what is a *fundamental unit of dimension* is arbitrary. Generally, mass, length, time, and temperature are adequate for geophysical purposes. However, there are cases in which combinations of these dimensions (such as force, with $[ML/t^2]$ dimensions) are convenient to use.

An example of where the number of fundamental units of dimensions must be changed can be found from the following argument. If we had n parameters, all with the same dimension ($m = 1$), we could form $n - m = n - 1$ independent ND parameters simply by dividing all parameters by any one of the others. This is evident if all parameters had dimensions of a single basic unit, say length. However, if all parameters had dimensions of velocity, L/t, one might expect $n - m = n - 2$ independent ND parameters. This is one less than before, and incorrect, since we can obtain the same number of independent ND parameters as in the previous example (with a single dimension). This example suggests that whenever the chosen fundamental units *always occur in the same combination* in all of the chosen parameters, then the combination should be viewed as one single basic dimension. In our example, we would then consider L/t, the units of velocity, as a fundamental dimension unit so that $m = 1$ and return to $n - m = n - 1$.

The same arguments are valid if we direct our attention to the chosen *parameters* for the problem. If two of them invariably occur only in a combination, then this combination should be viewed as a single parameter, with n reduced accordingly. These are unusual, but important, occurrences. For instance, the kinematic viscosity μ, with dimensions $[M/(tL)]$ often occurs only in combination with ρ as μ/ρ. Consequently this combination is defined as the kinematic viscosity ν, with dimensions $[L^2/t]$.

We see that intuition and flexibility are needed to do a dimensional analysis. The next examples will illustrate this procedure in familiar problems. Although the procedure can be (and will be) described formally, it is best learned by doing examples, as befits any process that relies heavily on intuition.

Example 3.6

The frequency T, of a pendulum (Fig. 3.6) is expected to be a function of its length, gravity, and possibly its mass. Develop a ND functional relationship for T.

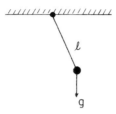

Figure 3.6 Pendulum swinging with frequency T, with mass M and force g.

Solution

Using the π theorem, we note that the number of parameters is $n = 4$, with basic dimensions mass, M; length, L; and time, t,

$$T[1/t],\ \ell[L],\ g[L/t^2],\ \text{and}\ M[M]$$

so that $m = 3$. Thus, we expect a minimum number of ND to be $4 - 3 = 1$.

Again, note that if we involve M, we cannot obtain a ND parameter since there are no other parameters with units of mass to cancel it. Hence we drop M. Our primary dependent variable T has dimensions $1/t$. We can get dimensions $1/t^2$ from g/ℓ, using ℓ to cancel the length scale in g. Thus we get the single ND parameter,

$$T/[g/\ell]^{1/2}$$

We could also have obtained $g/[\ell T^2]$; however, this is clearly related to $T/[g/\ell]^{1/2}$ as an inverse square and hence not independent.

Since a single ND parameter is a function of no others, it is a constant, and

$$T = C[g/\ell]^{1/2}$$

Note that if we hadn't included the possibility of M initially, we would still have had $n - m = 3 - 2 = 1$ ND parameters to be found. However, if we had included another parameter with dimensions including M, (e.g., the weight of the earth) then we would have had $n - m = 5 - 3 = 2$. The additional ND parameter would clearly be M/M_{earth}. If we took the pendulum to another planet, this would be a significant parameter, affecting T. We would then have to note that g is defined with respect to the different masses, the distance separating them, and a gravitational constant. This is the next example.

Example 3.7

Assume that there is a force of attraction between two masses, m_1 and m_2, and that it depends on only m_1, m_2, and the separation distance d. (See Fig. 3.7.) (a) See if you can obtain a relation for the force using dimensional analysis. (b) What if another parameter, G, with dimensions L^3/Mt^2 is involved? (c) What if acceleration $[L/t^2]$, or the newton, with dimensions of force ML/t^2 are used as basic dimensions? (d) Finally, Newton was aware, using arguments of symmetry, that F depended on the product, m_1m_2, rather than the individual masses. What effect has this?

m_1 m_2

d

Figure 3.7 Two masses, m_1 and m_2, separated by distance d.

Solution

(a) There are four parameters with dimensions,

$$F[ML/t^2], \quad m_1[M], \quad m_2[M], \quad \text{and} \quad d[L]$$

Note that the dimensions of force are the same as Ma, which can be obtained from Newton's law using dimensional homogeneity.

Our rule suggests $n - m = 4 - 3 = 1$ ND parameter can be found. This can only be m_1/m_2, since F alone has the time dimension and d cannot be combined with the masses nondimensionally. $m_1/m_2 = $ constant is not a satisfactory solution, since we have specified that $F(m_1,m_2,d)$. There are evidently some missing parameters. This illustrates that the rule does not work if one does not have a list of all pertinent parameters.

(b) If the gravitational constant, $G[L^3/Mt^2]$ were known to be an important parameter, then $n - m = 5 - 3 = 2$. We can obtain these two ND parameters by considering:

F is our primary dependent variable. We must involve G to nondimensionalize F, since only these two parameters have dimensions of t. Thus we divide F by G to eliminate the t-units, leaving dimensions

$$F/G[\text{M}^2/\text{L}^2]$$

The mass units can be removed by dividing by either m_1^2 or m_2^2; however, since we need to involve both, we'll use m_1m_2,

$$F/(m_1m_2G)\ [1/\text{L}^2]$$

Evidently, the length dimension is eliminated with d, and $Fd^2/(m_1m_2G)$ is dimensionless. The ratio m_1/m_2 is our other, independent, parameter. We can write,

$$Fd^2/(m_1m_2G) = \mathsf{F}(m_1/m_2)$$

or

$$F = Gm_1m_2/d^2\mathsf{F}(m_1/m_2)$$

This is not a satisfactory Newton's law. Experimental work would be needed to establish that $\mathsf{F}(m_1/m_2) \simeq 1$.

(c) We might have noted that t occurs only in combination with L as L/t^2, the units of acceleration. Call this a single unit of dimension A. We then have

$$F[\text{MA}], \qquad m_1[\text{M}], \qquad m_2[\text{M}], \qquad d[\text{L}], \quad \text{and} \quad G[\text{L}^2\text{A}/\text{M}]$$

with $n - m = 5 - 3 = 2$. There is no change, we have simply substituted A for t as a dimensional unit.

For force as a fundamental unit,

$$F[\text{N}], \qquad m_1[\text{M}], \qquad m_2[\text{M}], \qquad d[\text{L}], \quad \text{and} \quad G[\text{NL}^2/\text{M}^2]$$

with the same results, $n - m = 2$.

(d) Now we have

$$F[\text{N}], \qquad m_1m_2[\text{M}^2], \qquad d[\text{L}], \quad \text{and} \quad G[\text{L}^2\text{N}/\text{M}^2]$$

and $n - m = 4 - 3 = 1$. In a manner similar to above, first eliminate the dimension N, then M^2, then L^2, to obtain

$$Fd^2/(m_1m_2G) = C, \quad \text{a constant}$$

The constant can be absorbed in the constant G to obtain

$$F = Gm_1m_2/d^2$$

This is a correct expression of Newton's gravitational law. It is interesting that with dimensional analysis we can arrive at the inverse square law for this gravitational force. Newton did not use this procedure. Instead, he needed the observations that the accelerations of the moon and the apple in his orchard were in proportion to the ratio of the square of their distance from the earth's center. Thus, he set up a proportionality between F and $m_1 m_2 / d^2$.

It is evident that dimensional analysis provides valuable insights with minimal knowledge. However, it is also evident that it involves much trial and error.

We have determined that by simply considering the pertinent parameters in a problem and their dimensions, a unique formula for a functional relation between them can be obtained. The π theorem provides a basic rule for what can be done (obtaining a minimum number of ND parameters), but we need a systematic procedure to follow to obtain these parameters.

The procedure consists of three steps:

1. Determine *parameters*, independent and dependent, (I_n, D) that are important to the problem:

$$f\{I_1, I_2, \ldots I_{n-1}, D\}, \quad n \text{ parameters} \tag{3.7}$$

or

$$D = D\{I_1, I_2, \ldots I_{n-1}\}$$

For instance, these parameters can include the basic variables: all dimensional constants such as the gas constant R_g, the speed of sound c_0, the Coriolis parameter f, gravity g, etc.; and pure numbers such as π, $^1/_2$, and exponents.

2. Pick a fundamental *set* of dimensions (e.g., mass M, length L, time t, temperature T). Let m = the number of these fundamental *units* that occur in I_n (e.g., L and t only, $m = 2$).

3. Let π_n be the *nondimensional parameters*, then

$$f\{\pi_1, \pi_2, \ldots, \pi_{n-m}\} = 0$$

or

$$\pi_D = \pi_D\{\pi_1, \pi_2, \ldots, \pi_{n-m-1}\} \tag{3.8}$$

In other words, for n dimensional parameters with m units one can get a minimum of $n - m$ independent nondimensional parameters. (We nave noted that in some rare instances, the number of fundamental dimensions may be one less than the individual sum. This happens if two dimensions occur only in a fixed combination—such as L/t. In this case the dimensions are treated as one, m is one less, and $n - m$ is one greater).

Example 3.8

Find an expression for the speed of sound c_0 in water (at a given temperature). Assume it is a function of wavelength λ and gravity g.

Solution

Parameters: $c_0 = f\{g,\ \lambda\}$
    Units: $[L/t]$ $[L/t^2]$ $[L]$ (length and time)
    The number of parameters (c_0, g, λ) $n = 3$; the number of units (L,t) is $m = 2$; $n - m = 1$; and the minimum number of ND parameters is one. The one parameter is then a function of nothing, hence a constant.

$$\pi(0) = \text{Constant} = C \quad \text{(dimensionless)}$$

and by inspection,

$$\pi = c_0(g\lambda)^{-1/2} = C; \quad \text{or} \quad c_0 = C(g\lambda)^{1/2}$$

Note that this process yields only the form of the relationship. The constant C must be determined by experiment.

Whereas the π theorem states the *minimum* number of independent nondimensional terms that can be involved in a relation, the total number of *possible* dimensionless parameters is $n!/[(m + 1)!\ (n - m - 1)!]$. Thus, if $n - m = 6 - 3 = 3$ is the minimum number of nondimensional parameters that can be functionally related, the total number of possible nondimensional parameters is 15. To select the important parameters from the larger number of possibilities, experience suggests certain rules that can be used as guidelines for each nondimensional parameter.

3.2.3   *Rules for Nondimensionalizing*

When a large number of parameters are factors in a problem, there will generally be a selection of possible parameters with the same dimensions. Any of these are available for dividing out a particular dimension. Different choices of combinations of variables will yield different dimensionless parameters. If chosen in a haphazard fashion, complicated and interdependent parameters easily result. Thus it is advantageous to obtain rules for organizing the choice of parameters for the nondimensionalization.

In fluid dynamic problems, m is often three, and therefore three parameters can be chosen for nondimensionalizing the other parameters. To ensure independence, these three selected parameters should each be a characteristic of one of the following:

1. The flow
2. The flow geometry
3. A fluid characteristic

Typically, these would be (1) the flow velocity U, (2) a dimension scale d, and (3) for fluid characteristic, the density ρ or viscosity μ. Then each ND parameter is likely to require additional parameters to complete its non-dimensionalization. This step is the creative part of the process. When parameters are abundant, there are no unique choices, and the most we can obtain are some useful guidelines.

Some general rules have been developed to obtain the most suitable set of nondimensional parameters. For instance, one should involve the dependent variable only once, since you would like this ND parameter to be a function of the other parameters. Some guidelines for selecting the quantities to be used in nondimensionalizing are:

1. Don't select the quantity of most interest more than once (i.e., the primary dependent variable).

2. Prefer the "most important" quantities based on your expectations of dominant parameters. For example, form combinations first with ρ (or μ), then with V; then with L. If other variables are involved ($g, f, H_0, \ldots$), combine in order of the most complex dimension first and the simplest last.

3. Prefer quantities with "pure fundamental units" first, (such as $d[L]$ before a combination (v/U) $[L]$.

4. Investigate all choices.

The π theorem will yield the minimum number of independent ND parameters. More convenient parameters can sometimes be obtained with simple modifications:

1. Multiply by a constant (e.g., ρV^2 becomes $\frac{1}{2}\rho V^2$).

2. Raise to any power (e.g., $2K/fL^2$ replaces $\{2K/f\}^{\frac{1}{2}}/L)$.

3. Multiply any power of one parameter with any power of the others.

If the number of parameters is large, a systematic way of determining nondimensional parameters P is to *equate exponents,* sometimes called the *method of indices.* In this procedure, we can again select the primary dependent quantities P_0 as the nonrepeated ones, with an exponent of unity. The ith ND parameter can be written as a product of the repeated parameters $P_1, P_2, \ldots, P_m$, used to nondimensionalize it, raised to unknown powers,

$$\pi_i = P_0 \cdot P_1^a \cdot P_2^b \cdot P_3^c \ldots \qquad (3.9)$$

We can use the principle of dimensional homogeneity for this equation. Substitute the fundamental dimensions for each P_i and equate exponents of each dimension to zero to obtain the nondimensional parameter. That is, the total power of each dimension on each side of Eq. (3.9) must be zero. Note that the number of exponents can only equal the number of basic dimensions in P_0. The P_i must be chosen to include the dimensions contained in P_0.

Once again, the procedure is best learned by following an example that employs each of the steps described above.

Example 3.9

Assume a raindrop can be approximated as a sphere of diameter D falling with velocity W (Fig. 3.8). Use dimensional analysis to obtain an expression for the drag force F_D, as it falls through air of viscosity μ and density ρ.

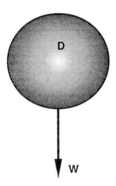

Figure 3.8 Raindrop falling with velocity W.

Solution

We first write the force as a function of the four parameters W, D, ρ, μ. Recall that force has dimensions the same as mass times acceleration and that weight of the drop is not a factor.

$$F_D = f\{W, \ D, \ \rho, \ \mu\}$$

$$\left[\frac{ML}{t^2}\right] = \left[\frac{L}{t}\right] \left[L\right] \left[\frac{M}{L^3}\right] \left[\frac{M}{Lt}\right]$$

There are five variables and three dimensions—therefore two ND parameters.

We nondimensionalize F_D with three of the other parameters characterizing the flow, the geometry, and the fluid. The flow parameter will be W, the geometry is D, and the fluid can be either ρ or μ. Choose ρ to eliminate the M-dimension from F_D. Since μ is then left over, it must be involved in the second ND parameter, and ρ is used to eliminate M from it, too.

$$\frac{F_D}{\rho} = f\{W, \qquad D, \qquad \mu/\rho\}$$

$$\left[\frac{L^4}{t^2}\right] = \quad \left[\frac{L}{t}\right] \qquad [L] \qquad \left[\frac{L^2}{t}\right]$$

Now use W to eliminate t.

$$\frac{F_D}{\rho W^2} = f\left\{D, \qquad \frac{\mu}{\rho W}\right\}$$

$$[L^2] = \qquad [L] \qquad [L]$$

Using D,

$$\frac{F_D}{\rho W^2 D^2} = f\left[\frac{\mu}{\rho WD}\right]$$

We can invert the last parameter to $\rho WD/\mu$ = Reynolds number. We can expect that the drag force plotted versus Re will yield a single curve (Fig. 3.9).

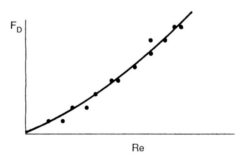

Figure 3.9 Drag force on a raindrop versus Reynolds number.

Example 3.10

Assume that you are interested in the stress force τ at the surface of a planetary boundary layer. If there were no equations to guide you, the value of this force would have to be determined by experiment. You would relate the force to the many parameters characterizing the planetary boundary layer flow. Assume these are the height H, the free stream velocity V, the fluid density ρ, eddy viscosity K, the Coriolis force parameter f, and some measure of the surface roughness z_o. Discuss the dimensional analysis to facilitate your experimentation and consider the benefits.

Solution

Without a dimensional analysis you would need to investigate the variation of τ with each parameter, holding all others constant. For example, see Fig. 3.10, where 10 experiments have been performed with H, ρ, K, z_o, and f held constant.

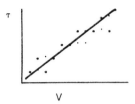

Figure 3.10   Variation of stress with velocity only.

One could then do this experiment for 10 different values for z_o (Fig. 3.11), where 100 experiments have been performed with H, ρ, f, and K held constant.

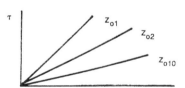

Figure 3.11   Variation of stress with velocity at various z_0.

It is clear that 10^6 runs of the experiment with 10^5 plots and perhaps 10 different fluids will be needed to provide different ρ and K. Perhaps we should give dimensional analysis a try.

Apply the π theorem. The variables are

stress	density	surface roughness
height	eddy-viscosity	Coriolis parameter
velocity		

$$\tau \quad (H, \quad V, \quad \rho, \quad K, \quad z_o, \quad f)$$

with dimensional units,

$$M/(Lt^2), \qquad L, \qquad L/t, \qquad M/L^3, \qquad L^2/t, \qquad L, \qquad 1/t$$

Hence, $n = 7$, $m = 3$, $n - m = 4$.

We can reduce the number of independent parameters to four nondimensional parameters. This will require 10^2 graphs, which are probably manageable.

Following the general rules for nondimensionalization, we choose three ($m = 3$) of the parameters with which to nondimensionalize the remainder. These should not include the primary dependent variable τ. They should include all units and should include a representative of the flow, the geometry, and the fluid characteristics. Choose V for flow, H for geometry, and ρ for fluid. Use the method of indices: π_1 involves the dependent variable τ plus an unknown combination of H, V, and ρ to make it dimensionless. Thus π_1 may be written, with τ the chosen variable to involve,

$$\pi_1 = \qquad \tau \qquad H^a \qquad V^b \qquad \rho^c$$

$$M^0 L^0 t^0 = [M/(Lt^2)]^1 \; [L]^a \qquad [L/t]^b \qquad [M/L^3]^c$$

Hence, equating coefficients of

$$M: 0 = 1 + c \rightarrow c = -1$$

$$t: 0 = -2 - b \rightarrow b = -2$$

$$L: 0 = -1 + a - 2 + 3 \rightarrow a = 0$$

Thus,

$$\pi_1 = \tau H^0 V^{-2} \rho^{-1} = \tau/(\rho V^2)$$

We have involved τ, H, V, and ρ. We still must incorporate K, z_0, and f. These will form the other π_i's to make up the expected minimum of four. Let π_2 involve K:

$$\pi_2 = \qquad K \quad H^a \quad V^b \quad \rho^c$$

$$M^0 L^0 t^0 = [L^2/t] \; L^a \; [L/t]^b \; [M/L^3]^c$$

From balancing coefficients,

$$M: 0 = c;$$

$$t: 0 = -1 - b \rightarrow b = -1:$$

$$L: 0 = 2 + a + b - 3c \rightarrow a = -1$$

Thus,

$$\pi_2 = K/(LV)$$

It follows in a similar manner that $\pi_3 = z_0/H$ and $\pi_4 = fV/H$.

{Note that often the nondimensionalization can be done by inspection: The units of τ are $[M/(Lt^2)]$. Only ρ is available to balance M so we must divide by ρ, leaving units of L^2/t^2. We can eliminate t with either V or K/H^2. V is simpler, and in fact also balances the L^2 dimension, yielding $\tau/(\rho V^2)$}. Thus, we may write

$$\tau/(\rho V^2) = f[K/(LV), fV/H, z_0/H]$$

and plot the ND τ parameter versus any of the other three ND parameters, holding the other two constant.

No physics has been used in these examples beyond the principle of dimensional homogeneity. Intuition or experience was used to divine the primary parameters, and there is no guarantee that all important ones have been considered. The relation found in Example 3.10 will apply only for neutral stratification. Otherwise temperature and gravity would have to enter as primary parameters.

In general, this process serves to reduce the number of parameters. In the special case when the number of original parameters is one greater than the number of basic units, the one ND dependent variable is then a function of nothing—a constant. In all cases the end product is a functional relation only. The explicit equation requires more knowledge, either from experiment or theory. One powerful source of additional knowledge is available when the governing equations are known. The application of dimensional analysis then takes a structure called similarity.

3.3 Similarity

Similarity is a powerful tool for both experimenters and theoreticians. It takes diverse forms, from flow around simple "look-alike" objects to the mathematical correlation between disparate fluid, mechanical, and electrical systems. Since the complete basic fluid-flow equations are generally too difficult to solve, various forms of similarity have been extremely valuable. As we have seen, experiments can be greatly simplified. Similarly, the equations that must be solved for a particular problem can be approximated in a systematic way.

3.3.1 Geometric Similarity

A geometrically similar model will have specified ratios of any length in the model to that in the actual configuration, often called the prototype. For perfect geometric similarity, all corresponding lengths will have the same ratio, called the scale factor. The model is like a three-dimensional picture of the object blown up or reduced by the scale factor.

In many cases, it is not possible to model all of the dimensions to the same scale factor. In other cases, one dimension may be effectively infinite. In these circumstances selected dimensions must be expanded or contracted, and a very distorted model can result. The model may better illustrate one aspect of its characteristics while losing its "actual" appearance. This is somewhat like presenting data on a log-linear plot, where the distortion of space on the log plot allows visualization of trends that could not be shown on a dual linear plot.

3.3.2 Flow or Kinematic Similarity

The sciences of hydraulics and aerodynamics have been largely based on the principles of similarity modeling. Generally, a geometrically similar model is placed in a laboratory flow field to simulate the actual object within the usually much larger flow regime. In this case it is necessary to have similarity in the flow regimes in addition to the geometric similarity. Two flow fields are said to be similar when the ratios between the velocities and the accelerations are constant.

Flow similarity is obtained when $U(x, t)$ and $a(x, t)$ of the model flow have constant ratios to the prototype values $U'(x', t')$ and $a'(x', t)$ at every point and all time. The flow model around a building is shown in Fig. 3.12.

When the fluids used in the model flow are identical to the prototype flow, then the flow streamlines around geometrically similar objects will look identical. A picture of one flow could be geometrically expanded to

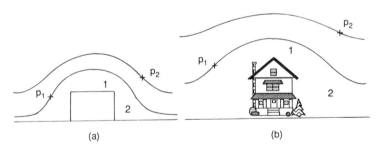

Figure 3.12 The air flow around (a) a house model in a wind tunnel and (b) in the atmosphere. The pressures and corresponding points are indicated in each (away from the poorly modeled surfaces).

identically overlay the other. In these cases, an actual large-scale flow field can be modeled in a laboratory. However, in many problems the need is to determine the forces on the objects. In this case, to model the inertial force on the small-scale object, we may consider substituting another fluid in the model flow. This generalization of the simple geometrically similar model leads us to a requirement that the ratio of the various forces in the model and prototype flow must have constant ratios. Since our basic equations of motion may be seen as a balance of forces, the resulting similarity criteria can be conveniently expressed in the equations. This leads to extremely valuable analysis procedures.

3.3.3 Dynamic Similarity

When the equations that govern the dynamics of a problem are known, substantial information about the solution for the flow is known. This can be extracted with the techniques of dimensional analysis without obtaining the analytic solution to the equations. This is because, once the appropriate equations have been derived, the intuitive step of the π theorem has been done. All of the important forces and parameters appear in the equations and the boundary conditions. In fact, availability of the equations plus the boundary conditions put certain constraints on how the parameters occur. For instance, only combinations of some parameters may appear, such as μ/ρ, or possibly only the gradients rather than the magnitudes of a parameter may be important for the problem. An inspection of the equations and the boundary conditions will reveal all pertinent parameters in the problem. These include dependent and independent variables, their boundary values, coefficients, and constants. The variables (dependent and independent) are the significant parameters, and the remaining parameters are to be used to nondimensionalize them. When a choice of the parameters with which to do the nondimensionalization are available, it is important to select one that is representative of the typical magnitude of the variable.

3.3.3.a Characteristic Values

When nondimensionalizing the variables in an equation, the constants chosen should have a *characteristic value* for each variable. This is a value, such as the value at a boundary, that is characteristic of the magnitude of the parameter in the domain. Then the nondimensional variables will have values near unity. When this is done for all of the variables in the equation, the nondimensional parameters that occur as coefficients in the equations will denote the relative importance of each term. This means that the selection of characteristic values must be done with care. For instance, in fluid-flow equations, pressure differences are more appropriate characteristic values than the total or mean pressures. They involve much smaller magnitudes. The smaller values are representative of the pressure gradients involved in the flow. Also, horizontal and vertical characteristic scales may be different. In the atmosphere, the variation of the characteristic scale in the vertical direction is typically very much less than that in the horizontal directions.

When we choose the nondimensionalizing parameters to be characteristic values of each variable in the problem, then the ND parameters are a ratio of each variable to its representative value and are each of order unity. When the nondimensional variables are substituted for the dimensional ones in the equations and boundary conditions, the characteristic scales will appear as coefficients of each variable. For instance, with $u = u'/U$, substituting for dimensional u', the contribution to the coefficient is U. Thus, each term will have a coefficient containing all of the characteristic values used in nondimensionalizing each variable in that term.

We can then divide through the equation by the coefficient of any selected term, leaving that term with no coefficient. Therefore that term is of order magnitude unity. This has the effect that all coefficients are now ND ratios. The magnitude of these ratios denote the importance of each term relative to unity. In fact, each dimensionless parameter is the ratio between the characteristic value of the term it multiplies and the one whose coefficient was chosen to divide through the equation. If we divided through by the coefficient of the pressure-gradient force term, the coefficient in front of the viscous term would denote the ratio of viscous to pressure-gradient forces. This is best illustrated in the following step-by-step development in an equation.

3.3.3.b Nondimensionalizing an Equation

We can nondimensionalize a "generic" equation and examine the possibilities of linearizing. Consider

$$d^2\varphi'/dz'^2 + A\varphi' d\varphi'/dz' + B\varphi' + C = 0 \qquad (3.10)$$

where the primes indicate dimensional parameters.

First we select characteristic values so that all nondimensional parameters are of order unity. Suppose that the problem involves investigating φ' in a layer of depth H, where $\varphi' = \Phi$ at $z' = H$. Using these boundary conditions for the nondimensionalization, φ and z will have the value unity, at least at and near the boundaries. Let

$$\varphi = \varphi'/\Phi, \quad z = z'/H$$

and

$$[\Phi/H^2]\, d^2\varphi/dz^2 + [A\Phi^2/H]\, \varphi\, d\varphi/dz + [B\Phi]\, \varphi + C = 0 \qquad (3.11)$$

The terms $d^2\varphi/dz^2$, $\varphi\, d\varphi/dz$, and φ each are of order unity. The characteristic values are in brackets and may have arbitrary magnitudes and dimensions. When we divide through the equation by the coefficient of the first term, we get

$$d^2\varphi/dz^2 + [AH\Phi]\, \varphi\, d\varphi/dz + [BH^2]\, \varphi + [CH^2/\Phi] = 0 \qquad (3.12)$$

Now, the variables are all dimensionless, and in particular, the first term is dimensionless. Thus, the coefficients in brackets are dimensionless. The parameter $AH\Phi$ is a ratio of the typical magnitude of the second term to that of the first term. The number BH^2 is the ratio of the size of φ to the first term, and CH^2/Φ evaluates the importance of the constant C. This equation is completely equivalent to Eq. (3.10). The only difference is that we have substituted three dimensionless parameters for A, B, and C. We will find that these nondimensional parameters are convenient for evaluating dynamic similarity. However, we can first look at their value in obtaining approximations of the complete equations.

If $[AH\Phi]$ in Eq. (3.12) is very small, the nonlinear term $\varphi\, d\varphi/dz$ can be dropped and the equation will be linearized.

$$d^2\varphi/dz^2 + [BH^2]\, \varphi + [CH^2/\Phi] = 0$$

If also $[BH^2]$ is small, then we have the simple equation

$$d^2\varphi/dz^2 = -[CH^2/\Phi] \approx \text{constant} \qquad (3.13)$$

If we continue in this vein and assume that $[CH^2/\Phi]$ is also very small, then we have

$$d^2\varphi/dz^2 = 0 \qquad (3.14)$$

This may or may not be a good approximation of Eq. (3.11). We chose the individual characteristic values to produce order unity for all of the nondimensional quantities, so the term in Eq. (3.14) should be of order unity. For it to equal zero seems to be a contradiction. Before using the solutions to this equation, we would have to check the characteristic values used for the nondimensionalization against those predicted in the solution. For instance, we used Φ to nondimensionalize φ—does the solution of Eq. (3.14) produce φ of order Φ? If they do not agree, then the process must be repeated using new characteristic values. In boundary layer problems, φ is often zero at $z = 0$, so that φ/Φ cannot be of order one at the surface. This is a fundamental difficulty for boundary layer problems, and constant vigilance is required to examine each solution.

Note that such careful attention to characteristic values is necessary only if we have dropped terms using scaling arguments, as in moving from Eq.

(3.11) to (3.14). If we merely used the nondimensionalization as an organizing tool, and proceeded to solve the complete Eq. (3.11), the solution would be valid regardless of characteristic values chosen. For instance, if they were poorly chosen, we might be plotting over a range of φ from 0.001 to 0.01 instead of 0 to 1.

Another interesting case occurs when $[AH\Phi]$ is very large.

$$[1/(AH\Phi)] \, d^2\varphi/dz^2 + \varphi \, d\varphi/dz + [BH/(A\Phi)] \, \varphi + C' = 0 \qquad (3.15)$$

We may choose to neglect the first term; however, incorrect solutions may be produced. This is because dropping this term produces a lower order equation. Thus fewer integrations are required to solve it, and fewer constants are produced. Therefore, the reduced equation cannot satisfy as many boundary conditions as the original. Nevertheless, the term is small in the domain characterized by H and Φ. The problem in satisfying the boundary conditions suggests that the chosen characteristic values of H and Φ are not valid in all regions. In particular they are poor values in the vicinity of the boundary condition that is not satisfied.

This quality of the approximate equations has particular importance in the boundary layer requirement to satisfy a zero velocity at a surface. If the velocity decrease takes place rapidly, as in a thin boundary layer, the characteristic height scale H must be small to satisfy this zero boundary condition. In this case $[1/(AH\Phi)]$ cannot be small, and the highest order term must be included in the governing equation. The neglect of the highest order term is called a *singular perturbation* to the equation. It is encountered when frictional forces are neglected. This is done in the derivation of the free-stream (inviscid) equations and in stability analyses of the governing equations.

We have seen that if a nondimensional coefficient multiplying a term in the equation is known to be extremely small, then that particular dimensional term is small with respect to the other terms. This term is a candidate to be neglected. This could be formally done in an asymptotic expansion with respect to the small parameter. In this way terms are dropped from the complete equations and more easily solvable equations may result. There is a danger in this process, however, that is associated with the characteristic values chosen. Because if the characteristic value is chosen incorrectly, the resulting equations may not be valid, as in a domain where the characteristic value chosen wasn't typical of the dependent variable.[2]

---

[2] See the theory of Kaplun, S. (1967). "Fluid Mechanics and Singular Perturbations," a collection of papers. Academic Press, New York; Van Dyke, M. (1964). "Perturbation Methods in Fluid Mechanics," Applied Mathematics and Mechanics series, Vol. 8, Academic Press, New York.)

Example 3.11

Consider the equation for one-dimensional flow along a streamline in the x-direction, which we can take at this point as given to be

$$p/\rho + gz + \tfrac{1}{2}u^2 = C = P_0/\rho \tag{3.16}$$

where the pressure $p = P_0$ at $z = 0$ and $u = 0$. Incompressible flow is assumed, with $P_0/\rho = 600 \text{ m}^2/\text{sec}^2$. The region of interest is

$$100 \le z \le 110 \text{ m} \quad \text{and} \quad 10 \le u \le U_{max} = 30 \text{ m/sec}$$

Nondimensionalize the equation and discuss it.

Solution

From the given equation and boundary conditions, choose P_0, $\Delta z \,(= 10 \text{ m})$, and $U_{max} \,(= 30 \text{ m/sec})$ for nondimensionalizing p, z, and u. We let $p' = p/P_0$, $z' = z/\Delta z$, and $u' = u/U_{max}$ and substitute (in this case, the primed variables are ND)

$$(P_0/\rho)p' + gHz' + \tfrac{1}{2}U_{max}^2 u'^2 = C = P_0/\rho \tag{3.17}$$

We are interested in the flow situation and therefore definitely want to retain the term involving the velocity. Divide through by the coefficient of the u'^2 term,

$$[2P_0/(\rho U_{max}^2)]\, p' + [2g\, \Delta z/U_{max}^2]\, z' + u'^2$$

$$= [2P_0/(\rho U_{max}^2)] \tag{3.18}$$

All nondimensional quantities are now of order unity. The relative magnitude of each term is contained in the bracketed term made up of characteristic values. When we insert the characteristic values into the dimensionless coefficients,

$$[2P_0/(\rho U_{max}^2)] = [2 \cdot 600 \text{ m}^2/\text{sec}^2/30^2 \text{ m}^2/\text{sec}^2] = 1.33$$

and

$$[2g\Delta z/U_{max}^2] = 2 \cdot 9 \cdot 8 \text{ m/sec}^2 \cdot 10 \text{ m}/30^2 \text{ m}^2/\text{sec}^2 = 0.2$$

Remembering that we are simply working with order of magnitude estimates, it appears that the second term is considerably smaller than the others. Therefore we can approximate Eq. (3.16) with

$$p/\rho + \tfrac{1}{2}u^2 = P_0/\rho \tag{3.19}$$

We will see later that Eq. (3.16) relates the pressure force energy, the potential energy of height, and the kinetic energy of the flow. Our scaling has indicated that for the conditions of this problem, the potential energy derived from the given height change is insignificant compared to the other energies.

When the dimensionless parameters for a given equation have been obtained, they provide valuable information about the solutions to the equation even without considering limits and approximations. For instance, all problems that have a combination of characteristic values that yield the same nondimensional parameters in the equations must be governed by the same equations. Then they have the same solutions. They are called *dynamically similar*.

3.3.4   Dynamic Similarity in the Equations of Fluid Dynamics

When the ND process is carried out in the equations for fluid flow, the equations remain unchanged except that new coefficients of the terms appear. These are combinations of the characteristic parameters used in the nondimensionalization. We nondimensionalize with V, L, ρ, μ, and with temperature T if thermodynamics is important. For instance, when u is nondimensionalized with its characteristic value V, then $u_{nd} = u/V$. When substitution is made for $u = Vu_{nd}$, the equation then contains the nondimensional velocity and the coefficient V. Some examples of the nondimensional coefficients that will occur in the fluid dynamics equations are

Reynolds number, Re $\quad = \rho VL/\mu$
Richardson number, Ri $\quad = (g/T)[dT/dz/(dV/dz)^2]$
Rossby number, Ro $\quad = V/(fL)$
Mach number, M $\quad = V/c_0$ (c_0 = speed of sound)
Drag coefficient, C_D $\quad = \tau/(\rho V^2)$

When these numbers appear in the force balance equations, each ND coefficient is a ratio of force magnitudes. This is a consequence of the nondimensionalization with typical values followed by division by one coefficient.

3.4   Some Similarity Concepts

Similarity exists between two quantities when they can be related by a constant. *Dynamic similarity* exists for two processes when they can be described by equations that can be cast into the same form by some transformation. The solutions to the equations then apply to both processes. The process of stretching one coordinate by nondimensionalizing it with another parametric scale in place of a characteristic value is called an *affine transformation*.

The transformation from height to pressure coordinates is common in atmospheric problems. This is a simple geometrical transformation. In general, one or all coordinates could be stretched, leading to a very distorted model for the actual flow. In other cases, one can find a transformation that takes a distorted real field and relates it to a simple flow in another frame of reference that has a theoretical solution. The transformed solution will then apply to the distorted field.

If an equation contains no coefficients, *then self-similarity* exists in the solutions. This means that the solutions are not dependent on any scale (length, velocity, temperature, etc.), and the one solution of the equations is valid for all flows. The scale of the solution is simply expanded or contracted depending on the characteristic parameters. For instance, if there doesn't exist a characteristic length in the problem, then there can be no dependence of the flow on the length scale. A flow moving around a corner is an example of such a flow. The picture in Fig. 3.13 would be the same regardless of the scale, since there is no characteristic scale for r. [This concept is generalized in the theory of fractals.] Since the flow can be broken into its components, there are cases where the flow depends on only one or two coordinates. It is then self-similar with respect to the rest.

There is dynamic similarity in two sets of equations if the nondimensional coefficients of the same terms can be matched. That is, the individual characteristic values of all parameters needn't be the same, only the *ratios* that occur in the dimensionless parameters must be the same. For instance, there can be Reynolds number ($\rho UL/\mu$) similarity for widely different values of ρ, U, L, and μ. When the nondimensional ratios are the same, the governing

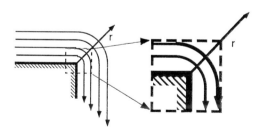

Figure 3.13 Flow around a corner with two looks at very different scales of radius. The different scale pictures will be identical since there is no characteristic scale at the corner.

equations are identical for the two problems and the solutions are geometrically similar.

Dynamic similarity requires that the forces have the same ratio. For instance, for the same Reynolds number, the ratio of the inertia to the viscous force must be the same. This can be true even if the characteristic lengths, densities, velocities and viscosities are quite different for each flow.

The result of a systematic nondimensionalization of the equations is that the dimensionless parameters which govern the problem are grouped into a smaller set called *similarity parameters*.

Example 3.12

Consider the equation

$$u'_{t'} = v u'_{z'z'}$$ (3.20)

where kinematic viscosity v is a constant, the prime denotes a dimensional variable, and subscripts denote partial derivatives. Nondimensionalize this equation with characteristic values L, V, and T, and discuss the results. Plot $u(z)$ at a given $t = T_0$.

Solution

Divide the dependent and independent variables in the equation by the characteristic scales L, V, and time T.

$$u = u'/V, \quad z = z'/L, \quad t = t'/T$$

and

$$(V/T)u_t = (vV/L^2)\, u_{zz}$$

or

$$u_t = [vT/L^2]\, u_{zz}$$ (3.21)

There is one nondimensional parameter in the brackets. It is a ratio of the viscous force to the inertial force. The experimental results could then be plotted versus the parameter $[vT/L^2]$ at $t = T_0$. (see Fig. 3.14).

If $[vT/L^2]$ is the same for two flow realizations, then the $u(z, t)$ are the same (similar).

Note that if we substitute $\zeta = z/(vT)^{1/2}$ in Eq. (3.21) (i.e., select $L = (vT)^{1/2}$), we get a basic equation without parameters.

$$u_t = u_{\zeta\zeta}$$

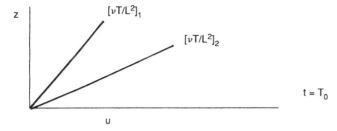

Figure 3.14 Data plots of velocity versus height for various parameters.

This solution is self-similar, and all solutions, for all scales, are given in one plot (Fig. 3.15).

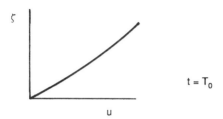

Figure 3.15 Data plots for self-similar conditions.

The application of similarity concepts has reduced the number of independent nondimensional variables from three ($\nu T/L^2$, z/L, t/T) to two ($z/(\nu T)^{1/2}$, t/T).

3.5 Summary

The concepts of scaling and similarity often seem nebulous and undisciplined when first encountered. However they provide great organization and systemization to both data presentation and analytic procedures. The presentation of data becomes particularly effective. Guidance for experimental planning can be optimized with dimensional analysis principles. Frustration is sometimes produced when trying to interpret data presented as one obscure ND variable versus another. But this liability is usually overcome by the advantages.

Evidently, dynamic similarity does more than simply effectively organize data presentation (although it does this very well). It provides a systematic way of evaluating the equations to help in obtaining valid approximate equations. And it allows us to investigate the possibilities of similarity solutions. In many circumstances, desired results can be obtained without ever needing the complete analytic solutions.

We have discussed only the concepts of scaling that are necessary to basic fluid dynamics analysis. Dimensional analysis can be put on a very formal level, as found in Ipsen (1960).[3] We will find the concepts essential to the develop ment of the stability equations and the boundary layer equations in later chapters.

METHODS OF ANALYSIS

Scaling
   Spectral Analysis
     Organized Waves
     Turbulence
Dimensional Analysis
   The π Theorem
Similarity
   Geometric
   Kinematic
   Dynamic
     Characteristic Values
     Nondimensional Equations
     Approximate Equations

Problems

1. Determine the dimensions of the following variables and combinations in terms of the length, mass, and time system of units: (a) $\rho U^2/2$; (b) $[\tau/\rho]^{1/2}$; (c) ζ (vorticity); (d) μ (viscosity); and (e) torque.

2. For very low velocities it is known that the drag force F_D of a small sphere falling through the air is a function of the velocity W, the diameter d, and the viscosity μ. Determine the dimensionless relationship for these parameters.

3. The velocity U of ripples on the ocean surface is a function of the

---

[3] Ipsen, E. C. "Units, Dimensions, and Dimensionless Numbers," McGraw-Hill, New York (1960).

wavelength λ, density ρ, and surface tension s. Derive an expression for U using dimensional analysis.

4. You wish to determine the viscosity of fluids, given the apparatus of a concentric cylinder viscometer, as shown in problem 12 in Chapter 1. Assume that μ (viscosity) $= f\{T(\text{torque}), R, \Omega, \text{ and } t/H\}$. Using dimensional analysis, obtain an expression for μ.

5. A barrage balloon 5 m in diameter and 30 m long is being studied as a possible suspension system for meteorology instruments in 20-m/sec winds. If the drag characteristics are to be studied in a wind tunnel with a 1/10-size model,

    (a) what windspeed must be used in the tunnel?
    (b) If the result in (a) is too high or too low, suggest alternatives.

6. The energy source driving the atmospheric and oceanic flows is ultimately the solar radiation. Approximately 30% of this radiation (including reflection) is absorbed and dispersed in the lower 20-km air layer. The remainder is incident on the surface and absorbed in the soil or water. This energy is then transported vertically. Measurements of the temperature fluxuations related to the *diurnal* forcing indicate active layers of temperature variation occur in about *2 km* of the atmosphere, *20 m* of the ocean, and *5 mm* of the soil.

    (a) Use dimensional analysis to get an approximate value for the thermal diffusivity K [m²/sec] for each medium.
    (b) Compare these with molecular diffusivities for air and water.
    (c) A similar analysis can be done for the planetary boundary layer wind or water regions. Here the Coriolis parameter f enters, and the observed characteristic heights are 500 m in the atmosphere and 50 m in the ocean. Calculate the eddy viscosity K for water and air.

7. A model of a high-rise building at 1:250 scale was tested in a wind tunnel to estimate the pressures and forces on the full-scale structure. The wind-tunnel airspeed was 20 m/sec at 20°C. The extreme values of the pressure coefficient on the windward wall, the side wall, and the leeward wall are measured at 1.0, -2.5 and -0.9, respectively. The full-scale structure is exposed to 120-mph winds at 20°C. What are the corresponding full-scale pressures? The lateral force (wind-induced force normal to the wind direction) was measured at 20 N in the model. The building can withstand a maximum force of $5 \cdot 10^6$ N. Will it collapse? $C_p = \Delta p/(\frac{1}{2}\rho U^2)$, $\Delta p = p - p_0$ where p_0 is gage pressure $= 0$.

8. Consider the pendulum problem. Use dimensional analysis to get the

period. (*Hint:* Possible parameters are length L, density ρ, gravity g, distance to center of earth R.)

9. Consider the design of a small submarine for oceanographic research, and use dimensional analysis to discover a relation for the drag D of the sub (to determine how much power it would need, how long energy would last). Discuss the size of a test model required and how much data is required to evaluate the design.

10. Assume the wind stress on the surface of the ocean is a function of the air density and the windspeed only. Use dimensional analysis to derive a relation.

11. Nondimensionalize the following equation, obtaining a self-similar equation.

$$fu + K \, d^2u/dz^2 = 0$$

where K is kinematic eddy viscosity, f is the Coriolis parameter, $u(0) = 0$, and $u(\infty) = U_G$.

12. A design for a bridge across a river will be tested in a wind tunnel. The size of the tunnel allows a $1/10$ scale model. Since the last bridge blew away in 100-mph winds, we must test for these.

(a) What will be the tunnel velocity for air at the same pressure and temperature?

(b) What is a potential problem with this project (related to compressibility)? Suggest ways to avoid this problem.

13. The vertical pressure gradient $(\delta P/\delta z)$ in the atmosphere depends on density and gravity. Use dimensional analysis procedures to obtain the relation.

14. Given that the phase velocity $u = f\{g, \lambda, v\}$, use nondimensionalization to get an expression relating u as a function of v (by inspection).

Chapter 4 | Tensors and Relative Motion

4.1 Tensors

The use of tensors has greatly facilitated work in many fields. These include the study of the space–time continuum, general physical concepts of proper frame of reference, and operational calculus. It is a tool that is a little abstract at first look (and at the last look in its most extensive form). However, in the relatively simple applications that we use, the benefits in compactness and elegance are sufficient to justify the work of learning some techniques. Although tensors are a general concept with applications up to infinite dimensions, we will deal with only a specific small subset of tensor theory.

    Our motivation for studying tensor notation is the need to express and manipulate the array of velocity gradients across our parcel. First, we wish to extend the one-dimensional Newton stress-rate-of-strain relation, τ (a scalar) $= \mu \, du/dz$, to a three-dimensional version. Here, τ has many components that are proportional to $\partial u/\partial x$, $\partial u/\partial y$, $\partial u/\partial z$, $\partial v/\partial x$, $\partial v/\partial y$, $\partial v/\partial z$, $\partial w/\partial x$, $\partial w/\partial y$ and $\partial w/\partial z$. Second, in our expression for the total derivative we have the change in velocity, a vector, that varies with respect to the change in position, also a vector. That is, each of the three components of

u change in each of the three directions of **x**, which takes a string of nine scalars to describe. When we group these nine components into an ordered set it is often possible, and very convenient, to treat them as one entity.

Let us call the ordered set of numbers that describes some thing a *tensor*. A tensor is a generalization of the familiar concepts of scalars and vectors. It includes these as special classes of tensor. A *zero order tensor* is a *scalar*—simply a number, such as temperature. A *first order tensor* is an *n*-D ordered array of scalars. We consider only those describing 3-D space, the Cartesian tensors. These three-dimensional arrays of three scalars are called *vectors* and are frequently used in dynamics. They can denote displacement, velocity, acceleration, forces, temperature gradient—anything that has a 3-D perspective. Time is a separate independent variable that we will treat individually.

In the spirit of proceeding from the simplest to the complex, consider the two-dimensional city map. Time and the third dimension, height, are constant. The map may change with time, but (in most cases) we can assume this is very slow. Location is represented by $P(i, j)$, where i is an index that designates a number on the horizontal axis and j, on the vertical. We could write this as P_{ij}. P can be considered as a two-dimensional vector from some origin where i and $j = 0$ to any spot on the map with specific values of i and j. Sometimes, for clarity, i ranges over the numbers while j uses the alphabet, as in Fig. 4.1.

In our three-dimensional plus time world at any given time, the location, velocity, and various forces can each be defined by three numbers. These denote their respective component magnitudes in the three orthogonal spatial directions. Any given point in space can be represented by several different three-number combinations depending on the reference system used. The common ones are Cartesian (x, y, z); cylindrical (r, θ, z); and spherical (r, θ, φ). (See Fig. 4.2.)

Finally, one can describe location in *n*-dimensional space, motion, and

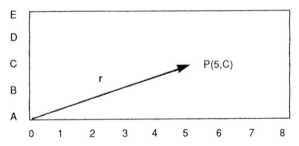

Figure 4.1 A two-dimensional map with locating numbers and letters and vector **r** locating P with respect to the origin.

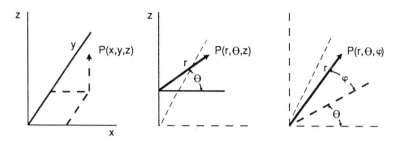

Figure 4.2   The point P with respect to origin O in (a) Cartesian, (b) cylindrical, and (c) spherical coordinates.

even color and texture, by using groups of numbers that perform together under given rules. For instance, a number that gives the number of dots per square centimeter in a picture will correspond to the resulting shade of gray. If one set of numbers represents the density of red dots, another of blue, and another yellow, then the distribution of the numbers and the relative values will produce color pictures. The quantity of numbers that are required depends on the complexity of the problem and the success one has in grouping the numbers. The use of number–picture relations is a very contemporary topic. It is important to numerical modeling techniques. It is used widely in data transmission for satellite and other remote sensing measurements.

The *second order tensor* will be used to represent the interaction between two vectors. For instance, a vector force can act on a surface where both the vector and the surface can have arbitrary orientations. Two examples are shown in Fig. 4.3, where the orientation of the area dA is given by a unit normal vector **n**.

The force on the surface dA needs to be expressed in terms of the normal component and the two components parallel to the sides of dA, since the effects on dA, compression and tangential stress, will depend on these components. In turn, these components will depend on the three components of **F** with respect to the three components of **n**. Since each of the three components of **F** must be decomposed to each of the three directions of **n**, there will be nine terms in all. The value of these nine terms for all possible orientations of **F** and **n** can be expressed as a second-order tensor. The tensor can be expressed in terms of the two vectors under an appropriate rule. One of our tasks in this chapter is to establish these rules.

In this text we will confine our attention to the 3-D variation of the 3-D vectors, involving "only" a nine-component second-order tensor. We will also employ only the simplest third order tensor, an identity-type operator (wherein 21 of the 27 terms are zero).

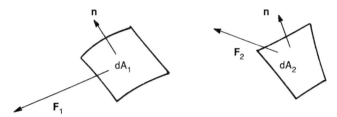

Figure 4.3 Examples of a force **F**, of area dA with unit normal **n**.

4.1.1   Tensor Notation

4.1.1.a   Symbolic Notation

Symbolic notation is the mathematical language represented in this text by boldface notation. In this language, scalars, vectors, and tensors are different entities. New definitions of sums and products are introduced. For instance, the sum of two scalars is just another number, $a + b = c$. However, we know that the vector sum, $\mathbf{a} + \mathbf{b}$, defines a specific ordered addition of the three components that make up each vector, $\mathbf{a} + \mathbf{b} = (a_1 + b_1, a_2 + b_2, a_3 + b_3)$. Thus the plus sign has a different meaning for vectors than it does in ordinary algebra. There are also different ways to define products.

Previously we introduced the symbol del (∇), which behaves like a vector in the multiplication rules. We use the descriptive terms gradient, divergence, and curl, or *grad*, *div*, and *curl*, to describe alternate derivative schemes. For instance, the gradient of a scalar is the sum of the partial derivatives in the three coordinate directions. In Cartesian coordinates,

$$\nabla\varphi \equiv \text{grad } \varphi \equiv \partial\varphi/\partial x + \partial\varphi/\partial y + \partial\varphi/\partial z$$

Recall that the divergence of a vector, div **a**, is the dot product and produces a scalar,

$$\text{div } \mathbf{a} \equiv \partial a_1/\partial x + \partial a_2/\partial y + \partial a_3/\partial z$$

The curl of a vector is the cross product, which produces a vector,

$$\text{curl } \mathbf{b} \equiv \nabla \times \mathbf{b} \equiv [\partial b_3/\partial y - \partial b_2/\partial z, \partial b_1/\partial z - \partial b_3/\partial x, \partial b_2/\partial x - \partial b_1/\partial y]$$

We will find that these operations have general definitions that apply to tensors of all orders.

In addition to this descriptive nomenclature there is great economy and mathematical purity in symbolic notation. The manipulation of large arrays of numbers can be represented in short statements. However, the practical application to geophysical problems inevitably requires calculation of the sums and products of the individual components. In contrast, the rules of index notation are advantageous in keeping track of the components. This method still gains much of the economy in notation of symbolic notation. We must learn both methods, symbolic and index, as both are valuable and in wide use in atmospheric science.

4.1.1.b Index Notation

Index notation has already been introduced as an alternative way of representing vectors. In our 3-D space, x_i exists. The index $i = 1, 2, 3$, and x_i represents the vector $\mathbf{x} = (x_1, x_2, x_3)$ in a chosen reference coordinate system. There will be occasions where we wish to refer to a specific component, say the ith component of $\mathbf{x}$. We will then have to use parentheses around the index to indicate the scalar $x_{(i)}$ to avoid confusion with the vector x_i. For instance, $x_{(i)}$ names only one of the scalars, x_1, x_2, or x_3, whereas x_i is the vector (x_1, x_2, x_3).

A scalar such as temperature will have a given value at a point in space regardless of the coordinate system used. A vector (or a higher-order tensor) will be made up of different components depending on the coordinate system. In fact, the definitions and functions of vectors and tensors are closely related to coordinate transformations. Let $\mathbf{x}$ $(x_1, x_2, x_3) \equiv x_i$, where $i = 1, 2, 3$ for the 3-D Cartesian space in which we deal. Consider another frame of reference with $\mathbf{x}' = (x_1', x_2', x_3')$ locating the same point. If the origin is the same and the new system is simply rotated about the origin, the length of the vector is unchanged.

$$x^2 = x_1^2 + x_2^2 + x_3^2 = x'^2_1 + x'^2_2 + x'^2_3 \tag{4.1}$$

The new coordinates can be written in terms of the old according to the following rules for rotation:

$$x_1' = x_1 e_{11} + x_2 e_{12} + x_3 e_{13}$$

$$x_2' = x_1 e_{21} + x_2 e_{22} + x_3 e_{23}$$

$$x_3' = x_1 e_{31} + x_2 e_{32} + x_3 e_{33}$$

where the e_{ij} can be any orthogonal coordinate set. For a transformation of coordinates, these conversion factors are the normalized direction cosines, $e_{ij} = \cos\{x_i', x_j\}$. Since $\cos\{x_i', x_j\} = \cos\{x_j, x_i'\}$, we have $e_{ij} = e_{ji}$.

It is valuable to use Einstein's summation convention: when an index is repeated *in a single term,* the term represents a sum over the repeated index. In other words, a short cut in notation is available by omitting the summation sign whenever there is repetition of an index within a single term. Thus,

$$x_i' = \sum_{j=1}^{3} x_j e_{ij} = x_j e_{ij} = x_1 e_{i1} + x_2 e_{i2} + x_3 e_{i3}$$

When the index appears in each term, it is called a *free* index (in contrast to a *repeated* index),

$$x'_i = 6x_i + 2x_i^2 - C_i$$

This represents three equations, one for each value of i,

$$x'_1 = 6x_1 + 2x_1^2 - C_1$$

$$x'_2 = 6x_2 + 2x_2^2 - C_2$$

$$x'_3 = 6x_3 + 2x_3^2 - C_3$$

Therefore, $x_j e_{ij}$ is a set of three equations denoted by the *free* index, $i = 1, 2, 3$. The free index occurs only once in each term of an equation and implies three equations as it takes on values 1, 2, and 3. Each of these equations will have three terms summed, denoted by the *repeated* index j, as it also takes on values 1, 2, and 3. The e_{ij} term represents nine scalars and is thus a second-order tensor. So this rule is our first example of a tensor vector product, $x_j e_{ij}$. We have also noted a dimensional homogeneity rule applied to vector equations: all terms must be of the same order tensor. When using index notation, this translates to the requirement that every term have the same free indices.

Example 4.1

Write out the terms, or equation, represented by the following, where e_{ij} are the normalized direction cosines: (a) $e_{ij} e_{ik}$, for arbitrary j and k; (b) the $i = j = 1$ case in (a); (c) the $i = 1, j = 2$ case of (a).

Solution

(a) We must sum over the i, whereas the j and k are free indices,

$$e_{ij} e_{ik} = \cos(x'_1, x_j)\cos(x'_1, x_k)$$

$$+ \cos(x'_2, x_j)\cos(x'_2, x_k)$$

$$+ \cos(x'_3, x_j)\cos(x'_3, x_k)$$

We could obtain the nine individual equations by first setting $i = 1, j = 1$; then $i = 1, j = 2$; $i = 1, j = 3$; $i = 2, j = 1$; $i = 2, j = 2$, and so on to obtain all possible combinations of i and j ranging over 1, 2, and 3.

(b) Since i and j are specified, this case is a single equation.

$$e_{i1}e_{i1} = \cos^2(x_1', x_1) + \cos^2(x_2', x_1)$$
$$+ \cos^2(x_3', x_1)$$

An inspection of the value of the $\cos(x_i', x_1)$ would reveal that this equation has a value of 1. This is also true for the cases $i = j = 2$ and $i = j = 3$.
(c)

$$e_{i1}e_{i2} = \cos(x_1', x_2)\cos(x_1', x_1)$$
$$+ \cos(x_2', x_2)\cos(x_2', x_1)$$
$$+ \cos(x_3', x_2)\cos(x_3', x_1)$$

The value of this equation is zero. This is true also for the cases where $i = 1, j = 3$; $i = 2, j = 1$; $i = 2, j = 3$; $i = 3, j = 1$; and $i = 3, j = 2$.

This example shows the property of the orthogonal Cartesian coordinate normalized cosines that the product of $e_{ij}e_{ik}$ is 1 whenever $i = j$, and 0 whenever $i \neq j$.

Index notation is in wide use in fluid dynamics. We will use it preferentially in our derivations.

4.1.1.c The Tensor Definition

Since we will find it easiest to define the second-order tensor rules in analogy to the first-order tensor rules, we must recall some basic definitions for vectors. First, a general definition of our three-dimensional first-order tensor (which is called a vector) is: a three-component object **a** under the transportation (rotation) rule

$$a_i' = e_{ij}a_j \tag{4.2}$$

Thus, the same *vector* is defined as an *array of three scalars, with index j = 1, 2, 3, which transform to another array of three scalars, with index i = 1, 2, 3, under the rule given in Eq. (4.2).*

In a similar manner, we define the second-order, three-dimensional tensor (the principal kind we deal with, henceforth simply called a *tensor*) as a nine-component object *under the transportation rule*

$$C_{ij}' = C_{k\ell}e_{ik}e_{j\ell} \tag{4.3}$$

The second-order tensor can be written in *matrix* form where

$$C_{\text{row, column}} = \begin{pmatrix} C_{11} & C_{12} & C_{13} \\ C_{21} & C_{22} & C_{23} \\ C_{31} & C_{32} & C_{33} \end{pmatrix} \tag{4.4}$$

In the term on the right side of Eq. (4.3), a double sum on k and ℓ are both repeated, leaving as "free" indices i and j. Thus, the right-hand side of Eq. (4.3) is a second-order tensor, as is the left side. The tensor with indices k and ℓ is transformed into a tensor with indices i and j. Here, we are summing over k and ℓ such that for each k there is a sum on $\ell = 1, 2, 3$, with an overall sum on $k = 1, 2, 3$, a total of nine terms. There are nine equations as i and j range from 1 to 3 independently. Each equation yields a component of the tensor. The components of the tensor are displayed as a matrix, with i designating the row and j, the column.

Example 4.2

Write out the terms for A_{13} in the expression

$$A_{ij} = B_{k\ell}e_{ik}e_{j\ell}$$

Solution

There is a double sum on the right side, over j and ℓ. Hence we simply replace i and j with 1 and 3, then sum over j and ℓ.

$$A_{13} = B_{11}e_{11}e_{31} + B_{12}e_{11}e_{32} + B_{13}e_{11}e_{33}$$

$$+ B_{21}e_{12}e_{31} + B_{22}e_{12}e_{32} + B_{23}e_{12}e_{33}$$

$$+ B_{31}e_{13}e_{31} + B_{32}e_{12}e_{32} + B_{33}e_{13}e_{33}$$

There are nine terms. This scalar is one of the nine terms making up the tensor A_{ij}, which occurs in the first row and the third column of the matrix display.

Example 4.3

The equations

$$u = ax + by + cz$$

$$v = dx + ey + fz$$

$$w = gx + hy + kz$$

represent a linear transformation from (x, y, z) space to (u, v, w) space. Write the equations in matrix form.

Solution

We can consider the three components $\mathbf{u} = (u, v, w)$ as a vector, which is a function of another vector, $\mathbf{x} = (x, y, z)$. If we let $\mathbf{u}$ and $\mathbf{x}$ be column vectors, we can arrange a matrix to pre-multiply $\mathbf{x}$ to get $\mathbf{u}$.

$$\begin{pmatrix} u \\ v \\ w \end{pmatrix} = \begin{pmatrix} a & b & c \\ d & e & f \\ g & h & k \end{pmatrix} \begin{pmatrix} x \\ y \\ z \end{pmatrix}$$

This could be written $\mathbf{u} = \mathbf{A}\mathbf{x}$ or $u_i = A_{ik}x_k$.

Since each increase in the rank of a tensor increases the number of terms by a factor of three, we see that an n order tensor has 3^n components. A vector can be written in either row or column matrix format. The three-dimensional matrix of the tensor can be considered to be made up of three-column or three-row vectors. This relationship between matrices and vectors is important in the process of matrix multiplication. We are already familiar with two vector products, the dot product,

$$\mathbf{a} \cdot \mathbf{b} = a_1 b_1 + a_2 b_2 + a_3 b_3 = a_i b_i$$

and the cross product,

$$\mathbf{a} \times \mathbf{b} = (a_2 b_3 - a_3 b_2, \, a_3 b_1 - a_1 b_3, \, a_1 b_2 - a_2 b_1)$$

There is also a vector-vector = tensor product, which will be defined in terms of index notation below.

The rule for multiplying two matrices, $\mathbf{A}$ and $\mathbf{B}$ to get a matrix $\mathbf{C}$ is

$$C_{ij} = A_{ik}B_{kj}$$

where

$$i = 1, 2, \ldots I_n, \quad j = 1, 2, \ldots J_n, \quad \text{and} \quad k = 1, 2, \ldots K_n$$

We can consider $\mathbf{A}$ as made up of i row vectors with k elements and $\mathbf{B}$ made up of j column vectors with k elements. In general i and j can be arbitrary, while k must be the same for both matrices. Thus, the number of elements in each row of $\mathbf{A}$ must equal the number of elements in each column of $\mathbf{B}$. In a square matrix, $I_n = J_n = K_n = n$, the number of elements in each row or column. Note that each element of C_{ij} is a scalar formed as the scalar product of the ith row of $\mathbf{A}$ with the jth column of $\mathbf{B}$. For instance, in a square matrix of dimension three, the scalar denoted by C_{23} occurs in the second row and the third column of Eq. (4.8). If $\mathbf{C} = \mathbf{AB}$, then the second row of $\mathbf{A}$ is a three-element vector that forms a dot product with the vector occupying the third column of $\mathbf{B}$ to yield

$$C_{23} = A_{2k} \cdot B_{k3}$$

This illustrates that when one index of a tensor is specified, a three-element unit remains, representing a vector part of the tensor.

Example 4.4

Calculate the tensor product.

$$\begin{pmatrix} 2 & 4 & 1 \\ 3 & 5 & 2 \\ 6 & 1 & 4 \end{pmatrix} \begin{pmatrix} 7 & 4 & 2 \\ 3 & 1 & 5 \\ 2 & 6 & 9 \end{pmatrix} \equiv \mathbf{AB}$$

Solution

We will create a tensor, C_{ij}.

$$\begin{pmatrix} C_{11} & C_{12} & C_{13} \\ C_{21} & C_{22} & C_{23} \\ C_{31} & C_{32} & C_{33} \end{pmatrix} = \begin{pmatrix} 2 & 4 & 1 \\ 3 & 5 & 2 \\ 6 & 1 & 4 \end{pmatrix} \begin{pmatrix} 7 & 4 & 2 \\ 3 & 1 & 5 \\ 2 & 6 & 9 \end{pmatrix}$$

The first element, C_{11}, is the scalar product of the first row vector of A with the first column vector of B,

$$C_{11} = (2, 4, 1) \cdot (7, 3, 2) = 14 + 12 + 2 = 28$$

In similar calculations,

$$C_{12} = (2, 4, 1) \cdot (4, 1, 6) = 18$$

$$C_{13} = (2, 4, 1) \cdot (2, 5, 9) = 33$$

$$C_{21} = (3, 5, 2) \cdot (7, 3, 2) = 39$$

$$C_{22} = (3, 5, 2) \cdot (4, 1, 6) = 29$$

$$C_{23} = (3, 5, 2) \cdot (2, 5, 9) = 47$$

$$C_{31} = (6, 1, 4) \cdot (7, 3, 2) = 53$$

$$C_{32} = (6, 1, 4) \cdot (4, 1, 6) = 49$$

$$C_{33} = (6, 1, 4) \cdot (2, 5, 9) = 53$$

It is evident here that the number of elements in the rows of A must equal the number of elements in the columns of B, or the dot products would not be defined.

Example 4.5

Consider the specific cases of the general rule for multiplying two matrices, where (a) $I_n = 1$, $J_n = K_n = 3$; (b) $J_n = 1$, $I_n = K_n = 3$; (c) $K_n = 1$, $I_n = J_n = 3$. Give specific examples with arbitrary numbers for the components in (a) and (b).

Solution

(a) In this case, $i = 1$ and j and $k = 1, 2,$ or 3.

$$C_{1j} = A_{1k}B_{kj}$$

Since C_{1j} has only three elements, C_{11}, C_{12}, and C_{13}, it is a vector. Another clue indicating that it is a vector is that C_{1j} also contains only one free index. Likewise for A_{1k}. However, B_{kj}, which has two free indices, is a tensor. Thus, we have a vector that is a product of a vector-tensor. For instance,

$$(C_{11}C_{12}C_{13}) = (3\ 5\ 2)\begin{pmatrix} 7 & 4 & 2 \\ 3 & 1 & 5 \\ 2 & 6 & 9 \end{pmatrix}$$
$$= (3\cdot 7 + 5\cdot 3 + 2\cdot 2,\ 3\cdot 4 + 5\cdot 1 + 2\cdot 6,\ 3\cdot 2 + 5\cdot 5 + 2\cdot 9)$$
$$= (40,\ 29,\ 49).$$

(b) Here, $C_{i1} = A_{ik}B_{k1}$. For instance,

$$\begin{pmatrix} C_{11} \\ C_{21} \\ C_{31} \end{pmatrix} = \begin{pmatrix} 2 & 4 & 1 \\ 3 & 5 & 2 \\ 6 & 1 & 4 \end{pmatrix}\begin{pmatrix} 2 \\ 3 \\ 6 \end{pmatrix} = \begin{pmatrix} 2\cdot 2 + 4\cdot 3 + 1\cdot 6 \\ 3\cdot 2 + 5\cdot 3 + 2\cdot 6 \\ 6\cdot 2 + 1\cdot 3 + 4\cdot 6 \end{pmatrix} = \begin{pmatrix} 22 \\ 33 \\ 39 \end{pmatrix}$$

(c) In this case, $C_{ij} = A_{i1}B_{1j}$. We have a tensor which is a product of two vectors.

$$\begin{pmatrix} C_{11} & C_{12} & C_{13} \\ C_{21} & C_{22} & C_{23} \\ C_{31} & C_{32} & C_{33} \end{pmatrix} = \begin{pmatrix} A_{11} \\ A_{21} \\ A_{31} \end{pmatrix}(B_{11},\ B_{12},\ B_{13})$$
$$= \begin{pmatrix} A_{11}B_{11} & A_{11}B_{12} & A_{11}B_{13} \\ A_{21}B_{11} & A_{21}B_{12} & A_{21}B_{13} \\ A_{31}B_{11} & A_{31}B_{12} & A_{31}B_{13} \end{pmatrix}$$

In all of these cases, it is clear that the index notation serves as a guide to the form of the vector or matrix. Each of the operations illustrated will appear in our derivations.

A few specialized tensors are convenient to use. One second-order unit tensor is called the *Kronecker delta,* defined as

$$\delta_{ij} \equiv \begin{cases} 1 & \text{if } i = j \\ 0 & \text{if } i \neq j \end{cases}$$

This can be written in matrix form.

$$\delta_{ij} = \begin{bmatrix} 1 & 0 & 0 & 0 & \cdots \\ 0 & 1 & 0 & 0 & \cdots \\ 0 & 0 & 1 & 0 & \cdots \\ \cdots & \cdots & \cdots & \cdots & \cdots \\ 0 & 0 & 0 & 0 & 1 \end{bmatrix}$$

We have seen in Example 4.1 that $e_{ij}e_{ik} \equiv \delta_{jk}$.

We also use one third-order tensor. This is the *alternating unit tensor* and contains 27 terms, of which only six are nonzero. It is used in index notation to indicate cross-multiplication. The terms are

$$\epsilon_{ijk} \begin{cases} = & 1 & \text{if} & ijk = 123,\ 231,\ 312 & \text{(even permutation)} \\ = & -1 & \text{if} & ijk = 321,\ 213,\ 132 & \text{(odd permutation)} \\ = & 0 & \text{if} & ijk & \text{has any two indices the same.} \end{cases}$$

When the tensor ϵ_{ijk} multiplies two vectors, it is equivalent to that of the cross product of those two vectors in symbolic notation. Thus,

$$\epsilon_{ijk}a_jb_k = \mathbf{a} \times \mathbf{b}$$

Note that the free index i denotes the product vector, the second index j, sums on the first vector, and the third index k, sums on the second vector. Reversing the order of j and k in either a_jb_k or ϵ_{ijk} would change the sign of the product, just as does

$$\mathbf{b} \times \mathbf{a} = -\mathbf{a} \times \mathbf{b}$$

Example 4.6

Write the following components, A_{12} and A_{33}, in the expression:

$$A_{ij} = u_iB_{kj}v_k + w_k\epsilon_{ijk} + a\delta_{ij}$$

Solution

$$A_{12} = u_1(B_{12}v_1 + B_{22}v_2 + B_{32}v_3) + w_3(\epsilon_{123}) + a\delta_{12}$$

$$= u_1(B_{12}v_1 + B_{22}v_2 + B_{32}v_3) + w_3$$

$$A_{33} = u_3(B_{13}v_1 + B_{23}v_2 + B_{33}v_3) + w_k(\epsilon_{33k}) + a\delta_{33}$$

$$= u_3(B_{13}v_1 + B_{23}v_2 + B_{33}v_3) + a$$

Note that there are two summed terms on the right side, although most of the terms involving the alternating unit tensor are zero.

Our final vector product, which we encounter in deriving the equations, is called the *dyadic* product, written in symbolic and index notation as

$$\mathbf{A} = \mathbf{a} ; \mathbf{b}, \quad \text{and} \quad A_{ij} = a_i b_j$$

The primary motivation toward adopting index notation in atmospheric dynamics is the need to express the variable force (a vector) on a variable unit area (a vector). However, it is also very handy for manipulating vectors in general.

4.1.1.d   Some Basic Definitions

A tensor **A** may be written in matrix form.

$$\mathbf{A} = A_{ij} = \begin{pmatrix} A_{11} & A_{12} & A_{13} \\ A_{21} & A_{22} & A_{23} \\ A_{31} & A_{32} & A_{33} \end{pmatrix}$$

The transpose of **A** is **A***, where $\mathbf{A}^* = A_{ij}$.

$$\mathbf{A} = A_{ji} = \begin{pmatrix} A_{11} & A_{21} & A_{31} \\ A_{12} & A_{22} & A_{32} \\ A_{13} & A_{23} & A_{33} \end{pmatrix}$$

A is *symmetric* if $A_{ij} = A_{ji}$ in a square matrix; and *antisymmetric* (or *skew symmetric*) if $A_{ij} = -A_{ji}$.

This distinction becomes significant due to the special characteristics of these tensors. For a symmetric tensor three of the elements are repeated: $A_{12} = A_{21}$, $A_{13} = A_{31}$, and $A_{23} = A_{32}$. Therefore, in our 3 × 3 matrix there are only six independent scalars: A_{12}, A_{13}, A_{23}, A_{11}, A_{22}, and A_{33}.

The antisymmetric tensor has $A_{11} = A_{22} = A_{33} = 0$, and the three independent scalars A_{12}, A_{13}, and A_{23} occur with both plus and minus signs.

These three independent scalars can constitute a unique vector called the *dual* vector. This dual vector is defined by an inner product. It is the product of a second- and a third-order tensor, which are doubly summed over two of their indices (leaving the third index on the third-order tensor as a free index).

$$d_i \equiv \epsilon_{ijk} A_{jk}$$

These particular tensors are important because of their symmetry properties, which give them a smaller number of unknowns. In the case of an antisymmetric tensor, we can deal with the associated vector instead of the tensor. In addition, there is the important fact that *any tensor can be separated into a symmetric plus an antisymmetric tensor* according to

$$C_{ij} = \tfrac{1}{2}(C_{ij} + C_{ji}) + \tfrac{1}{2}(C_{ij} - C_{ji}) \equiv S_{k\ell} + A_{mn}$$

Other symbolic terms we will use include an *expansion* process (here, φ is a scalar):

$$\text{grad } \varphi = \nabla \varphi = \partial \varphi / \partial x_i \qquad \text{a vector}$$

$$\text{grad } \mathbf{u} = \nabla \mathbf{u} = \partial u_i / \partial x_j \qquad \text{a tensor (second order)}$$

$$\text{grad } \mathbf{A} = \nabla \mathbf{A} = \partial A_{ij} / \partial x_k \qquad \text{a third order tensor}$$

The *contraction* process (a dot product) is

$$\text{div } \mathbf{u} = \nabla \cdot \mathbf{u} = \partial u_j / \partial x_j \qquad \text{a scalar}$$

$$\text{div } \mathbf{A} = \nabla \cdot \mathbf{A} = \partial A_{ij} / \partial x_i \qquad \text{a vector}^1$$

A *selection* process (a cross product) is defined by

$$\text{curl } \mathbf{u} = \nabla \times \mathbf{u} = \epsilon_{ijk} \partial u_k / \partial x_j \qquad \text{a vector}$$

It is clear that when second-order tensors are considered, many new possibilities arise for product definitions. Fortunately, we need only a select few in fluid dynamics. In the derivation of the equations of fluid dynamics, we will encounter the following tensor and vector products.

1. The vector–vector or tensor product: $a_i b_j = A_{ij}$. This is also called the dyadic product, $\mathbf{a} ; \mathbf{b}$. The order of the terms is important in the symbolic notation, since reversing results in the transpose tensor.

2. The tensor–tensor inner (scalar) product: $\mathbf{AB}$ or $A_{ij} B_{ji}$. Another scalar

---

[1] The term contraction is also used elsewhere to describe the sum of the diagonal terms A_{ii}.

product is $A_{ij}B_{ij} = $ **AB***. Note that since both indices are summed, there are no free indices, indicating a scalar product.

3. The vector–tensor product: $a_iA_{ij} = A_{ij}a_i$ or $\mathbf{a}\cdot\mathbf{A}$. There is a sum on one index and the other is a free index, indicating the product is a vector. We can say that the tensor operates on a vector to beget another vector.

The order of the terms is not important in index notation, as long as the indices are kept the same. However, the order is important in symbolic notation, and $\mathbf{A}\cdot\mathbf{a}$ is a different vector from $\mathbf{a}\cdot\mathbf{A}$. For instance, $A_{ij}a_j = a_jA_{ij}$ represents a different array of components to $A_{ij}a_i$.

We will introduce two other operators, the *rotation tensor,* rot **u**, and the *deformation tensor,* def **u**, when they occur. However, they are simply combinations of the above operations. We will use both the index and the symbolic methods of manipulating many numbers with symbols. The conventional 3-D vector symbols and the index notation are interchangeable. Our emphasis is on the components, and we will resort to component verification of the small number of symbolic identities used in our derivations.

Example 4.7

Show that when one tensor is symmetric and one antisymmetric, **AB** $\equiv 0$.

Solution

We are to calculate $A_{ij}B_{ji}$, where $A_{ij} = A_{ji}$ and $B_{ji} = -B_{ij}$.

The antisymmetric tensor B_{ij} has $B_{ii} = 0$, so that we can exclude all terms where $i = j$. Then,

$$A_{ij}B_{ji} = A_{12}B_{21} + A_{13}B_{31} + A_{23}B_{32}$$
$$+ A_{21}B_{12} + A_{31}B_{13} + A_{32}B_{23}.$$

Replacing B_{ij} with $-B_{ji}$ and A_{ij} with A_{ji} in the last three terms,

$$A_{ij}B_{ji} = A_{12}B_{21} + A_{13}B_{31} + A_{23}B_{32}$$
$$- A_{12}B_{21} - A_{13}B_{13} - A_{23}B_{32} = 0$$

Example 4.8

Show that curl grad $\varphi \equiv 0$.

Solution

We note that grad φ is a vector, equal to $\nabla\varphi = (\partial\varphi/\partial x,\ \partial\varphi/\partial y,\ \partial\varphi/\partial z)$. Curl indicates we are taking the cross product, $\nabla \times (\ )$, of this vector, which can be written in matrix form,

$$
\nabla \times \nabla\varphi = \begin{pmatrix} \mathbf{i} & \mathbf{j} & \mathbf{k} \\ \partial/\partial x & \partial/\partial y & \partial/\partial z \\ \partial\varphi/\partial x & \partial\varphi/\partial y & \partial\varphi/\partial z \end{pmatrix}
$$

$$
= \mathbf{i}[\partial^2\varphi/\partial y\ \partial z - \partial^2\varphi/\partial z\ \partial y]
$$

$$
+ \mathbf{j}[\partial^2\varphi/\partial x\ \partial z - \partial^2\varphi/\partial z\ \partial x]
$$

$$
+ \mathbf{k}\,[\partial^2\varphi/\partial y\ \partial x - \partial^2\varphi/\partial x\ \partial y] = 0
$$

A general summary of tensor relations can be found in the summary chart near the end of this chapter.

4.1.2   Applications to Atmospheric Variables

The main variables in atmospheric dynamics can be separated into several categories—fluid properties, body forces, and surface forces.

Fluid properties are scalar or vector fields, such as pressure—$P(x, y, z, t)$ $= P(x_1, x_2, x_3, t) = P(\mathbf{x}, t)$, density ρ, and temperature T. The *flow property* is usually velocity $\mathbf{V}$, although divergence and vorticity can also serve this purpose.

Body forces operate on each element of matter but may be represented as a force acting at the center of mass of the uniform parcel. This force may continuously vary in space and time but still have a specific value at a "point." In special areas of study forces such as the electromagnetic force are significant, but gravity is the main atmospheric body force.

Surface forces include the stress and pressure acting on the surface of our parcel. The expression of these forces generates the most difficult calculations in our derivation of the force balance on the parcel. However, we are now prepared to discuss the three-dimensional variation of this force vector.

We denote the force per unit area at a point as the stress at that point, τ.

$$
\tau = \lim_{\delta A \to 0} \frac{\delta\mathbf{F}}{\delta A}
$$

Consider the parcel at $t = 0$. It is a cube, but immediately before and after, its shape is different, varying with time as shown in Fig. 4.4. Let $\delta \mathbf{A}$ represent the normal to the incremental area on the surface of the parcel and $\delta \mathbf{F}$, the force on this area.

The stress is the ratio of two vectors representing the incremental force and area. They are shown on the face of the y-z plane in Fig. 4.5. Each component of both $\delta \mathbf{F}$ and $\delta \mathbf{A}$ can vary independently (there is a component of $\delta \mathbf{A}$ in the direction of each coordinate, and three components of force in each of these directions). Thus, there are 3^2 or 9 scalar values to describe the total force field. This can be organized in a second-order tensor.

Each component of δF_i acts on each δA_j. For example, the forces on δA_3 $= \delta A_z$ are shown in Fig. 4.6.

We can relate the vectors directly and compactly using tensor analysis in an efficient bookkeeping procedure. The indices i, j, k, etc., represent the entire array of a tensor, (e.g., τ_{ij}). When specific coordinates are used, such

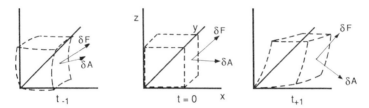

Figure 4.4 The force on the face of a parcel as the parcel distorts.

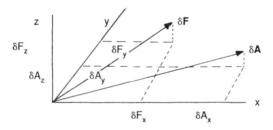

Figure 4.5 Resolution of force and area into components.

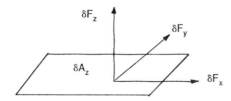

Figure 4.6 The three force components on δA_z.

as x, y, z, 1, 2, or 3, then a specific scalar component is indicated (e.g., τ_{xy}).

Thus, τ_{xy} (a scalar) is the component of stress on the face normal to the x-axis and parallel to the y-axis, whereas τ_{ij} (a tensor) represents all possible components.

$$\tau_{ij} = \lim_{\delta A_i \to 0} \frac{\delta F_j}{\delta A_i} = \begin{pmatrix} \tau_{xx} & \tau_{xy} & \tau_{xz} \\ \tau_{yx} & \tau_{yy} & \tau_{yz} \\ \tau_{zx} & \tau_{zy} & \tau_{zz} \end{pmatrix}$$

This is a nine-component array of the three stress components that act on the three faces of the parcel.

4.2   Relative Motion near a Point

We are interested in obtaining the Eulerian equations for the flow parameters at each point in a field. To write the force balance at a point in the field, we need to investigate the incremental velocity change at the point, or across our parcel. We can then relate the stress forces to the local velocity gradients, which will be the rate of strain. There are several ways to decompose the incremental velocity change into components, and the one presented below is simply the most useful. This section will only define and describe the relative motion in preparation for the later mathematical derivation.

Near a point where the velocity is $u(x)$ and the nearby velocity is $u(x + \delta x)$ or $u(x) + \delta u$ we can manipulate the vector incremental velocity change so that it is expressed in terms of an operator acting on the displacement vector. In a *Lagrangian* description the individual parcel position would be given by the vector $x(t)$, and the operator that gives the velocity of the parcel (at position x) is the derivative with respect to time. However, in our *Eulerian* frame of reference we know that the velocity change experienced by the parcel passing through any point is made up of the time-dependent velocity changes in the field plus a change due to the velocity variation in space. We must employ the total derivative as discussed in Chapter 2.

In the Eulerian description the combinations of the spatial derivatives that form the divergence and the curl yield important depictions of the nature of the field. We later consider cases where either or both of these parameters (divergence and curl of u) vanish to allow special flow solutions. Thus the capability to separate the velocity derivatives into the descriptive arrangements of derivatives called the curl and divergence can give a different perspective on the flow field and is useful in arriving at simplified equations and solutions.

Since we will frequently find it convenient to use indicial notation and the summation convention in this section, we will replace (x, y, z) with (x_1, x_2, x_3) and (u, v, w) with (u_1, u_2, u_3).

$$\mathbf{u}(\mathbf{x}) = u_i(x_1, x_2, x_3), \quad i = 1, 2, 3$$

or

$$\mathbf{u} = u_1(x_1, x_2, x_3)\,\mathbf{i} + u_2(x_1, x_2, x_3)\,\mathbf{j} + u_3(x_1, x_2, x_3)\,\mathbf{k}$$

One can think of δu in two different ways:

1. As we move successively in δx_1, δx_2, δx_3 directions,

$$\delta\mathbf{u} = (\partial u_1/\partial x_1\,\mathbf{i} + \partial u_2/\partial x_1\,\mathbf{j} + \partial u_3/\partial x_1\,\mathbf{k})\,\delta x_1$$
$$+ (\partial u_1/\partial x_2\,\mathbf{i} + \partial u_2/\partial x_2\,\mathbf{j} + \partial u_3/\partial x_2\,\mathbf{k})\,\delta x_2$$
$$+ (\partial u_1/\partial x_3\,\mathbf{i} + \partial u_2/\partial x_3\,\mathbf{j} + \partial u_3/\partial x_3\,\mathbf{k})\,\delta x_3$$

or

2. As the successive change in velocity components,

$$\delta u_1 = \partial u_1/\partial x_1\,\delta x_1 + \partial u_1/\partial x_2\,\delta x_2 + \partial u_1/\partial x_3\,\delta x_3$$
$$\delta u_2 = \partial u_2/\partial x_1\,\delta x_1 + \partial u_2/\partial x_2\,\delta x_2 + \partial u_2/\partial x_3\,\delta x_3$$
$$\delta u_3 = \partial u_3/\partial x_1\,\delta x_1 + \partial u_3/\partial x_2\,\delta x_2 + \partial u_3/\partial x_3\,\delta x_3$$

These three components can also be written as a vector that is the product of a tensor operating on the displacement vector. Shown here in matrix and indicial form:

$$(\delta u_1, \delta u_2, \delta u_3) = \begin{pmatrix} \partial u_1/\partial x_1 & \partial u_1/\partial x_2 & \partial u_1/\partial x_3 \\ \partial u_2/\partial x_1 & \partial u_2/\partial x_2 & \partial u_2/\partial x_3 \\ \partial u_3/\partial x_1 & \partial u_3/\partial x_2 & \partial u_3/\partial x_3 \end{pmatrix} \begin{pmatrix} \delta x_1 \\ \delta x_2 \\ \delta x_3 \end{pmatrix}$$

or

$$\delta u_i = \partial u_i/\partial x_j\,\delta x_j \tag{4.5}$$

Here, $\partial u_i/\partial x_j$ represents all of the possible rate-of-strain components and is called the *basic deformation tensor*. Now some of these components distort, some stretch, and some rotate the parcel. We can use the principal that *all tensors can be split into a symmetric plus an antisymmetric tensor* to separate the actions of the shearing forces on the parcel. The shears represented in the symmetrical tensor cause pure straining motion on the flow parcel, while those in the antisymmetrical tensor result in pure rotation.

Hence the elements of the total shearing action $\delta u_i/\delta x_j$ can be arbitrarily separated into two parts related to the relative velocities.

$$\delta u_i = [\tfrac{1}{2}\{\partial u_i/\partial x_j + \partial u_j/\partial x_i\} + \tfrac{1}{2}\{\partial u_i/\partial x_j - \partial u_j/\partial x_i\}] \, \delta x_j$$

$$= [\qquad \delta u_{ij}^s \qquad + \qquad \delta u_{ij}^a \qquad] \, \delta x_j$$

or

$$\frac{\delta u_i}{\delta x_j} = [\qquad e_{ij} \qquad + \qquad \cap_{ij} \qquad] \tag{4.6}$$

symmetrical tensor; + antisymmetrical tensor;
pure straining motion;   pure rigid body;
deformation tensor;   rotation tensor;
Def u   Rot u

where

$$e_{ij} = \tfrac{1}{2}[\partial u_i/\partial x_j + \partial u_j/\partial x_i]$$

and

$$\cap_{ij} = \tfrac{1}{2}[\partial u_i/\partial x_j - \partial u_j/\partial x_i].$$

The *basic deformation tensor* $\partial u_i/\partial x_j$ can be looked upon as a nine-component second order tensor that operates on the displacement δx_j to produce the vector velocity δu_i. It is also known as the *velocity gradient tensor*. The $\delta u_{(i)}$ are the three components of the velocity change across the parcel due to its advection through the velocity field u_i.

Above, we have separated the basic deformation tensor into two. One has properties of being symmetric ($e_{ij} = e_{ji}$) and is called the *deformation tensor*, and one is antisymmetric ($\cap_{ij} = -\cap_{ji}$) and is called the *rotation tensor*. These may be written in matrix form.

$$e_{ij} = \frac{1}{2} \begin{pmatrix} 2\,\partial u_1/\partial x_1 & \partial u_1/\partial x_2 + \partial u_2/\partial x_1 & \partial u_1/\partial x_3 + \partial u_3/\partial x_1 \\ \partial u_2/\partial x_1 + \partial u_1/\partial x_2 & 2\,\partial u_2/\partial x_2 & \partial u_2/\partial x_3 + \partial u_3/\partial x_2 \\ \partial u_3/\partial x_1 + \partial u_1/\partial x_3 & \partial u_3/\partial x_2 + \partial u_2/\partial x_3 & 2\,\partial u_3/\partial x_3 \end{pmatrix}$$

and

$$\cap_{ij} = \frac{1}{2} \begin{pmatrix} 0 & \partial u_1/\partial x_2 - \partial u_2/\partial x_1 & \partial u_1/\partial x_3 - \partial u_3/\partial x_1 \\ \partial u_2/\partial x_1 - \partial u_1/\partial x_2 & 0 & \partial u_2/\partial x_3 - \partial u_3/\partial x_2 \\ \partial u_3/\partial x_1 - \partial u_1/\partial x_3 & \partial u_3/\partial x_2 - \partial u_2/\partial x_3 & 0 \end{pmatrix}$$

4.2.1　Deformation Tensor

The deformation tensor e_{ij} represents straining motion of the parcel, without rotation, and is therefore also called the rate-of-strain tensor. The scalar terms $\partial u_{(i)}/\partial x_{(i)}$ represent the extension of two adjacent points along the axis $x_{(i)}$. The sum of the diagonal terms of e_{ij} is $\partial u_i/\partial x_i$, the divergence.

The *principal axes* theorem states that there is always a rotation of the axes possible such that the new e_{ij} has nondiagonal terms all equal zero. Thus there always exists some orientation of axes such that the strain is represented as pure extension or contraction of the parcel. The principal axes are not used for our general derivation of the field equations. This is because our coordinate system is most often set by the geometry of the problem, and the principal axes may be different at every point in the fluid. However, they come in handy when a local flow character is being discussed, and the coordinate system can be arbitrarily oriented with respect to that of the mean flow. One practical example of this is discussed in Example 4.9, where the development of atmospheric fronts is a local phenomenon embedded in the large-scale mean flow.

There will generally be six independent components to the deformation tensor. The three diagonal terms equal the divergence of **u**, which is a measure of the *elongation* (or compression) of lines along the axes at a point.

$$\partial u_i/\partial x_i \equiv \nabla \cdot \mathbf{u}$$

The other three independent terms represent *shearing* strain. This measures the angular change between perpendicular lines that coincide with the axes.

Example 4.9

A condition that characterizes a developing atmospheric front within an air mass is that the temperature gradient increases across the front. (See Fig. 4.7.) By looking at the effect of the local velocity gradients on the total change in the temperature gradient, derive a condition for frontogenesis.

Solution

Our parameter that determines whether a front is developing or decaying will be

$$\mathbf{F} \equiv D/Dt|\nabla T| > 0 \quad \text{is frontogenetic}$$

$$< 0 \quad \text{is frontolytic (decaying)}$$

Temperature is a conservative property, so its total change is zero.

$$DT/Dt = \partial T/\partial t + u \cdot \nabla T = 0$$

Hence, if we take the gradient of this equation,

$$\nabla DT/Dt = \partial \nabla T/\partial t + u \cdot \nabla \nabla T + \nabla u \cdot \nabla T = 0$$

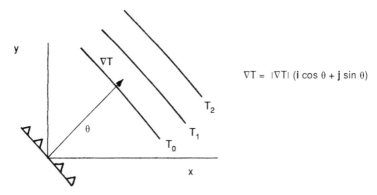

Figure 4.7   A surface air mass front and isotherms (constant T_1, T_2, etc.) and temperature gradient in the region.

or, combining the first two terms,

$$D \nabla T/Dt + \nabla \mathbf{u} \cdot \nabla T = 0$$

Since we can write $D(\nabla T)^2/Dt = 2|\nabla T| \, D|\nabla T|/Dt$

$$\mathbf{F} = 1/[2|\nabla T|] \, D(\nabla T)^2/Dt$$

Substituting for $\nabla T = |\nabla T|(\mathbf{i} \cos \theta + \mathbf{j} \sin \theta)$,

$$\mathbf{F} = -|\nabla T|[\cos^2 \theta \, \partial u/\partial x + \sin^2 \theta \, \partial v/\partial y + \sin \theta \cos \theta(\partial u/\partial y + \partial v/\partial x)]$$

When the flow is viewed in principal axes, the off-diagonal terms of the deformation tensor are zero.

$$\partial u/\partial y + \partial v/\partial x = 0$$

This leaves

$$\mathbf{F} = -|\nabla T|[(\partial u/\partial x - \partial v/\partial y) \cos 2\theta + (\partial u/\partial x + \partial v/\partial y)]$$

Generally, in atmospheric flow, the terms in the last parentheses, the divergence terms, are much less than the first term in this equation. Thus,

$$\mathbf{F} = -|\nabla T|(\partial u/\partial x - \partial v/\partial y) \cos 2\theta$$

Hence the criteria for whether the velocity field increases or decreases the strength of the front depends on the sign of the velocity gradients and the angle θ.

4.2.2   Rotation Tensor

The rotation tensor $\cap_{ij}$ is antisymmetric. Thus it has only three independent components, which can be used to define an associated vector. This vector has fundamental conservation properties that make it important in the description of the flow field dynamics. It is called the *vorticity vector*, and it is a measure of the parcel rotation. Vorticity will be the subject of Chapter 8. The vorticity vector may be written:

$$\bigcap_{ij} = -\tfrac{1}{2}\epsilon_{ijk}\,\zeta_k \tag{4.7}$$

where ζ = local vorticity, $\zeta = \nabla \times \mathbf{u}$, or $\zeta_i = \epsilon_{ijk}\,\partial u_k/\partial x_j$.

Consider the contribution to the relative velocity $\delta\mathbf{u}$ at a distance $\delta\mathbf{x}$ due to $\delta\mathbf{u}^a$. This is the velocity that would be produced by a rigid-body rotation with angular velocity $d\theta/dt \equiv \zeta/2$ about the point $\mathbf{x}$. Note that in general the fluid is not rotating about $\mathbf{x}$ as a rigid body. This rotation concept is strictly applicable only at the point represented by the parcel.

4.2.3 Representation in Cylindrical Coordinates

When the flow is basically in a plane, rotation is often expressed more easily in circular or cylindrical coordinates. We can examine the definitions and description of the three-part effect of the local velocity gradient on our parcel by looking at the relative velocity at a point P′ separated from a point P by $\delta\mathbf{r}$. There is a velocity $\mathbf{u}$ at P that is changing in the direction of $\delta\mathbf{r}$. (See Fig. 4.8.)

The relative motion can now be decomposed into the component along $d\mathbf{r}$ and that perpendicular to $d\mathbf{r}$.

1. The component δu_r will stretch or compress $\delta\mathbf{r}$ (a strain);
2. The component δu_θ has two contributions;

    (a) One part will cause solid-body rotation around P with a value equal to $\delta\mathbf{r}\, d\theta/dt$ (no-strain turning);
    (b) The remaining part of δu_θ will bring about angular deformation of any parcel at P (a strain).

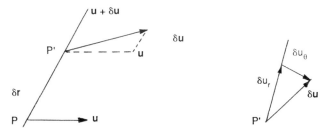

Figure 4.8 The vector velocity change along and normal to a curving path. The parcel moves from point P to P′, experiencing velocity change $\delta\mathbf{u}$ over the distance $\delta\mathbf{r}$. The velocity change can be separated into a part along $\mathbf{r}$ and a part normal to $\mathbf{r}$.

The complete effect on a parcel passing through P to P' can be imagined as a superposition of four effects, illustrated in Fig. 4.9.

1. A translation,
2. plus an elongation without a change in volume,
3. plus a rotation without distortion,
4. plus an angular shearing without change in volume. (The principal axis theorem states that δr can be chosen such that this term is zero.)

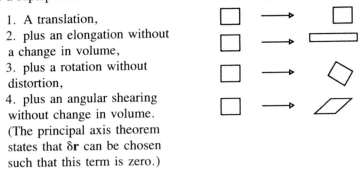

Figure 4.9   The total change in a flowing parcel broken into four parts.

In relating stress (forces acting on the parcel) to rate of strains (distortion caused by forces), we will group the components 1 and 2b together. The action of 2a does not involve strain or internal resistance. However, it is related to an important characteristic of the flow field called the vorticity.

Example 4.10

Consider pure shear in one direction, with $u_1(x_2)$ such that $\partial u_1/\partial x_2 = C_1$ as shown in Fig. 4.10.
  (a) Discuss the decomposition of this shear into strains and rotation.
  (b) Do the same for the case where $u_2(x_1)$ has simple shear.
  (c) Discuss the $\pm$ combinations of (a) and (b).

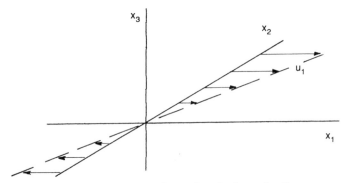

Figure 4.10   Example of pure shear in the x_2 direction.

Solution

(a) Substituting in Eq. (4.6), there is only one nonzero shear, which is split into contributions to strain and rotation. The deformation tensor reduces to

$$\frac{\delta \mathbf{u}}{\delta x_2} = \frac{\delta u_1}{\delta x_2} = \underbrace{\frac{1}{2}\frac{\partial u_1}{\partial x_2}}_{\text{shear strain}} + \underbrace{\frac{1}{2}\frac{\partial u_1}{\partial x_2}}_{\text{rotation}}$$

The shear can be separated into two equal parts, one contributing to strain and one to rotation.

(b) Similarly, if the u_2 component is uniformly varying in the x_1 direction (see Fig. 4.11.):

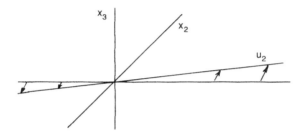

Figure 4.11   Example of pure shear in the x_1 direction.

$$u_2(x_1) = C_2 x_1$$

then

$$\frac{\delta \mathbf{u}}{\delta x_1} = \frac{\delta u_2}{\delta x_1} = \frac{1}{2}\frac{\partial u_2}{\partial x_1} + \frac{1}{2}\frac{\partial u_2}{\partial x_1} = \frac{1}{2}C_2 + \frac{1}{2}C_2$$

(c) Consider the sum and difference of these shear conditions.
   In this case, $\partial u_i/\partial x_j$ has two nonzero components:

$$\partial u_1/\partial x_2 = \tfrac{1}{2}(\partial u_1/\partial x_2 + \partial u_2/\partial x_1) + \tfrac{1}{2}(\partial u_1/\partial x_2 - \partial u_2/\partial x_1)$$

and

$$\partial u_2/\partial x_1 = \tfrac{1}{2}(\partial u_2/\partial x_1 + \partial u_1/\partial x_2) + \tfrac{1}{2}(\partial u_2/\partial x_1 - \partial u_1/\partial x_2)$$

The decomposition into symmetric and antisymmetric tensors becomes

$$e_{ij} = \frac{1}{2}\begin{pmatrix} 0 & \partial u_1/\partial x_2 + \partial u_2/\partial x_1 & 0 \\ \partial u_2/\partial x_1 + \partial u_1/\partial x_2 & 0 & 0 \\ 0 & 0 & 0 \end{pmatrix}$$

and

$$\cap_{ij} = \frac{1}{2}\begin{pmatrix} 0 & \partial u_1/\partial x_2 - \partial u_2/\partial x_1 & 0 \\ \partial u_2/\partial x_1 - \partial u_1/\partial x_2 & 0 & 0 \\ 0 & 0 & 0 \end{pmatrix}$$

Consider the special case where $\partial u_1/\partial x_2 = \partial u_2/\partial x_1$, as shown in Fig. 4.12.

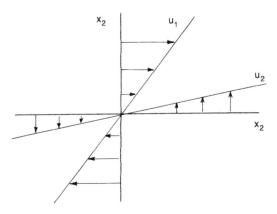

Figure 4.12   Example of two shears opposing each other in the $x_1 - x_2$ plane.

Then $e_{12} = e_{21} = \partial u_1/\partial x_2$ and $\cap_{12} = \cap_{21} = 0$. There is shear distortion only with no rotation.

Finally, in the case where $\partial u_1/\partial x_2 = -\partial u_2/\partial x_1$ as shown in Fig. 4.13.

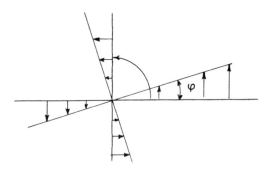

Figure 4.13   Example of two shears in the same direction.

Then $e_{ij} = 0$ for all i, j; while

$$\cap_{12} = \partial u_1/\partial x_2, \quad \text{and} \quad \cap_{21} = -\partial u_1/\partial x_2 = \partial u_2/\partial x_1$$

This case has pure rotation with no distortion. Although these cases consider the simplest of shear, they can be used as a first approximation to many atmospheric flows. We will explore some of these approximations in Chapter 9.

4.3 An Arbitrary Division of the Velocity

It is always possible to superimpose a constant velocity field on the flow without changing the terms involving velocity derivatives. Thus, including the mean flow, the velocity near a point may be separated into three components. These correspond to a uniform translation $\mathbf{u} = \mathbf{V}$ plus the pure strain associated with e_{ij} plus the pure rotation associated with $\cap_{ij}$. Using the definition of vorticity given in Eq. (4.7) the incremental velocity changes may be written

$$\delta u_i = \delta V_i + \tfrac{1}{2} \delta x_j e_{ij} + \tfrac{1}{2} \epsilon_{ijk} \zeta_j \, \delta x_k$$

The divergence and curl ($\mathbf{\nabla} \cdot \boldsymbol{\varphi}$) and ($\mathbf{\nabla} \times \boldsymbol{\varphi}$) of a vector function of position are basic operations yielding quantities that are independent of the choice of coordinate system. Thus, we can always write a general velocity field as (D is the divergence; ζ, the vorticity)

$$\mathbf{u} = \mathbf{V} + \mathbf{u}^s + \mathbf{u}^a \tag{4.8}$$

with

$$\mathbf{\nabla} \cdot \mathbf{u} = D, \quad \mathbf{\nabla} \times \mathbf{u} = \zeta$$

and

$\mathbf{\nabla} \cdot \mathbf{u}^s = D$	and $\mathbf{\nabla} \times \mathbf{u}^s = 0$
(divergent)	(irrotational)
$\mathbf{\nabla} \cdot \mathbf{u}^a = 0$	and $\mathbf{\nabla} \times \mathbf{u}^a = \zeta$
(incompressible)	(rotational)
$\mathbf{\nabla} \cdot \mathbf{V} = 0$	and $\mathbf{\nabla} \times \mathbf{V} = 0$
(V constant)	(V constant)

When there is divergence (or convergence) at any point in a field as denoted by the value of D at that point, there will be an associated velocity $\mathbf{u}^s$ directed away or toward the source of divergence. Similarly, if there is vorticity at a point given by the value of $\zeta(\mathbf{x})$, then there is an associated velocity at each point $\mathbf{x}'$ of the field tangential to the circle with radius $\mathbf{x}'$ $-$ $\mathbf{x}$. Specific atmospheric phenomena can be related directly to the local divergence or vorticity. Consequently, these properties of the flow field have practical value. This decomposition of the velocity field will be employed again in connection with the vorticity discussion in Chapter 8.

Summary

TENSORS

  Order

  Notation

    Symbolic   grad(∇), div($\nabla \cdot$), curl($\nabla \times$)

    Index $\partial a_i/\partial x_j \ldots$

    Repeated Index $a_1 b_1 = a, b, + a_2 b_2 + \ldots$

    Matrix

  Kronecker Delta $\delta_{ij} \equiv \begin{cases} 1 \text{ if } i = j \\ 0 \text{ if } i \neq j \end{cases}$

  Alternating Unit Tensor ϵ_{ijk}

  Dyadic Product $\mathbf{A} = a; b$

  Symmetric Tensors $A_{ij} = A_{ji}$

  Antisymmetric Tensors $A_{ij} = -A_{ji}$

RELATIVE MOTION NEAR A POINT $\dfrac{\delta u_i}{\delta x_j} = [e_{ij} + \cap_{ij}]$

  Rotation Tensor e_{ij}

  Deformation Tensor $\cap_{ij}$

  Principal Axes

ARBITRARY DIVISION OF A VELOCITY $\mathbf{u} = \mathbf{V} + \mathbf{u}^s + \mathbf{u}^a$

Problems

1. Write grad **a** in vector, indicial, and matrix forms.
2. Show the following identities by writing out terms in component form and rearranging.

(a) div $(f\mathbf{v}) = f\nabla \cdot v + \nabla f\, \nabla v$
(b) div grad $f = \nabla^2 f$
(c) curl $(f\mathbf{u}) = f\nabla \times \mathbf{u} + \nabla \mathbf{f} \times \mathbf{u}$
(d) $\mathbf{a} \times \mathbf{b} = \epsilon_{ijk} a_i b_j e_k$ (e_k is a unit vector in k-direction)
(e) $(\mathbf{Au}) \cdot \mathbf{w} = (\mathbf{A}^*\mathbf{w}) \cdot \mathbf{u}$
(f) $(\text{grad } \mathbf{a})\, \mathbf{a} = \text{grad } a^2/2 + [\text{grad } \mathbf{a} - (\text{grad } \mathbf{a})^*]\, \mathbf{a}$
 $= \text{grad } a^2/2 + (\text{curl } \mathbf{a}) \times \mathbf{a}$

3. Find, by expanding into coordinates,

(a) curl (grad f) $\equiv$?
(b) div curl $\mathbf{u}$ = ?
(c) curl curl $\mathbf{u} = \nabla \times \nabla \times \mathbf{u} = \text{grad div } \mathbf{u} - \nabla^2 \mathbf{u}$

4. State whether the following tensors are examples of a symmetric or an antisymmetric matrix.

(a) (b) (c)

$$\begin{pmatrix} 1 & 2 & 3 \\ 2 & 5 & 4 \\ 3 & 4 & 6 \end{pmatrix} \begin{pmatrix} 0 & -2 & -3 \\ 2 & 0 & -4 \\ 3 & 4 & 0 \end{pmatrix} \begin{pmatrix} 0 & 2 & 3 \\ -2 & 0 & 4 \\ -3 & -4 & 0 \end{pmatrix}$$

5. Show by expanding:

(a) Does the vector product of a vector and a tensor satisfy

$$\mathbf{u} \cdot \mathbf{A} = \mathbf{A} \cdot \mathbf{u} ?$$

(b) Does $u_j A_{ij} = A_{ji} u_j$?
(c) Does the dyadic (tensor) product of two vectors satisfy

$$\mathbf{uv} = \mathbf{vu} ?$$

(d) Does $u_i v_j = u_j v_i$?

6. Consider the following equation and discuss whether the terms are consistent dimensionally and with respect to tensor order:

$$\partial u/\partial t + u\, \partial u/\partial x + v\, \partial u/\partial y + w\, \partial u/\partial z$$
$$= -(\partial p/\partial x)/\rho + fv + v[\partial^2 u/(\partial x_i\, \partial x_i)]$$

7. How many elements are there in each of the following:

(a) Zero-order tensor   ———
(b) First-order tensor   ———
(c) Second-order tensor ———
(d) Third-order tensor   ———
(e) Fourth-order tensor ———

8. In our decomposition of the velocity gradient tensor, consider one-dimensional shear, $U = Cz$. What is the strain (extension and deformation)? What is the vorticity?

9. Write the deformation tensor in matrix form for (a) Couette flow, (b) flow in the x-direction with shear in the y- and z-directions, (c) vortex flow.

Part II | The Governing Equations for Fluid Flow

In this part we obtain the basic equations governing the flow of fluids. The concepts and techniques learned in the first four chapters will all come into play in the derivations. Part I was intended to provide a first introduction to the new and complex concepts and techniques of fluid dynamics. This part will provide a second, in-depth exposure. We will apply the methods towards deriving the basic equations and also applying them to geophysical phenomena. Experience has shown that this material really begins to get assimilated by the student at about the third exposure, which can be obtained from working the examples and problems.

In geophysical flows we need to describe the flow fields so that, when given certain boundary value measurements of field variables, we can describe the variation of any parameter in the entire domain. The unknown parameter might be velocity, density, moisture, temperature, or a pollutant. In general, the distribution of any parameter will be determined by the velocity field. Thus our first goal is to solve the equations for the velocity field. When another parameter is of interest, such as precipitation or pollution, the task is to find some relation between the velocity and this parameter at each point.

We start with statements that relate a parameter value at one point to its value elsewhere. They are generally derived from some principle of conservation for that parameter. These principles provide us with the basic physical laws. The fluid motion equations are obtained from the statements of conservation in physics. In these classical axioms, the principle is stated for a specific solid body. We have seen that in the dynamics of fluid flow, there is no solid body of fluid. We designate a proxy for this body, the *parcel*. To establish guidelines for the parcel, the concept of a *continuum* was elucidated. Also, since we are not generally concerned with a *specific* parcel, the *field description* and the concept of a *total derivative* at any point in the field was introduced.

The principles of conservation of mass, momentum, and energy evolved in connection with solid-body mechanics. Using the concepts of the parcel and the total derivative, we can apply these principles to a fluid flow to obtain field equations for the flow. Conservation of mass and energy of the parcel deal with scalars—the density and the temperature. The methods of tensor analysis are handy for providing compact derivations and expressions for these two equations in most applications, but they are not essential. They are essential for a general derivation of the momentum equation.

The other basic principle we employ is that defined by Newton's second law, which we can call the conservation of momentum. There are some fundamental differences between the conservation of momentum applied to a particle and that applied to a fluid parcel. Some authors prefer to refer to it simply as the momentum principle for fluid flow. Regardless of terminology (we will use "conservation of momentum"), the result is the basic equation for the velocity field in an Eulerian frame of reference.

The velocity is a vector. Consequently the interactions expressed in the total derivative become quite complicated. The methods of *tensors,* and *symbolic* and/or *index* notations come in handy in writing this equation. These techniques of tensor analysis and compact notation are essential, however, in expressing the forces that depend on the internal stresses. We "remove" the parcel from the surrounding fluid in order to apply the conservation principles to it. Thus, we must replace the effects of the surrounding fluid on the sides of the parcel with "viscous" forces. The description of these forces is the "tour de force" in deriving the equations, to the extent that the momentum equation is sometimes called *the* equation of motion. We will see, using our skills in *tensor notation,* that this vector equation fits in the same format as the other scalar conservation equations.

The conservation equations can be derived as *integral equations* for a specific volume, called a *control volume*. The integral form is most useful when the problem involves a confined flow volume. But when the equations are derived for a general field point, they are *differential equations*. Geophysical flows typically are unconfined, and the differential form of the equations is most valuable in their description. All of the conservation equations are of similar form, and we note the general form for these equations at the end of the study of conservation of mass in Chapter 5. This similarity of structure is an interesting generalization of the conservation equations. It could be used to provide a brief, concise, mathematical derivation of all of the equations. However, we will separately derive each of the equations to illustrate the individual terms and the underlying physics.

In the governing set of equations for each problem there must be as many relations as there are parameters. When considering geophysical flows we are generally interested in the velocity, pressure, density, and temperature—four variables. Thus we need to find four relations between these variables. Three of the relations are single equations for the scalars; density, pressure, and temperature. Only one of the relations will be a vector equation—for the velocity. Although the energy equation is simply a scalar relation, it does involve some complicated sources and sinks of energy.

In addition to these basic equations, several others are obtained. These equations are needed for specific classes of flows. The first supplemental equation, the *vorticity equation,* is valuable because of the special properties of *vorticity* and *vortices.* Also, vortices are frequently present in geophysical flows. The development of the vorticity equation uses concepts introduced in Chapters 2 and 4. Many phenomena in the atmosphere can be directly related to the vorticity field. Vorticity involves a specific combination of the velocity derivatives. It is closely related to the velocity equations (that is, the momentum equations), and we will derive it directly from them.

The second set of special equations are called the *potential flow* equations. They are special cases for the flow of inviscid fluids. They are related to the basic equations using the concepts of inviscid and irrotational flow. These assumptions reduce the basic equations to a particularly simple form, with important analytic solutions.

The third set of equations are the *perturbation equations.* They provide the governing equations for problems involving *waves, instabilities,* and *turbulence.* The nonlinear terms in the equations of motion are important when spacial gradients of the velocity are significant. Therefore dynamically active regions such as frontal zones, thin layers, and in general smaller than synoptic-scale dynamics require solution of the nonlinear equations. The starting point for such analyses, and for stability analyses in general, is the perturbation equations. They describe the linear and finite nonlinear perturbations. A wave in the flow can be considered as a perturbation on some basic mean flow. The equations determining the behavior of this perturbation can be obtained directly from the basic set of equations. The *stability* of a particular mean flow can be checked by examining whether an *infinitesimal perturbation* tends to grow explosively or to decay. *Finite perturbation waves* can be examined by letting the perturbation grow to the point that the nonlinear terms interact with the mean flow, perhaps to establish an equilibrium. Finally, these equations also provide insight into a basic geophysical tool for handling turbulence—the *eddy viscosity* concept.

In the last chapter, we derive the *boundary layer equations.* They provide an example of specialized equations obtained from the basic set of equations using the rules of dimensional analysis and similarity. The study of these equations is particularly revealing. There has been found an analytic solution to this version of the Navier–Stokes equations (momentum equations) that retains the highest-order terms, the viscous terms. In addition, this solution is unstable to infinitesimal perturbations, yet stable to nonlinear finite perturbations. All of the concepts developed in the first 10 chapters are brought to bear on this solution. And the solutions suggest ideas about the concepts of attractor solutions, coherent structures, and turbulence analysis by randomly occurring organized structures—chaos.

The following flow chart for the next seven chapters provides an overall picture of the equations and concepts.

GOVERNING EQUATIONS OF ATMOSPHERIC FLOW

EQUATION OF STATE—CHAPTER 1

CONTINUITY—CHAPTER 5

MOMENTUM EQUATION—CHAPTER 6
  Coriolis Term
  Stress Term

ENERGY EQUATIONS—CHAPTER 7
  First Law of Thermodynamics
  Thermal Energy Equation
  Mechanical Energy Equation
  Bernoulli Equation

VORTICITY TRANSPORT EQUATION—CHAPTER 8
  Line Vortex

POTENTIAL FLOW EQUATIONS—CHAPTER 9
  Ideal Vortex
  Velocity Potential
  Stream Function

PERTURBATION EQUATIONS—CHAPTER 10
  Reynolds Stress
  Eddy Viscosity

BOUNDARY LAYER EQUATIONS—CHAPTER 11

Chapter 5 | Conservation of Mass—Continuity

5.1 The Parcel Derivation

We are all familiar with the concept that mass cannot be created or destroyed (except under exceptional circumstances which do not apply in our domain of interest). However, in flow problems where we are approximating the flow as two-dimensional with a concentrated vertical flow, we might try to model this small area of flow as a point sink or source. An example of a point sink might be a drain hole, or the vertical flow in the eye of a storm. When the region of vertical flow is very small, it can be considered an exceptional (called singular) point of the flow. But in general, the idea that the mass of the fluid is conserved in the flow domain is a valid and intuitively comfortable assumption.

We begin the derivation of the conservation statement as it applies for every point in a field by considering the hypothetical cubical parcel with volume δV at a point **x**, time t (Fig. 5.1).

The mass of the parcel is simply the density times the volume. Density is the field variable, which may be changing. The parcel may be in a region of variable flow, so that more flux of mass is coming in from one direction than is leaving from another. For instance, *along a streamline* we might expect density to be increasing in a region that has converging flow and

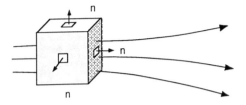

Figure 5.1   The parcel at (x, y, z) in a flow field U.

decreasing volume, and decreasing in a region that has diverging flow and increasing volume. This result would depend on the compressibility of the fluid. At any instant, the net change in density *at a point* will depend on the total change in mass due to flow in and out of the parcel. Thus the parcel changes density due to the characteristics of the flow. If the flow field is changing with time, then so will the density at each point.

We can also allow for the fact that the density may be changing with time at all points in the field, independent of any flow characteristics. This fluctuation or continuous change might require some distribution of sinks or sources of fluid. An example might be in a cloud domain where the condensation or evaporation of water affects the density. For instance, condensation of water takes place throughout the air mass due to the ascending and cooling of air that flows over a mountain range. Also, there is evaporation due to heating of a descending air flow. Thus, the parcel that is momentarily at point **x** may be changing density due to a "global" change with time in addition to the change due to its advection through a variable density field. When we express the total change in density as a field variable, there will be two parts—the local rate of change with time at each point, and the advective change due to the flow through the point.

The total change in the field variable density at a point—the time rate of change plus the advective change—can be recognized as the total derivative introduced in Chapter 2. In the absence of sinks or sources within the parcel, the time rate of change of mass of a parcel must equal the net mass flow through the surfaces of the parcel. Since there are usually no sinks/sources of mass in the atmospheric flow, this is a useful assumption.

We can express the advective change of mass with respect to the net velocity directed normal to the surface at every point on the surface of the parcel. Consider a unit vector normal to the surface as **n** so that the outward-directed component of flow is $\mathbf{u} \cdot \mathbf{n}$. The mass change will be positive if the flow rate is directed inward. Integrating over the volume to obtain the net volumetric change with respect to time, and the surface of the parcel to obtain the net inflow/outflow, we can set them equal,

$$\iiint \partial\rho/\partial t \, dV = -\iint \rho \, \mathbf{u} \cdot \mathbf{n} \, dA \tag{5.1}$$

We are interested in the changes at a *point* in the field. This means we would like an expression for the conservation over the volume of the parcel at the instant it occupies the point. We can then let the parcel shrink to a point within our continuum concept developed in Section 1.9. (We recall that this means that the parcel is small enough to allow the derivatives of the flow parameters to be defined across it, yet large enough to have uniform properties.) To get the expression (5.1) completely in terms of the volume, we can use the divergence theorem to relate the flow through the surface of a region to the corresponding changes within the volume,

$$\iiint \partial\rho/\partial t \, dV = -\iiint \operatorname{div} \rho \, \mathbf{u} \, dV = -\iiint \partial(\rho u_j)/\partial x_j \, dV$$

or

$$\iiint \{\partial\rho/\partial t + \partial(\rho u_j)/\partial x_j\} \, dV = 0$$

Now, in the limit as $\delta V \to 0$ at the point $\mathbf{x}$, we get

$$\partial\rho/\partial t + \partial(\rho u_j)/\partial x_j = 0 \tag{5.2}$$

This describes the change in density at a point in the Eulerian sense. It states that the rate of accumulation of mass per unit volume at $\mathbf{x}$ equals the net flow rate of mass per unit volume into that point.

Equation (5.2) may also be written

$$\partial\rho/\partial t + u_j \, \partial\rho/\partial x_j + \rho \, \partial u_j/\partial x_j = 0 \tag{5.3}$$

or

$$\underset{\substack{\text{rate of change of} \\ \text{density of a} \\ \text{fluid parcel}}}{D\rho/Dt} + \underset{\substack{\text{mass/unit volume times} \\ \text{parcel volume expansion} \\ \text{rate}}}{\rho \, \boldsymbol{\nabla} \cdot \mathbf{u}} = 0 \tag{5.4}$$

Equation (5.4) describes the change in density of a parcel as it moves through a velocity field in a Lagrangian perspective. The rate of change of density of the parcel with time depends only on the divergence field through which it moves. From an Eulerian view as stated in Eq. (5.3), the local change in density with respect to time depends on the divergence at a point, and also on the density gradient at the point.

If the divergence of $\mathbf{u}$, $\boldsymbol{\nabla} \cdot \mathbf{u} = 0$, we have a nondivergent flow (also called a solenoidal vector field) and conservation of mass is simply, from Eq. (5.4),

$$D\rho/Dt = 0 \qquad (5.5)$$

In the ocean, and often in the atmosphere, the density in the domain of concern (e.g., the PBL, or a thin horizontal region) can be considered approximately constant in both time and space. Then the conservation of mass equation is simply the expression of nondivergence.

$$\mathbf{\nabla} \cdot \mathbf{u} = 0 \qquad (5.6)$$

The condition of nondivergence is often called incompressible flow. However we should note that this is not necessarily the same as constant-density flow. The divergence in Eq. (5.3) might be zero while the temporal and advective changes in density balanced. We will examine the conditions of incompressibility more closely in Section 5.4.

These equations (5.2–5.6) which are expressions of the principle of conservation of mass applied to a fluid parcel, are also called *continuity* equations.

Example 5.1

Consider a constant-density fluid flow in a converging channel (Fig. 5.2). Find $v(x, y)$ when there is two-dimensional, steady-state flow and the channel width and along-stream velocity are given by

$$Y = Y_0/(1 + x/\ell)$$

and

$$u(x, y) = u_0(1 + x/\ell)[1 - (y/Y)^2]$$

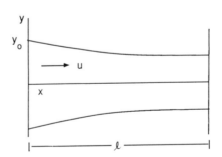

Figure 5.2 Flow through a converging nozzle.

Solution

We are given $u(x, y)$. Since the flow is steady state and continuous, conservation of mass will relate u and v. We start with the complete continuity equation (5.3).

$$\partial\rho/\partial t + \rho(\partial u/\partial x + \partial v/\partial y) + u\,\partial\rho/\partial x + v\,\partial\rho/\partial y = 0$$

The condition of constant density simplifies this equation considerably, since the time and spatial derivatives of ρ are zero, leaving

$$\partial u/\partial x + \partial v/\partial v = 0$$

We use the given $u(x, y)$ to get $\partial u/\partial x$ and then integrate with respect to y to get v.

$$u = u_0(1 + x/\ell)[1 - (y^2/Y_0^2)(1 + x/\ell)^2]$$

$$\partial u/\partial x = (u_0/\ell)\{1 - 3(y^2/Y_0^2)(1 + x/\ell)^2\} = -\partial v/\partial y$$

Hence, on integration,

$$v = -u_0/\ell \int [1 - 3(y^2/Y_0^2)(1 + x/\ell)^2]\,dy$$

$$= -(u_0/\ell)\,y + [u_0/(\ell Y_0^2)](1 + x/\ell)^2\,y^3 + C(x)$$

We can use the no-slip boundary condition at the boundary, $v(Y) = 0$, to evaluate $C(x)$.

$$-(u_0/\ell)[Y_0/(1 + x/\ell)] + u_0/(\ell Y_0^2)(1 + x/\ell)^2\,[Y_0^3/(1 + x/\ell)^3] + C(x) = 0$$

and therefore $C(x) = 0$.
Consequently,

$$v(x, y) = -(u_0/\ell)y + [u_0/(\ell Y_0^2)](1 + x/\ell)^2\,y^3$$

5.2 A Lagrangian Perspective

5.2.1 An Alternate Derivation of Continuity

We can derive the continuity equation in a slightly different manner, by considering a specific infinitesimal parcel in a Lagrangian sense. The derivation will illustrate the close connection between Lagrangian and Eulerian perspectives and we will end up with the familiar Eulerian expressions.

Starting with the Lagrangian perspective we consider a very small parcel such that $\delta V \to 0$, with no sinks or sources. We then follow the particular parcel that experiences volume and density changes with respect to time only. The field variables will vary infinitesimally across the small dimensions of the parcel. Then, the statement for the constant mass of this parcel, $\rho \, \delta V$, is completely expressed in the time derivative, $D(\rho \, \delta V)/Dt = 0$. However, when the parcel moves through the fluid, its volume must distort and change due to the changing forces in the flow field. The derivative is separated into density and volume changes by using the chain rule for differentiation. In the end, the derivative can then be converted to the Eulerian expression.

We start with the mass of the moving parcel expressed in terms of the density,

$$\{mass\} = \rho \, \delta V \tag{5.7}$$

The change in mass of the parcel as it moves through the field is

$$D\{mass\}/Dt = 0 = D(\rho \, \delta V)/Dt = \rho \, D \, \delta V/Dt + \delta V \, D\rho/Dt \tag{5.8}$$

We can expand the last term of this equation using the definition of the total derivative of the density.

$$D\rho/Dt = \partial \rho/\partial t + \mathbf{u} \cdot \text{grad } \rho \tag{5.9}$$

We can express the total rate of change in δV as the expansion or contraction due to the variable velocity field. This can be related to the divergence of the velocity over the volume using the divergence theorem. We then consider the value at any instant, where the change in δV equals the net outward/inward flow through the surface area:

$$D \, \delta V/Dt = \iint \mathbf{u} \cdot \mathbf{n} \, dA \tag{5.10}$$

This equals the divergence over the volume using the divergence theorem.

$$D \, \delta V/Dt = \iiint \nabla \cdot \mathbf{u} \, dV \tag{5.11}$$

To obtain the equation for an infinitesimally small parcel, let $\delta V \to 0$, to get

$$D \, \delta V/Dt = \nabla \cdot \mathbf{u} \, \delta V \tag{5.12}$$

Equations (5.9) and (5.12) express the time rate of change of mass of the parcel in terms of the spatial distribution of velocity and density. Substituting these expressions into Eq. (5.8) gives density as a function of time and space.

$$D(\rho\ \delta V)/Dt = \rho \text{ div } \mathbf{u}\ \delta V + (\partial\rho/\partial t + \mathbf{u}\cdot \text{ grad } \rho)\ \delta V = 0$$

or

$$\partial\rho/\partial t + \rho \text{ div } \mathbf{u} + \mathbf{u}\cdot \text{grad } \rho = 0 \tag{5.13}$$

Once again, this equation may be written in many ways, all of them called the continuity equation.

$$D\rho/Dt + \rho \text{ div } \mathbf{u} = 0$$

$$\partial\rho/\partial t + \boldsymbol{\nabla}\cdot\rho\mathbf{u} = 0$$

$$\partial\rho/\partial t + \boldsymbol{\nabla}\rho\cdot\mathbf{u} + \rho\boldsymbol{\nabla}\cdot\mathbf{u} = 0$$

$$\partial\rho/\partial t + \rho\ \partial u_j/\partial x_j + u_j\ \partial\rho/\partial x_j = 0$$

and

$$\partial\rho/\partial t + \partial(\rho u_j)/\partial x_j = 0 \tag{5.14}$$

The form of the equation used in a specific application will depend on the convenience of notation. The equations are all equivalent. The expressions are in a Eulerian sense since density is a function of time and position. Note that this is a single scalar equation with up to four unknowns (u, v, w, ρ). Three more equations are needed, and will be derived in the next chapter.

5.2.2   The Parcel Material Expansion

In the first of Eqs. (5.14) we have written

$$D\rho/Dt = -\rho\ \partial u_i/\partial x_i$$

This states that the rate of change of the density of a fluid parcel equals the mass per unit volume times the divergence. We can further explore the physics of this relation by inverting the process followed in Section 5.2.1. Writing the volume of the parcel in a Lagrangian sense, we have

$$\delta V = \iiint dV$$

Take the derivative with respect to time, applying Leibnitz's theorem.

$$D\delta V/Dt = d/dt\left\{\iiint dV\right\} = \iiint \partial(1)/\partial t\ dV + \iint \mathbf{u}\cdot\mathbf{n}\ dA$$

$$= \iint \mathbf{u}\cdot\mathbf{n}\ dA$$

Then, in applying the divergence theorem, we get

$$D \, \delta V/Dt = \iiint \nabla \cdot \mathbf{u} \, dV \qquad (5.15)$$

Now, to obtain our field relations we must consider the parcel volume at a point as $\delta V \to 0$.

$$\lim_{\delta V \to 0} \frac{D \, \delta V}{Dt} = \frac{\partial u_j}{\partial x_j} \delta V \qquad (5.16)$$

{Note: we could use the mean-value theorem for integrals on Eq. (5.15), which states that

$$D \, \delta V/Dt = \partial u_j/\partial x_j|_{P'} \, \delta V$$

for $\partial u_j/\partial x_j$ at some point P' in δV. In the limit $\delta V \to 0$, this must be at the point $P' \to P$.}

Thus, rewriting Eq. (5.16),

$$1/\delta V \, D \, \delta V/Dt = \partial u_j/\partial x_j \qquad (5.17)$$

The *rate of expansion* of a material parcel at a point equals the *divergence* at that point. This is also called the *dilation rate* at the point.

Finally, from (5.4) and (5.17) we can write

$$(1/\rho)(D\rho/Dt) \;\; = \;\; -(1/\delta V) \, D \, \delta V/Dt \qquad (5.18)$$

fractional rate change of $\;\;=\;\;$ fractional rate change of the
the density of the parcel $\qquad\qquad$ volume of the parcel

This states that, for no sources or sinks in the volume, the changes in density of a volume are due entirely to changes in its volume.

The change in mass may be written

$$\delta V \, D\rho/Dt + \rho \, D \, \delta V/Dt = D(\rho \, \delta V)/Dt = 0$$

and we have merely worked back to the statement that the mass of the parcel, $\rho \, \delta V$, is constant.

Example 5.2

Develop the equation for the dilation rate of a volume with sides δx_i in a flow field δu_i by considering the one-dimensional expansion rate of an individual side of the small volume (e.g., a parcel).

Solution

First, consider the side δx_1 at time t, and write the incremental velocity change across this distance.

At t,

$$\overline{\delta x_1}$$

$$\overrightarrow{u_1} \qquad\qquad \overrightarrow{u_1} + (\partial u_1 /\partial x_1) x_1 + \text{higher-order terms}$$

Then, we consider how this line stretches (or contracts) due to the difference in velocities at each end, at $t + \Delta t$.

$$\overline{\delta x_1}$$

$$|u_1 \Delta t| \qquad |(u_1 + \partial u_1/\partial x_1\, \delta x_1)\,\Delta t|$$

and the change in δx_1 is

$$\Delta(\delta x_1) = (u_1 + \partial u_1/\partial x_1\, \delta x_1)\Delta t - u_1\, \Delta t$$

This may be written

$$\frac{\Delta(\delta x_1)}{\Delta t} = \frac{\partial u_1}{\partial x_1} \delta x_1$$

and in the limit $\Delta t \to 0$,

$$D(\delta x_1)/Dt = (\partial u_1,/\partial x_1)\, \delta x_1$$

Similar expressions are obtained in the other directions x_2 and x_3, yielding three equations.

$$\frac{D(\delta x_i)}{Dt} = \frac{\partial u_{(i)}}{\partial x_{(i)}}\,(\delta x_i) \qquad\qquad (5.19)$$

Now we can write the volume and its change,

$$\delta V = \delta x_1 \delta x_2 \delta x_3$$

$$\frac{D\,\delta V}{Dt} = \frac{D\,\delta x_1}{Dx_1}\delta x_2\,\delta x_3 + \delta x_1\frac{D\,\delta x_2}{Dx_2}\delta x_3 + \delta x_1\,\delta x_2\frac{D\,\delta x_3}{Dx_3}$$

Substituting from Eq. (5.19) and dividing through by the volume,

$$\frac{1}{\delta V}\frac{D(\delta V)}{Dt} = \frac{\partial u_1}{\partial x_1} + \frac{\partial u_2}{\partial x_2} + \frac{\partial u_3}{\partial x_3} = \nabla\cdot\mathbf{u} \qquad\qquad (5.20)$$

This is the dilation rate = the rate of expansion = the volumetric strain.

It will be used in the energy derivations since it is associated with the work done on a parcel.

5.3 Two-Dimensional Version of Continuity

A more detailed view of the fluxes across the parcel can be obtained within a reasonable space of text if we restrict our attention to two dimensions. We can then write the equations for each component and look closely at the changes in these components.

Consider the planar view of a parcel with unit depth. Assume ρ is constant across the parcel, so we can write for the mass of the parcel,

$$\delta M = \rho \, \delta V = \rho \, \delta x \, \delta y \, \delta z \qquad (5.21)$$

In the two-dimensional flow, each component of velocity can vary in both the x and the y directions. We can approximate these velocity changes across our incremental parcel by a Taylor expansion. In this case we will consider the base values of quantities such as pressure and velocity to be the values at the center of the parcel and expand around these values. (Note that values at the corner, $x = y = z = 0$, could also be assumed as base values. Since the parcel is infinitesimal with respect to mean flow scales, the *magnitudes* of these base values are uniform across the parcel in the limit $\delta V \to 0$. We are writing the *incremental changes* in these quantities across the finite parcel, since we are calculating their *rate of change* at a point, which need not be zero.)

In Fig. 5.3, the higher-order terms in the change of the u-component in the $+x$-direction have been carried in this one term. This is to show that in the limit $\delta V = \delta x \, \delta y \, \delta z \to 0$, these terms vanish (since they retain terms in δx after division by δx).

Conservation of mass in this parcel is

$$D \, \delta M / Dt = (D/Dt)(\rho \, \delta x \, \delta y \, \delta z) = 0$$

or

$$\delta x \, \delta y \, \delta z \, D\rho/Dt + \rho(\delta y \, \delta z \, D \, \delta x/Dt + \delta x \, \delta z \, D \, \delta y/Dt$$

$$+ \, \delta x \, \delta y \, D \, \delta z/Dt) = 0 \qquad (5.22)$$

Once again we look at the total change in the density and the shape of the parcel as it instantaneously occupies the point (x, y). To evaluate the total derivative in Eq. (5.22) in a Eulerian frame of reference, we must express the density as a function of time and position.

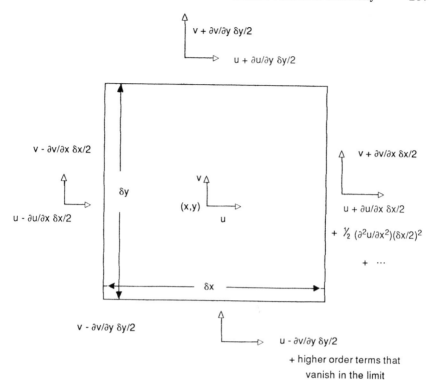

Figure 5.3   2-D parcel of unit height with incremental velocities from the origin at the center of the parcel.

$$Dρ/Dt = D/Dt[ρ\{x, y, z, t\}]$$

$$= ∂ρ/∂t + ∂ρ/∂x \, dx/dt + ∂ρ/∂y \, dy/dt + ∂ρ/∂z \, dz/dt$$

$$= ∂ρ/∂t + u \, ∂ρ/∂x + v \, ∂ρ/∂y + w \, ∂ρ/∂z \qquad (5.23)$$

Now $δx$ is the distance between faces and it is changing as the parcel distorts, one point moving faster than a nearby point due to the changing velocity.

$$D \, δx/Dt = u_{\text{right face}} - u_{\text{left face}}$$

$$= u + ∂u/∂x \, δx/2 - (u - ∂u/∂x \, δx/2) + \tfrac{1}{2}∂^2/∂x^2(δx/2)^2 + \ldots$$

$$= ∂u/∂x \, δx + ∂^2u/∂x^2(δx/2)^2 + \ldots$$

Similarly,

$$D \, δy/Dt ≈ ∂v/∂y \, δy; \qquad D \, δz/Dt ≈ ∂w/∂z \, δz \quad (= 0 \text{ in 2-D})$$

When these expressions are substituted in (5.20),

$$\delta x\,\delta y\,\delta z(\partial\rho/\partial t + u\,\partial\rho/\partial x + v\partial\rho/\partial y) + \rho\,\delta x\,\delta y\,\delta z(\partial u/\partial x + \partial v/\partial y)$$

$$+ \tfrac{1}{4}\delta x^2\,\delta y\,\delta z\,\partial^2 u/\partial x^2 + \ldots = 0$$

Divide by δV to get

$$\partial\rho/\partial t + u\,\partial\rho/\partial x + v\,\partial\rho/\partial y + \rho(\partial u/\partial x + \partial v/\partial y)$$

$$+ \tfrac{1}{4}\partial^2 u/\partial x^2\,\delta x + \ldots = 0$$

and for $\delta x \to 0$, we have the field equation for the density variation as a function of time and position.

$$\partial\rho/\partial t + u\,\partial\rho/\partial x + v\,\partial\rho/\partial y + \rho(\partial u/\partial x + \partial v/\partial y) = 0$$

or

$$\partial\rho/\partial t + \partial(\rho u)/\partial x + \partial(\rho v)/\partial y = 0 \tag{5.24}$$

The extension of this derivation to 3-D is straightforward, but with an attendant proliferation of terms. The economy of the five-line derivation in vector notation is now easily appreciated.

The Eulerian description of density depends on the location (x, y, z) and the time t. For example, in the atmosphere, $\rho(z)$ decreases with height but it is constant over relatively large changes in horizontal distances x or y. In addition, there may be a significant change in ρ with time—for example, due to diurnal heating near the ground $(z \to 0)$ or heat absorption at different levels in the atmosphere. However, we find that in many flow problems, the density does not change significantly, and air can be treated as an incompressible gas. The continuity equation will be used to investigate when this approximation is valid in Section 5.5.

5.4 The Integral Form of Continuity

If a problem involves flow in an enclosed region, such as a tank with several inlets and outlets, it is often simpler to apply the conservation of mass law in the integral form compared to the derivative form. For these cases, we will obtain the same two-part (the time rate-of-change over the volume plus the advection through the sides) integral for the total change in density as in the previous sections. However, instead of shrinking the volume to a point, we will expand it to a large domain called the *control volume*.

We can even allow for the possibility of an arbitrary motion of the selected region of material by using Leibnitz's rule, which we learned in Section 2.7.1 is associated with changing boundaries. We assume the velocity of the surface of the region is given by w_i. The rate of change of mass within the region is obtained by substituting density into Leibnitz's equation to get

$$d/dt \int \rho \, dV = \int \partial \rho / \partial t \, dV + \int \rho w_i n_i \, dA \qquad (5.25)$$

We then use continuity to change the second term,

$$d/dt \int \rho \, dV = - \int \partial(\rho u_i)/\partial x_i \, dV + \int \rho w_i n_i \, dA \qquad (5.26)$$

Finally, we obtain a relation between the total change of mass within the volume [the left side of (5.26)] and the mass flux through the surface. This is done by converting the second term to a surface integral using the divergence theorem and then combining the surface integrals to get

$$d/dt \int \rho \, dV = - \int \rho(u_i - w_i)n_i \, dA \qquad (5.27)$$

In Eq. (5.27), the mass flux through the boundary is equal to the density times the normal component of the flow relative to the boundary motion. If the control volume is a fixed region in space, then $w_i = 0$. If we consider a specific material volume as it moves along with the flow, then $w_i = u_i$; in other words, no material moves through the boundary.

Example 5.3

Use continuity on the steady flow in Fig. 5.4 to relate the velocity at one station to the velocity at another. Use a control volume approach.

Figure 5.4   Flow in a converging channel from area A_1 to A_2 with increase in velocity from u_1 to u_2.

Solution

We can consider the fixed volume between A_1 and A_2 as our finite control volume. In the case of steady-state flow, there is no change with respect to time, and $w_i = 0$. Equation (5.27) becomes simply

$$-\iint \rho u_j n_j \, dA = 0 \quad \text{(Integral continuity equation)}$$

where the integral is over the surface of the control volume. The sides of the channel can be treated as streamlines, for there is no flow across them. One can integrate around the surface of the control volume, which is comprised of the sides, end areas, and unit depth, to get

$$-\rho_1 u_1 A_1 + \rho_2 u_2 A_2 = 0$$

or

$$\rho u A = \text{constant} \tag{5.28}$$

If ρ is constant,

$$u_2 = u_1 A_1 / A_2 \tag{5.29}$$

Such a flow is called channel flow. It can provide rough estimates of wind variation in mountain passes, city streets and other corridors.

Example 5.4

Consider a piston moving in a chamber to eject fluid from a small hole, as shown in Fig. 5.5. Derive a relation for the exit velocity of an incompressible fluid.

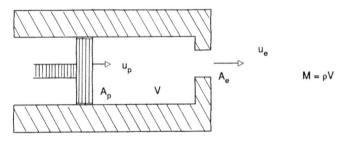

Figure 5.5 A piston moving fluid out of a small orifice.

Solution

In this example we choose the control volume V as that area with unit depth inside the chamber, and it is continually decreasing due to the action of the piston. From Eq. (5.27), the change in mass dM/dt within the chamber is equal to the relative velocity of the boundary and flow velocities. In this case, the left boundary is moving, so we must include the relative motion terms.

$$dM/dt = -\int \rho(u_i - w_i)\, n_i\, dA$$

The volume is decreasing as the surface at the piston moves inward with velocity $w_x = u_p$. The flow at the piston face also moves at u_p, so there is no relative flow through this surface. Thus the only flow across the bounding surface of the control volume is at the hole. At this point the velocity is u_e and the mass flux is $\rho u_e A_e$, so that

$$dM/dt = -\rho u_e A_e$$

We can substitute $M = \rho V$ and, for constant density flow,

$$dV/dt = -u_e A_e$$

If x is the distance from the piston to the end of the cylinder, the volume $V = A_p x$, changes according to

$$dV/dt = A_p\, dx/dt = -A_p u_p$$

Thus,

$$u_p = -1/A_p\, dV/dt = (A_e/A_p)\, u_e$$

This can be seen as an example of continuity written as $\rho U A = $ constant for constant density flow. The piston face merely establishes the flow velocity in the chamber. (In other words, we could discard the piston and imagine the flow as given at A_p, having been established by an upstream pressure head instead of the piston force.)

Example 5.5

Consider a steady-state convective plume of air—a vertically moving burst of air with a well defined expanding circular area—measured at height z_0. Given the parameters in Fig. 5.6, find a relation for velocity as a function

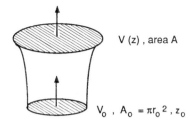

V (z) , area A

V_0 , $A_0 = \pi r_0{}^2$, z_0

Figure 5.6   Cross sections of a convective plume of air. As the plume ascends, the cross-sectional area changes and the vertical speed of the parcels changes.

of area, $V(A)$. (a) Assume no entrainment (i.e., no lateral mixing). (b) Assume lateral entrainment, $-u_e$, and linearly varying $r(z) = r_0 + C_1(z - z_0)$.

Solution

(a) From conservation of mass, we write

$$\text{mass flow in} = \text{mass flow out}$$

$$Q_{in} = Q_{out}$$

$$\rho_0 \,[\text{kg/m}^3]\{V_0 \,[\text{m/sec}] \cdot A_0 \,[\text{m}^2]\} = \rho VA$$

Assuming the vertical rise is not sufficient to cause significant changes in density, we have a simple relation between the velocity and the area.

$$V = V_0(A_0/A)$$

(b) We can allow for flow in through the sides of the control volume as $u_e A_e$, so that the integral over the sides of the control volume yields flow in from the bottom and sides and flow out the top.

$$\rho VA \equiv \rho_0 V_0 A_0 - \rho_e u_e A_e = 0$$

where

$$A_e = 2\pi \int_{z_0}^{z} r\,dz = 2\pi \int_{z_0}^{z} [r_0 + C_1(z - z_0)]\,dz$$

$$= 2\pi[r_0(z - z_0) + \tfrac{1}{2}C_1(z - z_0)^2]$$

We now have the velocity as a function of the given parameters and the area.

$$V(A) = V_0 A_0/A - 2\pi u_e/A\,[r_0(z - z_0) + \tfrac{1}{2}C_1(z - z_0)^2]$$

This example illustrates some of the very severe assumptions that must be made in approximating real geophysical flow events. In an actual cloud-forming plume one might have to consider entrainment of moist air at one level, detrainment of dry air at another level, and precipitation out the bottom.

5.5 Compressibility

Through common experience, we know that air is evidently quite compressible. This characteristic is an important feature in laboratory experiments that involve familiar uses of air in compressors and pistons. It is the essential characteristic in the transmission of sound. However, in the study of atmospheric flow, we are frequently able to treat the air as though it is incompressible. The actual approximation is that the effects due to the variation in air density are negligible compared to those of pressure and velocity variations. We can explore this approximation with the continuity equation. It is evident that when ρ in Eq. (5.4) is constant, $\nabla \cdot \mathbf{u} = 0$, so that we associate incompressibility with nondivergence. However, we can obtain more information if we first obtain the conservation of mass equation in terms of the *specific volume*, $V_s = 1/\rho$, the volume of a unit mass.

The expression for the fractional rate of change of specific volume, $\delta V/$ unit mass $= \delta x\, \delta y\, \delta z$, can be written

$$\frac{D(\delta x\, \delta y\, \delta z)/Dt}{\delta x\, \delta y\, \delta z} = \frac{1}{\delta x} D\, \delta x/Dt + \frac{1}{\delta y} D\, \delta y/Dt + \frac{1}{\delta z} D\, \delta z/Dt$$

or (with $D(\delta x_i)/Dt = \partial u_i/\partial x_i$)

$$\frac{1}{V_s} DV_s/Dt = \partial u/\partial x + \partial v/\partial y + \partial w/\partial z = \nabla \cdot \mathbf{u} \qquad (5.30)$$

This is continuity in terms of V_s. In this equation we see that nondivergence corresponds to negligible fractional rate of expansion, as introduced in Section 5.3. For large volumes, such as we see in the free atmosphere, we might expect the fractional rate of expansion to be small. However, for very small volumes such as those associated with sound waves, this term can be comparable to the divergence. (The scale of audible wavelengths is less than 30 cm; however, the displacement amplitudes are even smaller, $\approx 10^{-3}$ cm.)

Equation (5.30) can be changed to the familiar term with respect to density using

$$DV_s = -1/\rho^2\, D\rho$$

Thus, Eq. (5.30) may be written

$$-1/\rho \, D\rho/Dt = \mathbf{\nabla} \cdot \mathbf{u}$$

or

$$D\rho/Dt + \rho(\mathbf{\nabla} \cdot \mathbf{u}) = 0 \qquad (5.31)$$

We can also write the continuity equation in terms of pressure. To use the thermodynamic relation between density and pressure, we must assume negligible molecular transport effects—that is, set viscosity and thermal diffusivity equal to zero, so that we have isentropic flow with constant potential temperature. Then the perfect gas law applies and $p = \rho RT$ allows us to write

$$D\rho/Dt = \frac{1}{RT} Dp/Dt \bigg]_{\theta,S} = \frac{1}{c_S^2} Dp/Dt \bigg]_{\theta,S}$$

where

$$c_S = [\partial p/\partial \rho]_{\theta,S}^{1/2}$$

$$\equiv \text{speed of sound}$$

$$= (RT)^{1/2} \quad \text{for a perfect gas}$$

Since the speed of sound depends on compressibility, any flow velocities that approach this speed will also be affected by compressibility. The ratio of the flow speed to the speed of sound will yield some measure of the importance of compressibility. When we replace density with pressure in the continuity equation, c_S enters as a factor.

Continuity can be written in terms of pressure changes,

$$\frac{1}{\rho c_S^2} Dp/Dt + \mathbf{\nabla} \cdot \mathbf{u} = 0 \qquad (5.32)$$

When the first term is negligible in Eq. (5.32), the flow can be approximated as nondivergent and incompressible. This is likely to be true if c_S is very large compared to other velocities involved in a problem.

The general conditions necessary for $\mathbf{\nabla} \cdot \mathbf{u} = 0$ can be obtained from Eq. (5.32), a consideration of the perturbation equations (to be discussed in Chapter 10) and conditions that the vertical advection term $w \, \partial \rho/\partial z$ is negligible. They may be stated as

1. $u/c_S \ll 1$
2. Phase speed of perturbation waves/$c_S \ll 1$
3. Vertical scale of motion $\ll$ characteristic scale height of problem

The first condition is often violated in aerodynamics, the second in acoustic phenomena, and the third in dynamic meteorology. In the last case, meteorology, the characteristic atmospheric scale height is associated with the median of $\rho \, d\rho/dz$. In the atmosphere, this is obtained at a height of about 8 km, which is comfortably larger than planetary boundary-layer scales (about 1–2 km) but significantly less than tropospheric heights (about 12–20 km). Thus we can neglect density variations in isentropic PBL flows but must consider them in flows of depth greater than a few kilometers. This assumes that the vertical scale of the fluid motion can equal the height of the problem domain, but it may be considerably less. In the ocean, the characteristic scale height is an order of magnitude greater than the typical height scale of motion, making the incompressible assumption secure.

Example 5.6

We can obtain a dimensionless parameter for compressibility by nondimensionalizing the continuity equation written as Eq. (5.32). By replacing Dp/Dt with $p/\rho \, D\rho/Dt$ in Eq. (5.32), we get

$$\left[\frac{p}{\rho^2 c_s^2}\right] \frac{D\rho}{Dt} + \nabla \cdot \mathbf{u} = 0 \tag{5.33}$$

Then, by nondimensionalizing the terms in this equation with P_0, ρ_0, and U_0, L and t_0, where $U_0 = L/t_0$ and $P_0 = \rho U_0^2$, we get the dimensionless equation and use it to evaluate conditions where the dynamic effects on density are negligible.

Solution

Following the rules described in Chapter 3, substitute the nondimensional parameters.

With $u' = u/U_0$, $x' = x/L$, $t' = t/t_0$, etc., and

$$\partial/\partial x_i = \partial/\partial x_i' \, \partial x_i'/\partial x_i = 1/L \, \partial/\partial x_i'$$

so that

$$\nabla \cdot \mathbf{u} = U_0/L \, \nabla' \cdot \mathbf{u}'$$

we get

$$\left[\frac{P_0}{\rho c_s^2 t_0}\right] \frac{p'}{\rho'^2} \frac{D\rho'}{Dt'} + U_0/L \, \nabla' \cdot \mathbf{u}' = 0$$

If we divide through by U_0/L, we get a single coefficient in front of the density change term, which represents the ratio of the magnitude of the first term to that of the divergence term (recall that the dimensionless terms were dimensionalized to be of order unity). Thus,

$$\left[\frac{P_0}{\rho c_s^2}\frac{L}{U_0 t_0}\right]\frac{p'}{\rho'^2}\frac{D\rho'}{Dt'} + \nabla' \cdot \mathbf{u}' = 0$$

Since $P_0 = \rho U_0^2$ and $L/(U_0 t_0) = 1$, we have

$$\frac{U_0^2}{c_s^2}\frac{p'}{\rho'^2}\frac{D\rho'}{Dt'} + \nabla' \cdot \mathbf{u}' = M^2\frac{p'}{\rho'^2}\frac{D\rho'}{Dt'} + \nabla' \cdot \mathbf{u}' = 0 \qquad (5.34)$$

where $M \equiv U_0/c_s$ is the Mach number. We see that the condition of non-divergence is associated with low Mach number flow—when the Mach number is negligible, the flow is incompressible. The second factor, p'/ρ'^2, is significant only for large vertical motions.

The speed of sound in air is 341 m/sec, and in water, 1470 m/sec. Thus, for horizontal air speeds less than 50 m/sec ($M = 0.14$) the density-change term is negligible. In water, it is negligible for $U < 150$ m/sec. Atmospheric flows have $M \leq 0.2$. Thus the density term is important in the atmosphere only in association with large p'/ρ'^2. This criteria will be examined in the next example.

Example 5.7

Use the techniques of Section 3.3 to nondimensionalize the steady-state version of continuity

$$\nabla(\rho\mathbf{u}) = \mathbf{u} \cdot \nabla\rho + \rho\nabla \cdot \mathbf{u} = 0 \qquad (5.35)$$

to study the criteria for the incompressible approximation. We are interested in obtaining nondimensional parameters with magnitudes near unity. Thus, we must give consideration to the fact that characteristic scales for density changes in the horizontal are different (considerably greater) than those in the vertical. Also, velocity characteristic scales are considerably greater for horizontal winds compared to vertical winds. Thus, choose different characteristic scales for horizontal and vertical distances and velocities, $L \gg H$, $U, V \gg W$. Also, density changes a significant amount, with characteristic value $\Delta\rho_V$ over the vertical scale H but only a small amount in the horizontal $\Delta\rho_H$. When density appears alone, it is nondimensionalized with a base value ρ_0.

Solution

We let $u = u'/U$, $v = v'/U$, $w = w'/W$, $x = x'/L$, $y = y'/L$, $z = z'/H$, and $\rho = \rho'/\Delta\rho_V$ for vertical changes, $\rho'/\Delta\rho_H$ for horizontal.

$$\left[\frac{U\Delta\rho_H}{L}\right](u\,\partial\rho/\partial x + v\,\partial\rho/\partial y) + \left[\frac{W\,\Delta\rho_V}{H}\right](w\,\partial\rho/\partial z)$$

$$+ \left[\frac{\rho_0 U}{L}\right]\rho(\partial u/\partial x + \partial v/\partial y) + [\rho_0 W/H]\,\rho\,\partial w/\partial z = 0$$

When we divide by the coefficient $[\rho_0 U/L]$ of the third term, the horizontal divergence, we get, together with order of magnitude estimates,

$$\left[\frac{\Delta\rho_H}{\rho_0}\right]\left(\frac{u\,\partial\rho}{\partial x} + \frac{v\,\partial\rho}{\partial y}\right) + \left[\frac{WL\,\Delta\rho_V}{UH\,\rho_0}\right]\frac{w\,\partial\rho}{\partial z}$$

[0.01]   [1]   +   [0.1–0.7]   [1]

$$+ \rho\left(\frac{\partial u}{\partial x} + \frac{\partial v}{\partial y}\right) + \left[\frac{WL}{UH}\right]\frac{\rho\,\partial w}{\partial z} = 0$$

[1]   [1]   +   [≤1]   [1]

These coefficients permit us to determine which terms must be included in the continuity equation in addition to the horizontal divergence term.

The horizontal density advection term is negligible. If $WL/(UH) \leq 1$, then $w\,\partial\rho/\partial z$ and $\rho\,\partial w/\partial z$ must be considered. The magnitude of the first term also depends on $\Delta\rho_V/\rho_0$, which ranges from 0.1 for $H = 1$ km to 0.7 for $H = 10$ km. Thus there are cases in thin layers such as the PBL where only the $\rho\,\partial w/\partial z$ term must be included in addition to the horizontal divergence term. However, in many cases, W is assumed to be ≈ 0, so that $WL/UH \approx 0$, and the incompressible approximation is valid.

$$\nabla \cdot \mathbf{u} = 0 \tag{5.36}$$

This example illustrates (1) the complexities of nondimensionalizing with different characteristic scales, but also (2) the benefits of this process in determining whether or not certain approximations are valid.

5.6   Conservation Statement for the Quantity f

We have established a field equation for the mass parameter, density, in terms of flow velocity and the independent variables—the coordinate parameters and time. We can generalize the procedure we followed in order to apply it to the conservation of something other than mass. That is, for any conserved quantity, the temporal change integrated over the incremental volume equals the integrated net flux in and out through the surfaces of the volume. If we can then convert the surface integrals to volume integrals, the equation will be simply a volume integral of the differential rate of change of the quantity for the parcel. When we are seeking a field equation we would next consider the limit $\delta V \to 0$ to obtain an equation that governs the variable characteristics at all points in the flow field. To handle a wide range of variables, we can allow for possible sinks or sources of the quantity.

We will then have a very general statement of conservation for any parameter that describes some thermodynamic, state, or dynamic characteristic of the parcel (e.g., latent heat, moisture, NO_3, momentum, etc.). In each case, the parcel will experience changes in the chosen quantity due to the temporal change of the field plus the advective change as it moves through the flow field. At any particular point, the parcel may also encounter sources or sinks that produce or absorb the quantity. For instance, in the conservation of mass, these might be springs or drains of the fluid. And, in the case of momentum, we will find that forces provide the sinks and sources.

In each case we will write the equations as time-dependent integrals over the volume of the parcel plus an integral over the parcel surface. Following the procedure developed in this chapter, we will then proceed to use the divergence theorem on the area integrals. In this way we obtain a single integral over δV. We can then consider the limit for small volumes within the continuum to obtain the differential equation for the change in the considered quantity for our parcel at the point $\mathbf{x}$ at time t. Since the point is arbitrary and the source/sink field is assumed known, the field equation for the quantity is determined. This equation describes the variation of the quantity as a function of the flow field, the sink/source field, and the independent variables.

5.6.1   Conservation of an Arbitrary Quantity

Consider the conservation of $\mathbf{f}$, a quantity associated with our parcel, which may be a scalar, vector, or tensor. We know that the total change in $\mathbf{f}$ at a point in the field consists of a local time change plus a change due to the advection of $\mathbf{f}$ through the point. In addition, $\mathbf{f}$ may change if there is anything producing or eliminating it at the point. Thus we could write

$$D\mathbf{f}/Dt = \sum \mathbf{Q} \qquad (5.37)$$

where $\mathbf{Q}$ represents the sinks and sources of $\mathbf{f}$.

The total derivative of the quantity $\mathbf{f}$ in our parcel can be written in two parts. This is accomplished through an application of Leibnitz's theorem for the time rate of change of a function $\mathbf{f}$ integrated over a volume that may be changing due to its motion with velocity $\mathbf{u}$.

$$\frac{D}{Dt} \iiint \mathbf{f}\, dV = \iiint \frac{\partial \mathbf{f}}{\partial t}\, dV + \iint \mathbf{f}\,(\mathbf{u}\cdot\mathbf{n})\, dA \qquad (5.38)$$

The total change in the quantity $\mathbf{f}$ as expressed in Eq. (5.38) will be zero unless there are sources or sinks in the volume of the parcel. In other words, the local time change of $\mathbf{f}$ in the parcel plus the net $\mathbf{f}$ advected through the surface area [the right side of Eq. (5.38)], plus contributions from any sources or sinks in the parcel must add up to zero. Or

$$\iiint \frac{\partial \mathbf{f}}{\partial t}\, dV + \iint \mathbf{f}(\mathbf{u}\cdot\mathbf{n})\, dA = \sum Q \qquad (5.39)$$

For this derivation we also need the divergence theorem,

$$\iiint_V \operatorname{div} \mathbf{u}\, dV = \iint_A \mathbf{u}\cdot\mathbf{n}\, dA$$

applied to a tensor $\mathbf{A}$

$$\iiint \operatorname{div} \mathbf{A}\, dV = \iint \mathbf{A}\mathbf{n}\, dA$$

We can use the divergence theorem to convert the surface integral in (5.39) to a volume integral. First use an identity from the summary chart at the end of Chapter 4 to change $\mathbf{f}(\mathbf{u}\cdot\mathbf{n})$ to $(\mathbf{f}\,;\mathbf{u})\,\mathbf{n}$, then apply the divergence theorem.

$$\iint \mathbf{f}(\mathbf{u}\cdot\mathbf{n})\, dA = \iint (\mathbf{f}\,;\mathbf{u})\,\mathbf{n}\, dA$$

$$= \iiint \operatorname{div}(\mathbf{f}\,;\mathbf{u})\, dV$$

This states that the quantity $\mathbf{f}$ that is being transported through the surface area of the parcel by the velocity $\mathbf{u}$ can be related to the integral over the volume of $\operatorname{div}(\mathbf{f}\,;\mathbf{u})$. Since the divergence reduces the rank of the tensor on which it operates, the rank of the tensor $(\mathbf{f}\,;\mathbf{u})$ must be one greater than is $\mathbf{f}$. It must equal the order of the first term in Eq. (5.39).

Equation (5.39) becomes

$$\iiint \partial \mathbf{f}/\partial t \, dV + \iiint \text{div} \, (\mathbf{f} ; \mathbf{u}) \, dV = \sum \mathbf{Q} \qquad (5.40)$$

The term representing the sum of the sinks and sources, $\sum \mathbf{Q}$, can also be separated into a distributed part operating over the volume plus a part that operates on the surface of the parcel.

$$\mathbf{Q}_b + \mathbf{Q}_s = \iiint \mathbf{Q}_b' \, dV + \iint \boldsymbol{\sigma}_Q \, d\mathbf{A} \qquad (5.41)$$

where $\mathbf{Q}_b$ is the body contribution, $\mathbf{Q}_s$ is the surface contribution, and $\boldsymbol{\sigma}_Q$ *is defined as the operator that acts on the surface area to yield* $\mathbf{Q}_s$. Note that if $\boldsymbol{\sigma}_Q$ were the velocity vector, the dot product with $\delta \mathbf{A}$ would give only the outward-directed flow. The product $\boldsymbol{\sigma}_Q \, d\mathbf{A}$ must be a vector with components in the direction of the coordinate axes. Thus $\boldsymbol{\sigma}_Q$ must be a second-order tensor operating on $d\mathbf{A}$ to create the vector components of $\mathbf{Q}_s$. The $\boldsymbol{\sigma}_Q \, d\mathbf{A}$ term can be expressed as a vector operating on each face of the parcel, as discussed in Section 4.1. Both $d\mathbf{A} = \mathbf{n} \, dA$ and the surface force acting on $d\mathbf{A}$ can have arbitrary orientations. There are nine terms involved in the complete description of the three components of force acting in each of the three coordinate directions of $d\mathbf{A}$. The tensor operator can account for these terms.

The divergence theorem shows that the divergence operating on the tensor $\boldsymbol{\sigma}_Q$ produces the vector $\mathbf{Q}_s$. Therefore, we can express a surface sink/source of term $\mathbf{Q}_s$ in terms of a volume distribution.

$$\iint \boldsymbol{\sigma}_Q \, \mathbf{n} \, dA = \iiint \text{div} \, \boldsymbol{\sigma}_Q \, dV \equiv \mathbf{Q}_s \qquad (5.42)$$

In other words, $\boldsymbol{\sigma}_Q$ is an operator that operates on (vector-multiplies) the unit normal vector to beget the components of $\mathbf{Q}_s$ that are acting on the incremental area. There will be three components of $\mathbf{Q}_s$ acting on each of three components of $\mathbf{n} \, dA = d\mathbf{A}$, equaling nine terms.

We can now write

$$\iiint \{\partial \mathbf{f}/\partial t + \text{div} \, (\mathbf{f} ; \mathbf{U}) - \mathbf{Q}_b' - \text{div} \, \boldsymbol{\sigma}_Q\} \, dV = 0$$

or, in the limit as the incremental parcel shrinks to a point value,

$$\frac{\partial \mathbf{f}}{\partial t} + \text{div} \, (\mathbf{f} ; \mathbf{U}) = \mathbf{Q}_b + \text{div} \, \boldsymbol{\sigma}_Q \qquad (5.43)$$

The simplest expression will be obtained when $\mathbf{f}$ is a scalar, as in the conservation of mass in this chapter. In this case, there are generally no sinks or sources, so that the equation has only the terms of temporal and advective rate of change. We will find that in the case of energy, although the dependent variable is a scalar—either internal energy or temperature—there are frequently sinks and sources of energy located in the fluid flow domain. In the conservation of momentum discussed next, $\mathbf{f} = \rho\mathbf{u}$, so that we have a vector equation. In addition, there are sinks and sources of momentum arising from the various forces acting on the fluid and from those forces internal to the fluid flow. The latter requires the introduction of a tensor source/sink and $\boldsymbol{\sigma}_Q = \boldsymbol{\sigma}$, the stress tensor.

We will employ the basic procedures of this chapter, together with everything else we have learned, in developing the equations for the conservation of momentum.

Summary

CONSERVATION OF MASS—CONTINUITY

Integral Format $\displaystyle\iiint \frac{\partial\rho}{\partial t}\, dV = -\iint \rho\, \mathbf{u}\cdot\mathbf{n}\, dA$

Differential Format $D\rho/Dt + \rho\nabla\cdot\mathbf{U} = 0$
Incompressible $D\rho/Dt = 0$
Constant Density $\nabla\cdot\mathbf{u} = 0$
Two-Dimensional
$$\frac{\partial\rho}{\partial t} + \frac{\partial(\rho u)}{\partial x} + \frac{\partial(\rho v)}{\partial y} = 0$$
Control Volume
Compressibility

GENERAL STATEMENT OF CONSERVATION OF $\mathbf{f}$

$\partial\mathbf{f}/\partial t + \mathrm{div}(\mathbf{f}\,;\mathbf{U}) = \mathbf{Q}_b + \mathrm{div}\,\boldsymbol{\sigma}_Q$

where $\mathbf{f}$, $\mathbf{U}$, and $\mathbf{Q}$ are vectors and $\boldsymbol{\sigma}$ is a 2nd-order tensor.

Problems

1. Use continuity to simplify $\partial(\rho u_i)/\partial t + \partial(\rho u_i u_j)/\partial x_j$.
2. The wind is blowing down the main street with flow off of a side

street, as shown in the figure. What is an estimate of the velocity in the side street shown? Assume 2-D flow, constant density.

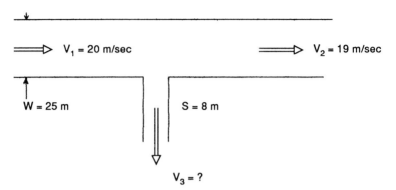

3. A cylindrical plunger moves downward into a conical receptacle, filled with liquid. At what height z in terms of d will the mean upward velocity of the liquid between the plunger and the wall equal the same magnitude as the downward plunger speed?

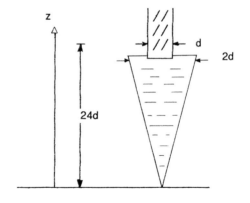

4. A tornado has velocity components,

$$(u_r, u_\theta) = (-C_r/r, -C_\theta/r)$$

Find an equation for the streamlines. If this velocity is valid for 30 m $<$ $r <$ 100 m and there is a constant upward velocity in the center, what is this velocity if $C_r = 300$ m^2/sec, $C_\theta = 800$ m^2/sec?

5. The formula for an incompressible flow is given as

$$u = U(x^3 + xy^2), \quad v = U(y^3 + yx^2), \quad w = 0$$

Is such a flow possible? What if $w = -4Ur^2z$ ($r^2 = x^2 + y^2$)?

6. A plume of wood smoke carcinogens moves from its source past you with the entrainment characteristics shown. Get a formula for the pollutant concentration (ρ_p/unit volume) as a function of distance. Assume incompressible, and a cylindrical plume with diameter $D = D_0 + 0.02 \times xm$.

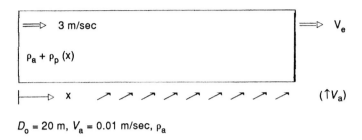

$D_0 = 20$ m, $V_a = 0.01$ m/sec, ρ_a

7. Use a version of conservation of mass to obtain V_3 for flow in and out of a chamber with characteristics as given in the sketch.

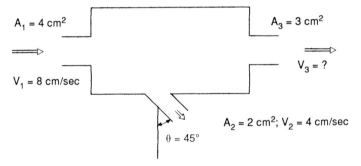

8. Consider a two-dimensional box with flow in and out as shown in the sketch.

For constant density ρ, what is V_3?

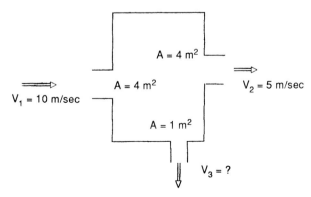

9. For problem 8, what are the forces on the box (with respect to ρ)?

10. Using $d(\delta x_i)/dt = \partial u_{(i)}/\partial x_{(i)} \, \delta x_{(i)}$ [(i) is not summed], get an expression for the rate of volumetric strain. What is its value in incompressible flow?

11. Water is flowing at a constant flow rate through a large pipe of diameter $D_1 = 600$ mm that branches into two smaller pipes of diameter $D_2 = 400$ mm and $D_3 = 250$ mm. A discharge of 5 m³/sec is maintained in the 600-mm diameter pipe. An average velocity of $V_2 = 24$ m/sec was measured in the 400-mm diameter pipe. Determine the discharge and the average velocity in the 250-mm diameter pipe.

12. Water enters a circular chamber of radius R through a circular pipe, radius r. It effluxes out the periphery of the chamber. The chamber height is h, the flow rate in is Q. What is the efflux velocity?

13. Write a general equation for the conservation of lawyers in Bellevue, WA.—that is, the conservation of Y in domain B.

14. In an atmospheric experiment, radiosonde measurements of the wind are taken vertically through the PBL to a height of 2 km at four corners of a rectangle. Average values are as shown. Use these values to determine the vertical velocities at the top of the box. Assume compressibility is negligible in the PBL.

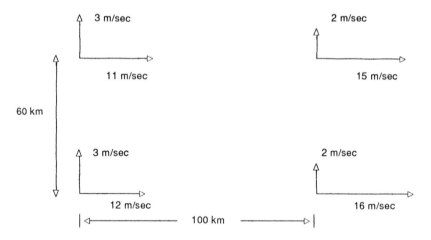

Chapter 6 | Momentum Dynamics

In the study of fluid flow, the velocity is our most important dependent variable. The product of velocity and mass, the momentum, has a fundamental statement of conservation. The principle of the conservation of momentum results in the basic equations for determining the velocity fields. These are the equations of motion.

Newton formulated the principle that the momentum is conserved unless acted upon by a force. Although he was concerned with a point mass, we can apply an analogous principle to the continuum parcel. Forces are the sources and sinks of momentum. In Chapter 5 we left the source and sink terms out of the conservation of mass equation since creation or destruction of mass does not generally arise in atmospheric problems. However, the forces that create and destroy momentum are very important to the flow dynamics. In fact we will find that often in our problems the acceleration is zero. Then the momentum equation reduces to a statement that the sum of the forces equals zero.

Atmospheric problems frequently involve rotation of domains from parcels to air masses. Thus, we might expect angular momentum to be an important factor, at least in some situations. However, in a great many flow problems, linear momentum dominates. We simply note that this section deals with a special case of momentum conservation, the conservation of linear momentum. We will return to examine rotation and angular momentum in Chapters 8 and 9.

6.1 Conservation of Momentum—Newton's Law

Newton's law applied to the continuum parcel states that the rate of change of momentum equals the sum of forces acting on the parcel, Σ **F**.

$$D(m\mathbf{u})/Dt = \sum \mathbf{F} \qquad (6.1)$$

However, the fluid parcel is not a constant aggregate of mass particles. To express this balance for a finite fluid volume, both sides of this equation need special attention. First consider the left side, expressing the rate of change of momentum as the total derivative of $m\mathbf{u}$. For the elemental parcel of fluid with uniform density we can write the mass as the mass per unit volume, hence density, and the total derivative can be written

$$D(m\mathbf{u})/Dt = D(\rho\mathbf{u})/Dt = \partial(\rho\mathbf{u})/\partial t + (\mathbf{u} \cdot \text{grad})(\rho\mathbf{u}) \qquad (6.2)$$

The local time change plus the advective change of the aggregate mass of fluid can be written in terms of the integral over a specific volume of fluid called the control volume. Since the control volume is at a specific point in space, there is continuous flow through the volume. However, at any instant, there is a specific mass of fluid in the control volume, to which we will apply the conservation of momentum principle. We must keep in mind that there is a rate change contribution from the net flow in and out of the control volume. This doesn't go to zero at $\delta t \to 0$ (or $\delta x \to 0$). Thus the total rate of change of the momentum of any small material volume of fluid must be written in two parts. One is the time-dependent change of momentum integrated over the volume. The other is the change due to flux of momentum through the surface.

$$\iiint \partial(\rho\mathbf{u})/\partial t \, dV + \iint \rho\mathbf{u}(\mathbf{u} \cdot \mathbf{n}) \, dA = \sum \mathbf{F} \qquad (6.3)$$

Here, we have a vector integral equation for the conservation of momentum. This equation results when $\rho\mathbf{u}$ replaces **f** in the general conservation statement, Eq. (5.39). These terms can be illustrated by examining the components with index notation, where $(u, v, w) \equiv (u_1, u_2, u_3)$. (See Fig. 6.1).

The quantity being transported across the incremental area δA [the third

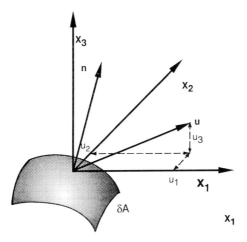

Figure 6.1   The incremental area δA with unit normal **n** and velocity **u**.

term in Eq. (5.38)] is the momentum, $\rho\mathbf{u}$, and it may be written in several ways using indicial, matrix, and symbolic notation.

Indicial:

$$\rho u_i(u_j n_j) = \rho u_i(u_1 n_1 + u_2 n_2 + u_3 n_3) = \rho(u_i u_j)\, n_j$$

Matrix:

$$= \rho \begin{pmatrix} u_1 u_1 n_1 & + u_1 u_2 n_2 & + u_1 u_3 n_3 \\ u_2 u_1 n_1 & + u_2 u_2 n_2 & + u_2 u_3 n_3 \\ u_3 u_1 n_1 & + u_3 u_2 n_2 & + u_3 u_3 n_3 \end{pmatrix}$$

$$= \rho \begin{pmatrix} u_1 u_1 & u_1 u_2 & u_1 u_3 \\ u_2 u_1 & u_2 u_2 & u_2 u_3 \\ u_3 u_1 & u_3 u_2 & u_3 u_3 \end{pmatrix} \begin{pmatrix} n_1 \\ n_2 \\ n_3 \end{pmatrix} =$$

Symbolic:

$$= \rho\mathbf{u}(\mathbf{u}\cdot\mathbf{n}) = \rho(\mathbf{u}\, ;\, \mathbf{u})\,\mathbf{n}$$

Thus, the left side of (6.1) may be written per unit mass:

$$\iiint \partial(\rho\mathbf{u})/\partial t\, dV + \iint (\rho\mathbf{u})(\mathbf{u}\cdot\mathbf{n})\, dA$$

$$= \iiint \partial(\rho\mathbf{u})/\partial t\, dV + \iint (\rho\mathbf{u}\, ;\, \mathbf{u})\cdot\mathbf{n}\, dA$$

Note that the last term consists of a tensor, $\rho(\mathbf{u} ; \mathbf{u})$, operating on a vector $\mathbf{n}$, to produce a vector. Employ the divergence theorem on this term to get

$$\iiint \partial(\rho\mathbf{u})/\partial t \; dV + \iiint \text{div}(\rho\mathbf{u} ; \mathbf{u}) \; dV$$

or

$$\iiint \partial(\rho u_i)/\partial t \; dV + \iiint \partial/\partial x_j(\rho u_i u_j) \; dV$$

or

$$\iiint [u_i \, \partial\rho/\partial t + \rho \, \partial u_i/\partial t + \rho u_j \, \partial u_i/\partial x_j + u_i \, \partial(\rho u_j)/\partial x_j] \; dV$$

From continuity, [Eq. (5.8)], $\partial(\rho u_j)/\partial x_j + \partial\rho/\partial t = 0$; and the $u_i \, \partial(\rho u_j)/\partial x_j$ terms cancel, leaving

$$\iiint \rho[\partial u_i/\partial t + u_j \, \partial u_i/\partial x_j] \; dV = D(\rho\mathbf{u})/Dt \tag{6.4}$$

We have now obtained the left-hand side of the momentum equation. The total derivative of momentum has been expressed as an integral over a specific volume of material in an inertial frame of reference. We will discuss the ramifications of the geophysical rotating frame of reference later. Now we need the forces.

When we speak of the imaginary parcel isolated from the surrounding fluid, we must account for the action of the surroundings on the parcel by using the concept of *internal* forces on the parcel surface. The fluid actions on the surface of the parcel are pressure and friction forces, represented by $\mathbf{F}_S$. The total fluid forces will consist of two parts, forces that act on the mass bulk of the material in the parcel, called *body* forces, and the $\mathbf{F}_S$, called *surface* forces.

$$\mathbf{F} = \rho\mathbf{F}_b + \mathbf{F}_S \tag{6.5}$$

The body forces act equally on each element of the parcel because the parcel characteristics are uniform. Thus, the net body force can be represented by a resultant force per unit volume acting at the center of the parcel in the direction of the vector $\mathbf{F}_b$. (Formally, we would have to invoke the mean-value theorem to state that the integral of the body forces over the parcel volume is equal to the force acting on a mean value of the integral, somewhere within the volume.) The density is included with $\mathbf{F}_b$ because our body force is gravity, $\mathbf{g}$, which is the weight force per unit mass. In the limit for an infinitesimal parcel at field point **x**, this resultant force acts at the point.

$$\rho \mathbf{F}_b = \int\!\!\int\!\!\int \rho \mathbf{F}_b' \, dV$$

where $\mathbf{F}_b'$ are the body forces acting on each element of the parcel.

Example 6.1

For an arbitrary control volume, the momentum balance from Eq. (6.1) is simply (Section 5.6)

$$D(\rho \mathbf{V})/Dt = \partial/\partial t \int\!\!\int\!\!\int (\rho \mathbf{V}) \, dV + \int\!\!\int (\rho \mathbf{V}) \, \mathbf{V} \cdot \mathbf{dA}$$

$$= \int\!\!\int (\rho \mathbf{V}) \, \mathbf{V} \cdot \mathbf{dA} = \sum F_i \quad \text{(for steady state).} \quad (6.6)$$

Apply this to a streamtube bending in space to get the net forces on the streamtube. Apply the result to get the forces on an elbow in a pipe with the configuration as shown in Fig. 6.2. Let $d_1 = 20$ cm; $d_2 = 10$ cm; $Q = 0.25$ m^3/s; and $\rho = 1000$ kg/m^3.

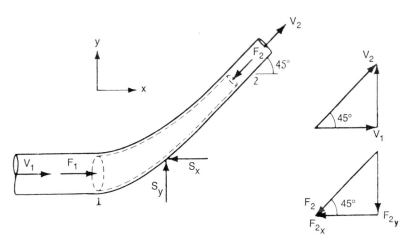

Figure 6.2   The flow through a reduced elbow.

Solution

Consider the control volume shown in Fig. 6.2 as representative of an arbitrary streamtube in space. Since there is no flow through the sides, Eq. (6.6) applies. At the inlet,

$$\iint (\rho \mathbf{V}) \, \mathbf{V} \cdot \mathbf{dA} = -\rho_1 V_1 V_1 A_1 = -\rho Q_1 V_1 = -DM_1/Dt$$

Similarly, at the outlet,

$$\iint (\rho \mathbf{V}) \, \mathbf{V} \cdot \mathbf{dA} = \rho_2 V_2 V_2 A_2 = \rho Q_2 V_2 = DM_2/Dt$$

Hence, in general,

$$\sum F = \Delta M / \Delta t \tag{6.7}$$

where M is the mass flow rate across the streamtube.

When this formula is applied to the flow through the elbow, with $Q = 0.25 \text{ m}^3/\text{sec}$; $d_1 = 20$ cm; $d_2 = 10$ cm and $\rho = 1000 \text{ kg/m}^3$,

$$A_1 = \pi(20)^2/4 = 0.0314 \text{ m}^2; A_2 = 0.00785$$

$$F_1 = p_1 A_1 = 160 \cdot 0.0314 = 5 \text{ kN}$$

$$F_2 = p_2 A_2 = 140 \cdot 0.00785 = 1.1 \text{ kN}$$

and

$$V_1 = 0.25/0.0314 = 7.96 \text{ m/sec}$$

$$V_2 = 0.25/0.00785 = 31.85 \text{ m/sec}$$

The balance in the x-direction yields

$$F_1 - F_{2(x)} - S_x = \rho Q V_{2(x)} - \rho Q V_1$$

In the y-direction,

$$-F_{2(y)} + S_y = \rho Q(V_{2(y)} - V_{1(y)})$$

Hence,

$$5{,}000 - 1{,}100 \cos 45° - S_x = 1{,}000 \cdot 0.25(31.85 \sin 45° - 7.96)$$

or

$$S_x = 5622 \text{ N}$$

and

$$S_y = 1,100 \sin 45° + 1,000 \cdot 0.25 \cdot 31.85 \cos 45°$$

$$= 6407 \text{ N}$$

The force on the elbow to hold it in place must have magnitude and direction:

$$S = (5622^2 + 6407^2)^{1/2} = 8524 \text{ N}$$

at

$$\theta = \tan^{-1} [5622/6407] = 41°$$

However, the internal fluid forces that act on the *surface* of the parcel may vary in space. Thus, there is a rate-of-change of the forces at any point in the fluid. We can obtain the net surface force on the parcel by considering the finite change in $\mathbf{F}_S$ across the infinitesimal width of the parcel. The net force on the parcel as it shrinks to a point will depend on the gradients of the force at the point. The surface force concept assumes instantaneous action. It applies to the material Lagrangian parcel moving with the fluid at any point in the fluid. Or it applies to the imaginary Eulerian parcel fixed at a point. In the limit $t \to 0$, the parcel moving through a point and the parcel fixed at the point coincide and the forces at that point are identical, regardless of the frame of reference used.

As discussed in Chapter 4, resolving the components of the vector force on the surface of the parcel into the component directions of $\mathbf{dA} = \mathbf{n} \, dA$ requires a nine-component array of numbers. The force has components along and normal to $\mathbf{n}$. The area vector is aligned with the coordinate axes.

Let us now define the *tensor* $\boldsymbol{\sigma}$, such that when it operates on the unit vector normal to a surface it produces the *stress vector force* on the surface. Thus, it is a nine-component set of numbers that will produce the stress force components on each of the surfaces of the parcel. Since the stress forces are proportional to rate of strain, $\boldsymbol{\sigma}$ will be made up of the various components of *velocity shear*.

The total force on the parcel due to the surface forces is the surface integral of the vector forces,

$$\mathbf{F}_s = \iint \boldsymbol{\sigma} \mathbf{n} \, dA$$

Using the divergence theorem, we get

$$\mathbf{F}_s = \iiint \text{div } \boldsymbol{\sigma} \, dV \tag{6.8}$$

Now we are ready to combine the expressions for the change of momentum and the forces. Both expressions are in terms of integrals over the incremental volume and are placed together into the basic conservation-of-momentum equation [substitute (6.2) and (6.8) into Eq. (6.1)] to get

$$\iiint \{\partial(\rho u_i)/\partial t + \partial/\partial x_j(\rho u_j u_i) - \rho F_{bi} - \partial \sigma_{ji}/\partial x_j\}\, dV = 0$$

In the limit $\delta V \to 0$ for a parcel at $\mathbf{x}$,

$$\partial(\rho u_i)/\partial t + \partial/\partial x_j(\rho u_j u_i) - \rho F_{bi} - \partial \sigma_{ji}/\partial x_j = 0 \qquad (6.9)$$

or

$$\rho\, \partial u_i/\partial t + \rho u_j\, \partial u_i/\partial x_j + u_i\, \partial\rho/\partial t + u_i\, \partial(\rho u_j)/\partial x_j$$

$$- \rho F_{bi} - \partial \sigma_{ji}/\partial x_j = 0$$

As before, we combine the first two terms in this equation into $\rho Du_i/Dt$, and the second two into $u_i[\partial\rho/\partial t + \partial(\rho u_j)/\partial x_j]$, which is equal to zero from the conservation of mass. This results in

$$\rho\, D\mathbf{u}/Dt = \rho \mathbf{F}_b + \text{div } \boldsymbol{\sigma} \qquad (6.10)$$

The total derivative is merely a shorthand for the time plus advective derivatives in the Eulerian frame of reference. This is a compact form of the statement of conservation of linear momentum. However, it leaves the form of the stress tensor, $\boldsymbol{\sigma}$, still to be determined.

Example 6.2

Consider a simple steady-state, one-direction, constant-density flow where only gravity, normal forces (pressure), and tangential surface stress forces exist. Use the method for an infinitesimal parcel to get the governing equation for the z-direction variation in velocity $\mathbf{u}$ along the channel shown in Fig. 6.3. Note that the vertical is denoted by Z here.

Gravity provides a body force. In this case it has a component along the flow and normal to the flow. The surface forces include pressures normal to the surfaces $\mathbf{F}_p$ and stress parallel to the surfaces $\mathbf{F}_\tau$.

Solution

We can start out by writing the basic law,

$$\delta m a = \delta \mathbf{F}_g + \delta \mathbf{F}_p + \delta \mathbf{F}_\tau$$

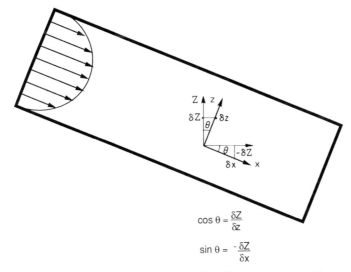

Figure 6.3   The downslope flow in a channel. Coordinates x, y, z are with respect to the channel. Z designates the vertical direction (aligned with gravity).

Then, we picture the parcel with the forces (Fig. 6.4). These consist of body forces acting at the center and surface forces.

The x-component force is

$$\delta ma_x = \rho\,\delta x\,\delta z(1)(\partial u/\partial t + u\,\partial u/\partial x + w\,\partial u/\partial z)$$

$$= [-\partial p/\partial x\,\delta x\,\delta z(1)] + g_x\rho\,\delta x\,\delta z(1)$$

$$+ [\tau + \partial\tau/\partial z\,\delta z/2 - (\tau - \partial\tau/\partial z\,\delta z/2)]\,\delta x$$

where $\delta\mathbf{W} = \delta m\mathbf{g}$.

Divide this equation by the mass, $\rho\,\delta x\,\delta z(1)$, so that, for a unit mass at the point (x, y, z, t) as $\delta V \to 0$ the equation is

$$\partial u/\partial t + u\,\partial u/\partial x + w\,\partial u/\partial z = -1/\rho\,\partial p/\partial x + g_x + 1/\rho\,\partial\tau/\partial z$$

The forces across the parcel can be expressed with respect to either the center of the parcel (Fig. 6.4) or the lower left corner (Fig. 6.5). They will yield the same force at the point in the limit of $\delta V \to 0$. In the future, we employ the latter choice because there are less terms to write.

If the flow is steady, $\partial u_i/\partial t = 0$, with parallel channel flow, and ρ can be considered constant, then $v = w = 0$, $u \neq 0$, and $\partial v/\partial y = \partial w/\partial z = 0$. Thus, $\partial u/\partial x = 0$ from continuity, and

$$\delta m\, a = \delta F_g + \delta F_p + \delta F_\tau$$

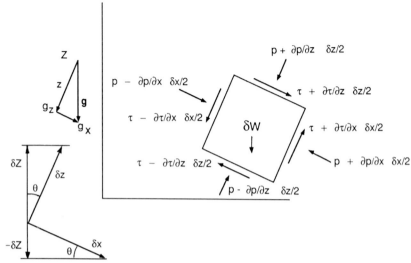

Figure 6.4 Forces on a two-dimensional parcel (unit depth) located at (x, y, z) in an inclined channel. The components of gravitation force must be considered.

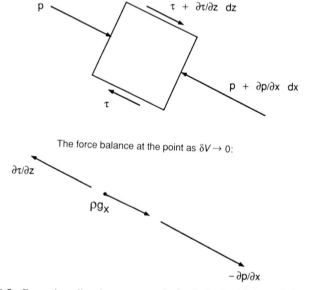

Figure 6.5 Forces in x-direction on a parcel of unit depth with lower left corner located at (x, y, z).

$$\rho[\partial u/\partial t + u\,\partial u/\partial x + w\,\partial u/\partial z] = -\partial p/\partial x + \rho g_x + \partial\tau/\partial z = 0$$

Similarly,

$$Dw/Dt = -\partial p/\partial z + \rho g_z + \partial\tau/\partial x = 0$$

Since $\partial\tau/\partial x$ depends on velocity gradients in the x-direction and these are zero, $\partial\tau/\partial x = 0$.

We can write

$$g_z = -g\cos\theta, \quad g_x = g\sin\theta$$

$$\cos\theta = \partial Z/\partial z, \quad \sin\theta = -\partial Z/\partial x$$

Hence,

$$\rho g_z = -\rho g\,\partial Z/\partial z$$

$$\rho g_x = -\rho g\,\partial Z/\partial x$$

And

$$-\partial p/\partial x - \rho g\,\partial Z/\partial x + \partial\tau/\partial z = 0$$

$$-\partial p/\partial z - \rho g\,\partial Z/\partial z = 0 = \partial[p + \rho gZ]/\partial z$$

$$\{\text{Or, } \partial p/\partial z = -\rho g\,\partial Z/\partial z = \partial p/\partial Z\,\partial Z/\partial z; \quad \partial p/\partial Z = -\rho g.\}$$

This states that $p + \rho gZ$ is a function of x only, and

$$d[p + \rho gZ]/dx - \partial\tau/\partial z = 0$$

or, using the one-dimensional viscosity relation,

$$d[p + \rho gZ]/dx - \mu\,d^2u/dz^2 = 0$$

This is the desired equation for $u(z)$. It is a balance between the pressure, viscous, and gravity forces on the parcel. This problem illustrates the derivation of the equation of motion for a specific arrangement where the body force contributes to both components of the momentum equation.

6.2 Derivation with Respect to the Infinitesimal Elementary Parcel—Lagrangian Perspective

We now consider the changing momentum of a particular parcel as it moves through an arbitrary point in the field. If we express the rate of change with respect to time only, we have a Lagrangian point of view. Thus, imagine moving with the parcel. We can write Newton's law balancing the force on the parcel with the rate of change of the momentum, $\rho \mathbf{u} \, \delta V$, where δV is the volume, as

$$\mathbf{F} = D(\rho \mathbf{u} \, \delta V)/Dt = \mathbf{F}_b + \mathbf{F}_s \qquad (6.11)$$

where all of the terms are to be expressed at time t and at a point x in the flow. For example, the surface forces can be written

$$\mathbf{F}_s = \lim_{\delta V \to 0} 1/\delta V \iiint \operatorname{div} \boldsymbol{\sigma} \, dV = \operatorname{div} \boldsymbol{\sigma} \text{ at a point}$$

For a small finite volume δV, Eq. (6.11) may be written

$$\rho \mathbf{u} \, D \, \delta V/Dt + \delta V \, D\rho \mathbf{u}/Dt = \rho \mathbf{F}_b \, \delta V + \operatorname{div} \boldsymbol{\sigma} \, \delta V \qquad (6.12)$$

Now recall that the volumetric change may be written [Eq. (5.36)]

$$D \, \delta V/Dt = \operatorname{div} \mathbf{u} \, \delta V$$

and by canceling the δV,

$$\rho \mathbf{u} \operatorname{div} \mathbf{u} + D\rho \mathbf{u}/Dt = \rho \mathbf{F}_b + \operatorname{div} \boldsymbol{\sigma} \qquad (6.13)$$

The conservation-of-mass equation is employed to write the left side of Eq. (6.13) in a much simpler form,

$$\rho \, D\mathbf{u}/Dt = \operatorname{div} \boldsymbol{\sigma} + \rho \mathbf{F}_b \qquad (6.14)$$

These equations are the same as those obtained in Section 6.1, giving a field description of $\mathbf{u}(x, y, z, t)$. The derivation is much briefer, because we employed the definitions of $\boldsymbol{\sigma}$ and the volumetric rate of change. However, the elucidation of the forces is less clear in this procedure.

Another option for a very quick yield of the momentum principle equation would have been to use the general conservation equation (5.43). In this case, simply substitute momentum for $\mathbf{f}$ and the force terms for the sinks and sources.

6.3   The Stress Term—Liquids and Gases

Three equations [three *component* equations, or one vector equation (6.14)] are now added to our one continuity equation for the description of the flow field. We can generally assume that the body force $\mathbf{F}_b$ is known (usually it is the force of gravity, although sometimes it includes a geomagnetic force). However, we have now replaced $\mathbf{F}_s$ with an unknown tensor $\boldsymbol{\sigma}$, which has nine new unknowns. This tensor hides the intricacies of the internal forces in the fluid. Its definition constitutes the premier challenge for the derivation of the general equations for fluid flow.

Although we have the conservation equation, it is in terms of the operator σ, which has no physical significance. If stress were as easy to measure as velocity, then we might be done, since the components of σ are stress terms. However, good stress meters do not exist at present, and stress is most often a term derived from velocity measurements. So, with the addition of σ, we have more unknowns than equations, and the set of equations cannot be solved for all unknowns—it is not closed.

To have a closed set, we must have a like number of unknowns and equations. Thus, we must relate the terms in σ to known constants, functions, or the other unknowns. We are going to employ the equation relating stress to velocity shear with viscosity as the factor of proportionality. The result will be a relation between the terms in σ and those in the velocity gradient tensor, $\partial u_i / \partial x_j$. This relation, with its large number of unknowns, presents a fundamental difficulty in solving the momentum equations. It is known as the closure problem. The closure problem was solved in this manner for classical laminar flow molecular stress by Stokes in 1845. He simply replaced shear with rate-of-shear in the existing law for elastic solid bodies (Hooke's law). However, we do not expect molecular stresses to be an important factor in atmospheric motions.

The viscous forces due to air viscosity are many orders of magnitude smaller than typical atmospheric flow forces. The viscous force term cannot be ignored by atmospheric scientists, however. The small viscous forces become important when an atmospheric problem depends on the microphysics. Examples are found in the calculation of precipitation growth rates or the flux mechanisms at the molecular-scale interface between the air and the sea. Another example arises in the modeling of turbulent flow.

In the atmosphere, the closure problem becomes the problem of accounting for the diffusive (flux) characteristics of the turbulent eddies. When scales become large, Reynolds numbers, $\rho UL / \mu$, become large, and transition to turbulence becomes inevitable. In geophysical flows, scales are very large, and different regimes of turbulence appear. The methods of modeling turbulence are inseparable from scaling analyses of the equations and the flow domain. Since the usual first approximation is an *ad hoc* analogy to the molecular law of friction, we must be familiar with the derivation in terms of molecular friction.

We recall the discussion on viscosity in Section 1.11. A one-dimensional scalar stress is related to the rate of strain (the velocity shear) in the equation

$$\tau = \mu \, du/dz \qquad (6.15)$$

Since the internal stress force is given by **σ**, we expect that the unknown elements in this tensor will involve the various component shears and the viscosity coefficient.

The one-dimensional parametrization of stress used in the definition of viscosity (Section 1.11.3) can be extended to three-dimensions. The definition must include the array of possible stress components. These stress components will be proportional to the array of the shears, represented in the velocity gradient tensor—nine possible shears. We learned in Chapter 4 that this tensor can be separated into parts. One part contributes to deformation and another part simply rotates the parcel without distortion. The first part can be further separated into two contributions:

1. Deformation

    (a) Volume strain (contraction, expansion)
    (b) Distortion (without volume change or rotation)

2. Rotation

The first component of the deformation (1a) is the *volumetric strain*. The rate of volumetric strain per unit volume is

$$\frac{1}{\delta V}\frac{d\,\delta V}{dt} \tag{6.16}$$

It is valuable to recall that in solid mechanics an elastic solid has volumetric strain

$$(D\,\delta V)/\delta V = \mathbf{\nabla}\cdot\mathbf{S} = \partial(\delta x_j)/\partial x_j$$

where **S** is the elongation vector,

$$\mathbf{S} = \mathbf{\delta x}_i$$

In the case of a fluid parcel, the shape and volume change as the surface bounding the volume is distorted. In Chapter 5 we found the expansion rate of a parcel to be

$$1/\delta V\, D\,\delta V/Dt = \partial u_j/\partial x_j$$

This is the sum of the *three diagonal elements* of the velocity gradient tensor. We see where the component velocities have replaced the incremental lengths of the parcel in the strain expression.

The second component of the shear deformation is the *distortion* that deforms the parcel without changing its volume. This is often the most important kinematic property, as it is directly related to the shearing stress force on the parcel. We can examine this distortion from a 2-D perspective, as shown in Fig. 6.6. The deformation of δx_1 includes the effects from the

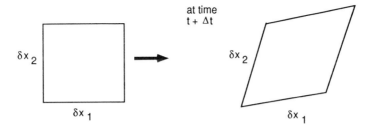

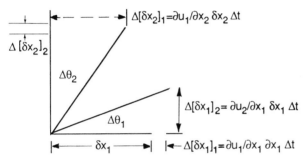

Figure 6.6 Distortion components with respect to $\Delta\theta_1$.

stretching produced by $\partial u_1/\partial x_1$ plus that produced by $\partial u_1/\partial x_2$. Similarly, the change in δx_2 is due to $\partial u_2/\partial x_2$ and $\partial u_2/\partial x_1$. The net distortion can be related to the change in the angle between δx_1 and δx_2, which is $\Delta\theta_1 + \Delta\theta_2$ in Fig. 6.6.

From the figure, we can write

$$\tan\Delta\theta_1 = \left[\frac{\partial u_2/\partial x_1 \, \delta x_1 \Delta t}{\delta x_1 + \partial u_1 \, \partial x_1 \, \delta x_1 \, \Delta t}\right]$$

$$\approx \partial u_2/\partial x_1 \, \Delta t \qquad [= \Delta(\delta x_2)/\delta x_1)]$$

$$\approx \Delta\theta_1.$$

In the limit, as $\Delta\theta, \Delta t \to 0$, $\Delta\theta_1/\Delta t = d\theta_1/dt = \partial u_2/\partial x_1$

Similarly,

$$\Delta\theta_2 \approx \Delta(\delta x_1)/\delta x_2, \qquad \text{and} \qquad d\theta_2/dt = \partial u_1/\partial x_2 \qquad (6.17)$$

The average shear deformation is proportional to the average angular change,

$$\Gamma_{21} \equiv \tfrac{1}{2}(\theta_1 + \theta_2)$$

$$d\,\Gamma_{21}/dt = \tfrac{1}{2}d(\theta_1 + \theta_2)/dt$$

and

$$d\Gamma_{21}/dt = \tfrac{1}{2}(\partial u_1/\partial x_2 + \partial u_2/\partial x_1)$$

In the same way,

$$d\Gamma_{31}/dt = \tfrac{1}{2}(\partial u_1/\partial x_3 + \partial u_3/\partial x_1); \qquad d\Gamma_{23}/dt = \tfrac{1}{2}(\partial u_3/\partial x_2 + \partial u_2/\partial x_3)$$

and

$$d\Gamma_{ij}/dt = d\Gamma_{ji}/dt \qquad (6.18)$$

This tensor consists of the *off-diagonal elements* of the 3-D deformation tensor. We have defined them as def **u**.

Now we will write the relation between the stress force and the deformations by drawing on the well-established stress-strain relationship from solid mechanics. Note that in elastic solid mechanics, the shear deformation is often simply one dimensional, with stress proportional to strain, as shown in Fig. 6.7.

$$\tau_{21} \equiv G\Gamma_{21}$$

and in general,

$$\tau_{ij} \equiv G\Gamma_{ij} \qquad (6.19)$$

where G is the shear modulus, a constant.

In dealing with fluids we will substitute the viscosity for the shear modulus, and the rate of strain for the strain in Eq. (6.19). The result is an expression for the stress-rate of strain of distortion:

$$\tau_{ij} = \mu\, d\Gamma_{ij}/dt \qquad (6.20)$$

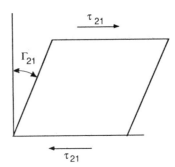

Figure 6.7 An elastic solid shear.

This is a tensor representing all of the internal distortion stresses on the surface of a parcel. It is part of the general surface force tensor $\boldsymbol{\sigma}$.

6.3.1   The General Three-Dimensional Stress Tensor

In this section we will write out the details of the individual stress tensor terms as a function of the velocity derivatives. To proceed with a manageable relationship between the individual elements of div $\boldsymbol{\sigma}$ and the various possible velocity gradients, we must make assumptions about the basic nature of the fluid. The following assumptions define a Newtonian fluid. They are based on laboratory observations of fluid behavior.

1. In a static fluid, there are only forces normal to the parcel surface, called the pressure forces.

2. σ_{ij} is independent of the heat flux, depending only on local kinematic and thermodynamic states.

3. There do not exist any characteristic (preferred) directions.

4. The stress is proportional to the velocity gradient.

These assumptions will allow us to write the stress tensor components in terms of the pressure and the rate-of-strains discussed in the previous section. We will write out the relations for the three-dimensional parcel.

In our sketch of the infinitesimal parcel (Fig. 6.8), we align the coordinate axes x_1, x_2, x_3 with the corner of the cubical parcel. The unit vectors in the coordinate directions are denoted by $\mathbf{i}$, $\mathbf{j}$, $\mathbf{k}$. The first subscript of the stress tensor component will indicate the axis to which the face is normal; the second will indicate the direction of the force on the face.

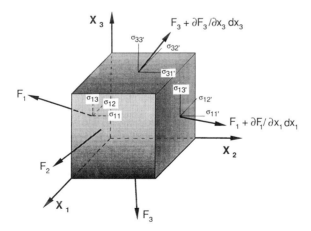

Figure 6.8   The parcel showing the forces on each face.

Now we can write the forces on each face touching the origin, which is located at the lower left corner of the parcel in Fig. 6.8 as

$$\mathbf{F}_1 = \mathbf{i}\sigma_{11} + \mathbf{j}\sigma_{12} + \mathbf{k}\sigma_{13}$$

$$\mathbf{F}_2 = \mathbf{i}\sigma_{21} + \mathbf{j}\sigma_{22} + \mathbf{k}\sigma_{23} \qquad (6.21)$$

$$\mathbf{F}_3 = \mathbf{i}\sigma_{31} + \mathbf{j}\sigma_{32} + \mathbf{k}\sigma_{33}$$

The net shearing force on the $x_2 x_3$ surface is $\partial \mathbf{F}_1/\partial x_1\ dx_1 (dx_2\ dx_3) = \partial \mathbf{F}_1/\partial x_1\ dV$. Adding the components, the total surface force on the element per unit volume is

$$\mathbf{F}_{tot} = \{\partial \mathbf{F}_1/\partial x_1 + \partial \mathbf{F}_2/\partial x_2 + \partial \mathbf{F}_3/\partial x_3\}$$

or

$$= \mathbf{i}\{\partial \sigma_{11}/\partial x_1 \quad + \partial \sigma_{21}/\partial x_2 + \partial \sigma_{31}/\partial x_3\} \dots X_1\text{-component}$$

$$+ \mathbf{j}\{\partial \sigma_{12}/\partial x_1 + \partial \sigma_{22}/\partial x_2 + \partial \sigma_{32}/\partial x_3\} \dots X_2\text{-component}$$

$$+ \mathbf{k}\{\partial \sigma_{13}/\partial x_1 + \partial \sigma_{23}/\partial x_2 + \partial \sigma_{33}/\partial x_3\} \dots X_3\text{-component}$$

$$\begin{array}{ccc} \text{on face 23} & \text{on face 31} & \text{on face 12} \\ \perp \text{ to } X_1 & \perp \text{ to } X_2 & \perp \text{ to } X_3 \end{array}$$

Therefore, Newton's law, $m\mathbf{a} = \mathbf{F}$, may be written

$$\rho\, D\mathbf{u}/Dt = \rho \mathbf{F}_b + \text{div } \boldsymbol{\sigma}$$

or, in component form,

$$\rho\, Du_1/Dt = \rho F_1 + \partial \sigma_{11}/\partial x_1 + \partial \sigma_{21}/\partial x_2 + \partial \sigma_{31}/\partial x_3$$

$$\rho\, Du_2/Dt = \rho F_2 + \partial \sigma_{12}/\partial x_1 + \partial \sigma_{22}/\partial x_2 + \partial \sigma_{32}/\partial x_3 \qquad (6.22)$$

$$\rho\, Du_3/Dt = \rho F_3 + \partial \sigma_{13}/\partial x_1 + \partial \sigma_{23}/\partial x_2 + \partial \sigma_{33}/\partial x_3$$

There are nine components of the stress force. If we assume that the parcel is in equilibrium with the forces, there are no unbalanced forces at the point represented by the parcel. Then by taking moments about any axis one finds that $\sigma_{ij} = \sigma_{ji}$ (problem 6.1). This reduces the unknowns by three. However, even if the body forces ($\mathbf{F}_b$) are known, there are still nine unknowns ($u_i + \sigma_{ij}$), and only three equations. We need a relation between σ_{ij} and the other flow parameters (u_i) to provide the other six equations.

6.3.1.a Pressure

We discussed pressure in Chapter 1 as a term in the equation of state where it appears as a thermodynamic variable. We defined a Newtonian fluid such that in a static state the only surface force is the normal pressure force,

$$\sigma_{ij} = -p\,\delta_{ij}; \qquad \sigma_{13} = \sigma_{12} = \sigma_{23} = 0 \qquad (6.23)$$

and only $\sigma_{11} = \sigma_{22} = \sigma_{33}$ are not zero.

In a moving fluid, the normal component of stress on an elemental face depends on the orientation of the face with respect to the flow. The orientation is denoted by the normal unit vector. The stress in any direction is related to the velocity gradient in the same direction. The internal stress creates the *tangential* shearing forces. In addition, there is an internal stress contribution to the *normal* force. Thus, there is no longer a "pressure" acting equally in all directions. However, we can obtain a scalar that acts like the pressure for the moving fluid by defining the average,

$$\sigma = [\sigma_{11} + \sigma_{22} + \sigma_{33}]/3 \equiv -p \qquad (6.24)$$

where σ is invariant under rotation of the axes. It will be defined as *pressure* for a fluid, static or moving.

In the general case of a moving fluid, normal forces due to the shearing action can come into play. It is convenient to separate the pressure and shearing contributions to the stress by defining a new tensor, τ, which is obtained from σ by removing the pressure forces. In matrix notation:

$$\sigma = \begin{pmatrix} -p & 0 & 0 \\ 0 & -p & 0 \\ 0 & 0 & -p \end{pmatrix} + \begin{pmatrix} \tau_{11} & \tau_{12} & \tau_{13} \\ \tau_{21} & \tau_{22} & \tau_{23} \\ \tau_{31} & \tau_{32} & \tau_{33} \end{pmatrix} \qquad (6.25)$$

Here, τ_{ij} is referred to as the shearing stress tensor, often shortened to stress tensor. Its elements are the shearing stress that arises due to the motion of the fluid. It is called the deviatoric stress tensor by some authors. From the definition of the pressure in Eq. (6.24),

$$\tau_{ii} = 0$$

The problem now is to relate τ to the velocity shear. As noted before, when this fluid problem arose, there already existed a classical treatment of stress in elastic solids. The law of friction for fluid flow borrows heavily from that relation.

6.3.1.b   Hooke's Law for Elastic Solids

For elastic solids, a coefficient called the *modulus of elasticity* relates the one-dimensional elongation to the normal stress. Poisson's ratio gives the relation between the effects of the elongation in one direction and the corresponding contraction strains in the other directions. These two constants can be related to the shear modulus, yielding the stress-strain relation in the matrix equation. This is Hooke's law. We realize that this brief synopsis of the theory for the strain of an elastic solid body cannot be satisfying from the standpoint of understanding the solid mechanism. However, this relation has been applied to the stress-rate-of-strain relation for laminar flow of a fluid. And we also extend it into the eddy-stress-rate-of-strain concept. Since the last hypothesis is often tenuous, a knowledge of its historical roots may be important toward evaluating the validity of its application in particular circumstances. We can write Hooke's law by using Eqs. (6.20–6.25) to define the *stress tensor for an elastic body,*

$$
\boldsymbol{\sigma} = \begin{pmatrix} \sigma_{11} & \sigma_{12} & \sigma_{13} \\ \sigma_{21} & \sigma_{22} & \sigma_{23} \\ \sigma_{31} & \sigma_{32} & \sigma_{33} \end{pmatrix} = \begin{pmatrix} \sigma & 0 & 0 \\ 0 & \sigma & 0 \\ 0 & 0 & \sigma \end{pmatrix}
$$

$$
+ G \begin{pmatrix} 2\epsilon_1 & \Gamma_{12} & \Gamma_{13} \\ \Gamma_{21} & 2\epsilon_2 & \Gamma_{23} \\ \Gamma_{31} & \Gamma_{32} & 2\epsilon_3 \end{pmatrix} + G^* \begin{pmatrix} 2\epsilon_1 & \Gamma_{21} & \Gamma_{31} \\ \Gamma_{12} & 2\epsilon_2 & \Gamma_{32} \\ \Gamma_{13} & \Gamma_{23} & 2\epsilon_3 \end{pmatrix}
$$

$$
- \frac{2}{3} G \begin{pmatrix} \text{div } \mathbf{S} & 0 & 0 \\ 0 & \text{div } \mathbf{S} & 0 \\ 0 & 0 & \text{div } \mathbf{S} \end{pmatrix} \tag{6.26}
$$

This is *Hooke's law* for elastic solid bodies. It results from the assumption that stress is proportional to magnitude of strain. It is a simple step from here to a law for fluids and gases.

6.3.2   Stokes' Law of Friction

We have noted that the only internal forces on the parcel in a static fluid are the pressure forces. These forces are therefore conveniently separated from the *general stress tensor* $\boldsymbol{\sigma}$, leaving the *stress tensor* $\boldsymbol{\tau}$. Now we use the observation that a fluid can support strain only when in motion and that the stress is proportional to the velocity gradient—hence the *rate of strain*. Assuming that stress is proportional to rate of change of strain instead of simply the strain, we will use an equation similar to Eq. (6.19) for fluids. Thus, replace the *displacement vector* $\mathbf{S}$ with

$$
d\mathbf{S}/dt = \mathbf{u} = \mathbf{i}u_1 + \mathbf{j}u_2 + \mathbf{k}u_3
$$

the proportionality constant G with viscosity μ, and finally, σ with $-p$.
Stokes's law is then written

$$\boldsymbol{\sigma} = \begin{pmatrix} \sigma_{11} & \sigma_{12} & \sigma_{13} \\ \sigma_{21} & \sigma_{22} & \sigma_{23} \\ \sigma_{31} & \sigma_{32} & \sigma_{33} \end{pmatrix} = \begin{pmatrix} -p & 0 & 0 \\ 0 & -p & 0 \\ 0 & 0 & -p \end{pmatrix} + \mu \begin{pmatrix} u_x & u_y & u_z \\ v_x & v_y & v_z \\ w_x & w_y & w_z \end{pmatrix}$$

$$+ \mu \begin{pmatrix} u_x & v_x & w_x \\ u_y & v_y & w_y \\ u_z & v_z & w_z \end{pmatrix} + \alpha \begin{pmatrix} \text{div } \mathbf{u} & 0 & 0 \\ 0 & \text{div } \mathbf{u} & 0 \\ 0 & 0 & \text{div } \mathbf{u} \end{pmatrix} \qquad (6.27)$$

This may be written in symbolic form,

$$\boldsymbol{\sigma} = -p\mathbf{I} + \mu\{\text{grad } \mathbf{u} + \text{grad* } \mathbf{u}\} + \alpha\mu \text{ div } \mathbf{uI} \qquad (6.28)$$

or

$$\boldsymbol{\sigma} = -p\mathbf{I} + \mu \text{ def } \mathbf{u} + \alpha \text{ div } \mathbf{uI} = -p\mathbf{I} + \boldsymbol{\tau} \qquad (6.29)$$

in close analogy to Eq. (6.26). I is the identity tensor, $I_{ij} = 1$ if $i = j$, 0 if $i \neq j$.

Here, α is sometimes called the "second viscosity." Since $\tau_{ii} = 0$, and

$$0 = (2\mu + 3\alpha) \text{ div } \mathbf{u}$$

$$\alpha = -\tfrac{2}{3}\mu$$

However, in atmospheric work we can often assume that div $\mathbf{u} \approx 0$ in Eq. (6.29). This must be true compared to the deformation term. In this case the second viscosity term is dropped and this parameter is not required.

The relation (6.29) provides nine equations for the nine unknown components of σ_{ij} (only six of which are independent). The stress must be empirically related to the mean velocity gradients with a coefficient of viscosity. When the viscosity (or eddy-viscosity) assumption is used to form these relations, the set of equations is closed—there are an equal number of equations and unknowns.

$$\rho \, D\mathbf{u}/Dt = \rho\mathbf{F}_b + \text{div } \boldsymbol{\sigma} = \rho\mathbf{F}_b + \text{div}\{-p\mathbf{I} + \boldsymbol{\tau}\}$$

$$= \rho\mathbf{F}_b - \text{grad } p + \text{div}\{\mu \text{ def } \mathbf{u} + \alpha \text{ div } \mathbf{uI}\} \qquad (6.30)$$

These equations expressing the conservation-of-momentum principle and providing a parametrization relation for the stress terms are attributed to Poisson (1831), St. Venant (1843), Navier (1827), and Stokes (1845).

6.3.3 An Alternate Derivation

The relation of stress tensor to rate of strain can be obtained directly from vector calculus.[1] To illustrate this technique, we can examine this derivation

---

[1] See Jeffereys, H. (1931). "Cartesian Tensors." Cambridge Univ. Press, London.

of our closure relation. We must accept a tensor identity or two as given, since their derivation is beyond the requirements of this text. However, the procedure is elegant, and it reveals a problem in the eddy-viscosity relation that has yet to be resolved.

Under Stokes's assumption that the stress is proportional to rate of strain or velocity, one could expand the stress tensor in a Taylor series with respect to the velocity gradient about a point. Here, the constants in the expansion are tensors.

$$\boldsymbol{\sigma} = \mathbf{A} + \mathbf{B} \text{ grad } \mathbf{U} + \mathbf{C} \cdots$$

or

$$\sigma_{ij} = A_{ij} + B_{ijkl} \, \partial u_k/\partial x_l + C_{ijklm} \, \partial^2 u_k/\partial x_l \, \partial x_m + \cdots$$

$$\approx A_{ij} + \{\alpha \, \delta_{ij} \, \delta_{kl} + \mu(\delta_{ik} \, \delta_{il} + \delta_{il} \, \delta_{jk}) + \beta(\delta_{ik} \, \delta_{il} - \delta_{il} \, \delta_{jk})\} \, \partial u_j/\partial x_l$$

general form of isotropic tensor   zero, from symmetry

From the no-flow boundary condition, $A_{ij} = -p \, \delta_{ij}$, and

$$\sigma_{ij} = -p \, \delta_{ij} + \alpha \, \partial u_k/\partial x_k \, \delta_{ij} + \mu(\partial u_i/\partial x_j + \partial u_j/\partial x_i)$$

or, in symbolic notation,

$$\boldsymbol{\sigma} = -p\mathbf{I} + \alpha \text{ div } \mathbf{u}\mathbf{I} + 2\mu \text{ def } \mathbf{u}$$

stress = pressure +   volume   +   deformation   (6.31)

dilation   tensor

This can be used to define $\boldsymbol{\tau}$.

$$\boldsymbol{\sigma} = -p\mathbf{I} + \boldsymbol{\tau} \tag{6.32}$$

Substituting for $\alpha = -\frac{2}{3}\mu$,

$$\boldsymbol{\tau} = 2\mu \text{ def } \mathbf{u} - \frac{2}{3}\mu \text{ div } \mathbf{u}\mathbf{I}$$

Then the divergence of $\boldsymbol{\sigma}$ is

$$\text{div } \boldsymbol{\sigma} = -\text{grad } p + \text{div}[2\mu \text{ def } \mathbf{u} - \frac{2}{3}\mu \text{ div } \mathbf{u}\mathbf{I}] \tag{6.33}$$

For constant μ and the approximation div $\mathbf{u} = 0$, use the identity 2 div def $\mathbf{u}$ = div grad $\mathbf{u}$ + grad div $\mathbf{u}$ to obtain

$$\text{div } \boldsymbol{\sigma} = -\text{grad } p + \mu \text{ div grad } \mathbf{u} \tag{6.34}$$

The stress term can also be written as

$$\partial\tau_{ji}/\partial x_j = \partial/\partial x_j\{\mu(\partial u_i/\partial x_j + \partial u_j/\partial x_i)\}$$

$$= \mu\,\partial^2 u_i/\partial x_j\,\partial x_j \qquad \text{for } \mu \text{ constant}$$

Placing Eq. (6.34) in the conservation of momentum equation yields

$$\rho\,D\mathbf{u}/Dt = \rho\mathbf{F}_b - \operatorname{grad} p + \mu\,\nabla^2\mathbf{u} \qquad (6.35)$$

or, in index notation,

$$\rho(\partial u_i/\partial t + u_j\,\partial u_i/\partial x_j) = \rho\sum F_{bi} - \partial p/\partial x_i + \mu\,\partial^2 u_i/\partial x_j\,\partial x_j$$

$$(6.36)$$

The change in velocity is a function of body forces, pressure gradient forces, and the internal stress forces. We have assumed that μ is a universal constant. This is a good estimate for molecular viscosity. However, when we are attempting to model turbulent diffusive fluxes we replace μ with an eddy viscosity K. Now K depends on the turbulence distribution, which depends on the velocity shear, and this coefficient could be different for each of the B_{ijkl}. If the eddy-viscosity was indeed a fourth-order tensor, it would probably not be a practical model for the turbulent stress versus rate-of-strain effect. Fortunately, simpler forms of K seem to suffice for many practical applications.

6.3.4   Eddy Viscosity

The equations developed by Navier and Stokes are strictly for laminar flow with a constant viscosity coefficient. Reynolds suggested that when the flow is turbulent, the equations apply to the instantaneous velocities. However, from a practical viewpoint, we must write the Navier–Stokes equations with respect to the mean velocities, which are usually measured. To do this, we must write the equations in terms of the turbulent velocities and then average the velocities. This will be done in Chapter 10. For now, we can simply assume that the random motion of the turbulent eddies transports momentum in a manner analogous to that of the molecules. The stresses due to turbulence are then expressed in a formula similar to the Newtonian friction law.

We have already remarked that in geophysical flows internal stress due to molecular forces is much smaller than the other forces. And we have noted that in atmospheric flows the huge scales mean very high Reynolds numbers and, consequently, unavoidable turbulence. Still, we use Eq. (6.36) in two ways.

1. In the inviscid approximation, where the viscous term is neglected with respect to both molecular stress and eddy turbulent fluctuations.

2. In highly turbulent regimes, where we assume that μ can be replaced by an eddy viscosity that accounts for the much larger mixing effects of the turbulence.

When we make the eddy viscous approximation, the eddy viscosity K simply replaces μ in Eq. (6.36). This implies that the flow is eddy-laminar; that is, laminar-like mean flows coexist with the small-scale turbulence.

A primary limitation in the K-theory approximation is that K represents the mixing action of the small-scale turbulence, and in the atmosphere this can vary in space. In general it is a fourth-order tensor; however, analytic solutions can result by assuming that it is constant or varies only in one direction, usually the vertical. These approximate equations have resulted in successful models for many real flow situations. When the eddy-viscosity assumption is made, the preceding development becomes

$$\boldsymbol{\sigma} = -p\mathbf{I} + K_1 \operatorname{div} \mathbf{u}\mathbf{I} + K_2\{\operatorname{grad} \mathbf{u} + \operatorname{grad}^* \mathbf{u}\}$$
$$= -p\mathbf{I} + \boldsymbol{\tau} \tag{6.37}$$

where K_1 and K_2 are scalar coefficients replacing α and μ. The eddy coefficients depend on the turbulence distribution, which may depend on the local state of the fluid. The stress term is

$$\operatorname{div} \boldsymbol{\tau} = \operatorname{grad}(K_1 \operatorname{div} \mathbf{u}) + \operatorname{div}\{K_2(\operatorname{grad} \mathbf{u} + \operatorname{grad}^* \mathbf{u})\} \tag{6.38}$$

One simplification that frequently applies in geophysical flows is obtained by assuming that the eddy viscosity varies only in the vertical. By means of vector identities (see the summary chart at the end of Chapter 4), Eq. (6.38) may be written ($\mathbf{k}$ is the unit vertical vector)

$$\operatorname{div} \boldsymbol{\tau} = (K_1 + K_2)\{\nabla^2\mathbf{u}\} + K_2 \nabla \times \nabla \times \mathbf{u}$$
$$+ \mathbf{k}\, dK_1/dz\, \nabla \cdot \mathbf{u} + \mathbf{k}\, dK_2/dz(\nabla\mathbf{u} + \nabla^*\mathbf{u}) \tag{6.39}$$

This expression is still complicated. However, since generally we are dealing with cases where $\nabla \cdot \mathbf{u} = 0$,

$$\operatorname{div} \boldsymbol{\tau} = K \nabla^2\mathbf{u} + dK_2/dz\, (w_x + u_z, w_y + v_z, 2w_z) \tag{6.40}$$

where $K = K_1 + K_2$. When K_2 is constant, the stress term is merely

$$\operatorname{div} \boldsymbol{\tau} = K \nabla^2\mathbf{u} \tag{6.41}$$

The development of the general closure scheme is complicated by the fact that the stress vector force varies in each of the three directions. We will find that in most applications, this generality is unnecessary and simple stress forms such as Eq. (6.41) will be sufficient.

6.4   The Coriolis Term

In Section 2.5 we found that since we are using a noninertial frame of reference it is necessary to add a virtual term, $-2\boldsymbol{\Omega} \times \mathbf{u}$, the Coriolis acceleration (or force per unit mass), to our equation of motion. The vector $\boldsymbol{\Omega}$ can be associated with a skew-symmetric tensor Ω. Therefore, the Coriolis acceleration can be expressed in the following ways:

$$2\boldsymbol{\Omega} \times \mathbf{u} = \mathbf{u}\Omega = u_j\Omega_{ji} = \epsilon_{ijk} k'_j u_k$$

and

$$\mathbf{k}' \equiv (0, f', f),$$

$$\Omega \equiv \begin{pmatrix} 0 & f & -f' \\ -f & 0 & 0 \\ f' & 0 & 0 \end{pmatrix} \tag{6.42}$$

and

$$f = 2\Omega \sin \theta, \qquad f' = 2\Omega \cos \theta \tag{6.43}$$

For our description in Cartesian coordinates, note that

$$u\Omega = \mathbf{k}' \times \mathbf{u} = (f'u_3 - fu_2, fu_1, -f'u_1)$$

In the atmosphere generally,

$$f'u_3 \ll fu_2, \qquad \text{(e.g., horizontal motion);}$$
$$f'u_1 \ll \text{other terms in the vertical momentum equation;}$$

and

$$2\boldsymbol{\Omega} \times \mathbf{u} = \mathbf{u}\,\Omega \approx \epsilon_{ijk}k'_j u_k f = \mathbf{k}' \times \mathbf{u} = (-fu_2, fu_1, 0)$$

neglecting $f'u_3$ and $f'u_1$.

In this case we can write $\mathbf{k}' \times \mathbf{u} = f\mathbf{k} \times \mathbf{u}$.

When we include the one-body force due to gravity, $\rho g\delta_{i3}$, the momentum equation may be written

$$\partial/\partial t(\rho u_i) + u_j \partial(\rho u_i)/\partial x_j = -\partial p/\partial x_i$$

$$+ \partial/\partial x_j\{\mu(\partial u_i/\partial x_j + \partial u_j/\partial x_i) + \alpha\, \partial u_k/\partial x_k\, \delta_{ij}\} - \rho g\delta_{i3} - \rho f\epsilon_{ijk}k_j u_k$$

Using continuity to simplify this equation,

$$\partial u_i/\partial t + u_j\, \partial u_i/\partial x_j = -1/\rho\, \partial p/\partial x_i + 1/\rho\, \partial/\partial x_j$$

$$\{\mu(\partial u_i/\partial x_j + \partial u_j/\partial x_i) + \alpha\, \partial u_k/\partial x_k\, \delta_{ij}\} - g\delta_{i3} - f\epsilon_{ijk}k_j u_k$$

For div $\mathbf{u} = 0$ and μ constant $= \rho \nu$, we get

$$Du_i/Dt = -1/\rho \ \partial p/\partial x_i + \nu \ \partial^2 u_i/\partial x_j \partial x_j$$

$$- g \ \delta_{i3} - f\epsilon_{ijk}k_j u_k \qquad (6.45)$$

Writing this equation out in component form,

$$\partial u/\partial t + u \ \partial u/\partial x + v \ \partial u/\partial y + w \ \partial u/\partial z = -1/\rho \ \partial p/\partial x + fv$$
$$+ \nu(\partial^2 u/\partial x^2 + \partial^2 u/\partial y^2 + \partial^2 u/\partial z^2),$$

$$\partial v/\partial t + u \ \partial v/\partial x + v \ \partial v/\partial y + w \ \partial v/\partial z = -1/\rho \ \partial p/\partial y - fu$$
$$+ \nu(\partial^2 v/\partial x^2 + \partial^2 v/\partial y^2 + \partial^2 v/\partial z^2),$$

$$\partial w/\partial t + u \ \partial w/\partial x + v \ \partial w/\partial y + w \ \partial w/\partial z = -1/\rho \ \partial p/\partial z - g$$
$$+ \nu(\partial^2 w/\partial x^2 + \partial^2 w/\partial y^2 + \partial^2 w/\partial z^2) \qquad (6.46)$$

These equations provide excellent approximations for many atmospheric applications. Yet one must remember the assumptions: constant viscosity; replacing div [μ 2def $\mathbf{u}$] with $\mu \ \nabla^2\mathbf{u}$ (exact for constant density, yielding div $\mathbf{u} = 0$); and a constant Coriolis parameter.

Example 6.3

Viscosity is often measured in an apparatus that consists of two rotating concentric cylinders. (See Fig. 6.9). Use the Navier–Stokes equations to determine the velocity between the cylinders. Show that this yields an approximate Couette flow across the short distance between the cylinders. The Navier–Stokes equations in cylindrical coordinates for incompressible flow are

$$Du_r/Dt = -1/\rho \ \partial p/\partial r + \nu[\nabla^2 u_r - u_r/r^2 - (2/r^2)\partial u_\theta/\partial\theta]$$

$$Du\theta/Dt = -(1/\rho r)\partial p/\partial\theta + \nu[\nabla^2 u_\theta + (2/r^2)\partial u_r/\partial\theta - u_\theta/r^2]$$

$$Dw/Dt = -1/\rho \ \partial p/\partial z + g + \nu \ \nabla^2 w$$

Here,

$$\nabla^2 \equiv (1/r)\partial/\partial r[r \ \partial/\partial r] + (1/r^2) \ \partial^2/\partial\theta^2 + \partial^2/\partial z^2 \qquad (6.47)$$

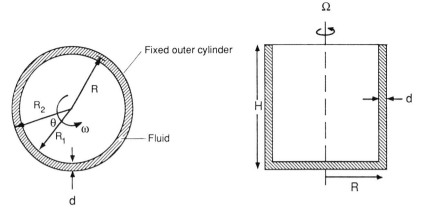

Figure 6.9 Coaxial cylinders with tangential laminar flow between them. The inner cylinder rotates at ω.

Neglect gravitational forces and assume separation between cylinders, $d \ll R_2$.

Solution

For this case, the terms remaining in the Navier–Stokes equations are

$$-u_\theta^2/r = -1/\rho \, dp/dr \qquad (6.48)$$

$$0 = v[d^2u_\theta/dr^2 + d/dr(u_\theta/r)]$$

The second of these equations integrates to

$$du_\theta/dr + u_\theta/r = C_1$$

A trial solution, $u_\theta = ar + b/r$, yields

$$u_\theta = (C_1/2)r + C_2/r \qquad (6.49)$$

The boundary conditions at R_1 and R_2 give

$$R_2\omega = C_1R_2/2 + C_2/R_2$$

$$0 = C_1R_1/2 + C_2/R_1$$

Hence,

$$C_1 = -2R_2^2\omega/(R_1^2 - R_2^2)$$

$$C_2 = R_1^2R_2^2\omega/(R_1^2 - R_2^2)$$

Substitution into Eq. (6.49) yields

$$u_\theta = [R_2^2\omega/(R_1^2 - R_2^2)][R_1^2/r - r] \tag{6.50}$$

The pressure distribution across the gap can be found by integrating the first of Eqs. (6.46).

The approximation for $d = R_1 - R_2 \ll R_2$ can be found by substituting for d to get

$$u_\theta \le R_2/(2d)[x - d]$$

where $x = r - R_2$. Thus the velocity across a thin gap approximates Couette flow.

Example 6.4

Write out the momentum equations in component form and discuss the approximations necessary to yield the following balance equations:

1. *Geostrophic flow:* Balance between Coriolis and pressure gradient forces only.
2. *Gradient flow:* Balance between Coriolis, pressure gradient and inertial (the advective terms) forces. (These apply in highly curved flows where the centrifugal forces are significant. They are most simply written in polar coordinates.)
3. *Isallobaric flow:* Same as 1. with time dependency.
4. *Planetary boundary layer flow:* Same as 1. with viscosity terms.

Solution

In component form, with eddy viscosity K,

$$\partial u/\partial t + u\,\partial u/\partial x + v\,\partial u/\partial y + w\,\partial u/\partial z = -\partial p/\partial x/\rho + fv + K(\partial^2 u/\partial x_i\,\partial x_i)$$

$$\partial v/\partial t + u\,\partial v/\partial x + v\,\partial v/\partial y + w\,\partial v/\partial z = -\partial p/\partial y/\rho - fu + K(\partial^2 v/\partial x_i\,\partial x_i)$$

$$\partial w/\partial t + u\,\partial w/\partial x + v\,\partial w/\partial y + w\,\partial w/\partial z = -\partial p/\partial z/\rho - g + K(\partial^2 w/\partial x_i\,\partial x_i)$$

1. Assume steady state; $w \ll u, v$; horizontal homogeneity; and inviscid, leaving

$$0 = -\partial p/\partial x/\rho + fv,$$
$$0 = -\partial p/\partial y/\rho - fu. \tag{6.51}$$

2. Assume steady state; $w \ll u, v$; and inviscid;

$$u\, \partial u/\partial x + v\, \partial u/\partial y = -\partial p/\partial x/\rho + fv \tag{6.52}$$
$$u\, \partial v/\partial x + v\, \partial v/\partial y = -\partial p/\partial y/\rho - fu$$

3. Assume $w \ll u, v$; inviscid;

$$\partial u/\partial t + u\, \partial u/\partial x + v\partial u/\partial y = -\partial p/\partial x/\rho + fv \tag{6.53}$$
$$\partial v/\partial t + u\, \partial v/\partial x + v\, \partial v/\partial y = -\partial p/\partial y/\rho - fu$$

4. Assume steady state; $w \ll u, v$; horizontal homogeneity;

$$0 = -\partial p/\partial x/\rho + fv + K(\partial^2 u/\partial x_i\, \partial x_i) \tag{6.54}$$
$$0 = -\partial p/\partial y/\rho - fu + K(\partial^2 v/\partial x_i\, \partial x_i)$$

In these equations, one can also assume that

$$\partial^2 u/\partial z^2 \gg \partial^2 u/\partial x^2 \qquad \text{or} \quad \partial^2 u/\partial y^2$$

and

$$\partial^2 v/\partial z^2 \gg \partial^2 v/\partial x^2 \qquad \text{or} \quad \partial^2 v/\partial y^2$$

leaving

$$0 = -\partial p/\partial x/\rho + fv + K\, \partial^2 u/\partial z^2 \tag{6.55}$$
$$0 = -\partial p/\partial y/\rho - fu + K\, \partial^2 v/\partial z^2$$

These equations also require the incompressible assumptions $\nabla \cdot \mathbf{u} \approx 0$. This may not be valid when problems involving significant vertical motion or large vertical extent are being considered. In these cases, the relative magnitude of each term can be checked using dynamic similarity.

6.5   The Dimensionless Navier–Stokes Equations

The equations derived in Section 6.3 can be made dimensionless using the methods of Chapter 3. This is most helpful in atmospheric dynamics, since we deal with a wide variety of scales.

We begin by selecting arbitrary characteristic values for each parameter. This will allow choices for particular problems to be inserted in the resulting nondimensional parameters. Select L and V for characteristic length and velocity. Assume density and viscosity are nearly constant. Thus, consider the dimensional parameters in Eqs. (6.46) as all primed, so that the nondimensional parameters are plain. Then,

$$u_i = u_i'/V;\ x_i = x_i'/L;\ t = (V/L)t';\ \text{and}\ p = p'/(\rho V^2)$$

When these are placed in Eqs. (6.46),

$$V^2/L[\partial u/\partial t + u\,\partial u/\partial x + v\,\partial u/\partial y + w\,\partial u/\partial z] =$$

$$-V^2/L\,\partial p/\partial x + fV\,v + \nu V/L^2\,(\partial^2 u/\partial x^2 + \partial^2 u/\partial y^2 + \partial^2 u/\partial z^2),$$

$$V^2/L[\partial v/\partial t + u\,\partial v/\partial x + v\,\partial v/\partial y + w\,\partial v/\partial z] =$$

$$-V^2/L\,\partial p/\partial y - fV\,u + \nu V/L^2\,(\partial^2 v/\partial x^2 + \partial^2 v/\partial y^2 + \partial^2 v/\partial z^2),$$

$$V^2/L[\partial w/\partial t + u\,\partial w/\partial x + v\,\partial w/\partial y + w\,\partial w/\partial z] =$$

$$-V^2/L\,\partial p/\partial z - g + \nu V/L^2\,(\partial^2 w/\partial x^2 + \partial^2 w/\partial y^2 + \partial^2 w/\partial z^2).$$
$$(6.56)$$

Multiplying these equations through by L/V^2 will leave two coefficients only.

$$\partial u/\partial t + u\,\partial u/\partial x + v\,\partial u/\partial y + w\,\partial u/\partial z =$$

$$-\partial p/\partial x + 1/\mathrm{Ro}\,v + 1/\mathrm{Re}(\partial^2 u/\partial x^2 + \partial^2 u/\partial y^2 + \partial^2 u/\partial z^2),$$

$$\partial v/\partial t + u\,\partial v/\partial x + v\,\partial v/\partial y + w\,\partial v/\partial z =$$

$$-\partial p/\partial y - 1/\mathrm{Ro}\,u + 1/\mathrm{Re}(\partial^2 v/\partial x^2 + \partial^2 v/\partial y^2 + \partial^2 v/\partial z^2), \quad (6.57)$$

$$\partial w/\partial t + u\,\partial w/\partial x + v\,\partial w/\partial y + w\,\partial w/\partial z =$$

$$-\partial p/\partial z - L/V^2 g + 1/\mathrm{Re}(\partial^2 w/\partial x^2 + \partial^2 w/\partial y^2 + \partial^2 w/\partial z^2).$$

The Rossby number $\mathrm{Ro} = V/(fL)$, the Reynolds number $\mathrm{Re} = \rho VL/\mu$, and a modified gravitational force are the parameters in these equations. Their values relative to unity will determine the importance of Coriolis, viscous, and gravitational forces. When $\mathrm{Re}_\tau = \rho VL/K$, then the turbulent viscous forces are considered.

From Eqs. (6.57) we see that when the distance scale is large, Ro is small and the Coriolis term is relatively large. For large scales the Coriolis term must be included in the equations. For these large scales the Re is large and hence the viscous term is small. However, for atmospheric scales, this also means that the critical Re for turbulence transition is passed. Then, Re_τ must be used. However, it is also large, and for synoptic scales the turbulent viscous term is often negligible. However, K is large enough in some domains that viscous terms must be retained. One such region is the PBL, where the vertical scale is about 1 km.

Example 6.5

A synoptic scale can be defined as that horizontal scale where Coriolis forces are significant and viscous forces are negligible. For air at mean density of 1.23 kg/m3 and velocity 10 m/sec, what scale length will this include? Approximate eddy viscosity as $K \leq 20$ m^2/sec.

Solution

We need the Ro = 0[1] and Re $\gg$ 1. At mid-latitudes, the Coriolis parameter is $f = 10^{-4}$/sec. Hence,

$$Ro = V/(fL) = 10 \text{ (m/sec)}/[10^{-4} (1/\text{sec}) \cdot L(\text{m})] \leq 1$$

Hence,

$$L \leq 10^5 \text{ m} = 100 \text{ km}$$

Note that Ro will still be near unity at smaller scales for lower velocities (50 km for 5 m/sec). As the equator is approached, $f \rightarrow 0$, Ro $\rightarrow \infty$, and the Coriolis force is nonexistent.

When there is ample source of perturbation to a laminar flow, as there is in the lower layers of the atmosphere, transition takes place at Re$_\tau \leq$ 2000. The scale at which this takes place is

$$L \leq 2000 \, K/V = 2000(20)/10 = 4 \text{ km}$$

This is a very short distance, and turbulence is assured.

On the 100-km scale,

$$Re \leq 10 \text{ (m/sec) } 10^5 \text{ (m)}/20 \text{ (m}^2/\text{sec)} = 50,000$$

Thus, in Eqs. 6.53, the eddy-viscous terms are about four orders of magnitude smaller than the other terms. In fact, only when the scale is on the order of kilometers and there is significant turbulence will the viscous terms be important. This happens with respect to the vertical scale in the PBL.

6.6 Summary

The Navier–Stokes equations is a phrase used principally to describe the momentum equations, which Navier and Stokes independently derived. But the phrase is often used to include the continuity equation and the equation of state. Even the energy equation may be included.

There are two aspects of these equations which make them extremely difficult to solve. First, they are nonlinear. The advective terms have the product of the velocity and the velocity derivatives. This has far-reaching consequences in terms of the uniqueness of the solutions. In uniform flows with the velocity spatial derivatives equal to zero, the number of terms is greatly reduced. These assumptions result in linear versions of the equations that have simple solutions. They provide quite successful approximations for many flow situations. However, in atmospheric flows, waves frequently are present, with resulting horizontal velocity gradients. The nonlinear terms must be addressed. The consequences of the nonlinear effects are large. These problems are dealt with under such topics as stability theory, turbulence, chaos, strange attractors, and coherent structures. These topics are at the cutting edge of applied mathematics, and the Navier–Stokes equations provide a paramount challenge for them.

The second great difficulty in solving the general Navier–Stokes equations arises in dealing with the viscous terms. These contribute by far the most terms when the equations are written out in component form. The equations are much easier to solve when the viscous terms can be neglected. Consequently, the study of inviscid flow is a popular separate field of study. It has proven very fruitful, resulting in excellent approximations for general atmospheric flow above the boundary layer regions.

Viscous forces are essential to the flow near a boundary, where the flow must come to a halt at the boundary. The viscous terms are the highest order terms. Thus, they require more constants of integration in their solutions and permit these solutions to satisfy more boundary conditions. In particular, the no-slip conditions at the boundary surface can be satisfied.

Furthermore, viscous effects must be included in many flows even though they are not dominant terms. The inviscid equations applied to weather forecasting are limited in the extent of their forward integration with time. The contribution of the viscous terms eventually becomes important. Also, in numerical integrations of the basic inviscid equations for some simple freestream flows, numerical instabilities may develop. These can grow to invalidate the solution unless one adds a small viscous term that has the effect of damping these waves. However, one must take care that the viscous terms are realistic, and that important real physical waves are not also eliminated.

Even when viscous effects are included, the complete stress term is generally not calculated, and various approximations are made. With an understanding of the complete equations, we are now in a good position to spell out the conditions under which the various approximations to the com-

stokes7l

plete equations are valid. However, first we must investigate temperature variations and the energy equation.

The NAVIER–STOKES EQUATIONS

The Momentum Equation:

Acceleration	=	Body force	+	Coriolis force	+	Static pressure and normal forces	+	Strain rate tensor, def **u** off-diagonal terms

$$\rho\, Du/Dt = F_x + \rho f V - \partial p/\partial x + \partial/\partial x[\mu(2\,\partial u/\partial x - 2/3\,\partial u_k/\partial x_k)]$$
$$+ \partial/\partial y[\mu(\partial u/\partial y + \partial v/\partial x)] + \partial/\partial z[\mu(\partial w/\partial x + \partial u/\partial z)]$$

$$\rho\, Dv/Dt = F_y - \rho f U - \partial p/\partial y + \partial/\partial y[\mu(2\,\partial v/\partial y - 2/3\,\partial u_k/\partial x_k)]$$
$$+ \partial/\partial z[\mu(\partial v/\partial z + \partial w/\partial y)] + \partial/\partial x[\mu(\partial u/\partial y + \partial v/\partial x)]$$

$$\rho\, Dw/Dt = F_z - \partial p/\partial z + \partial/\partial z[\mu(2\,\partial w/\partial z - 2/3\,\partial u_k/\partial x_k)]$$
$$+ \partial/\partial x[\mu(\partial w/\partial x + \partial u/\partial z)] + \partial/\partial y[\mu(\partial v/\partial z + \partial w/\partial y)]$$

Continuity:

$$\partial\rho/\partial t + \partial(\rho u_k)/\partial x_k = 0$$

State:

$$p = \rho R T$$
(5 equations in **u**, p, and ρ)

The Constant Density or Nondivergent Approximation:

$\nabla \cdot \mathbf{u} = 0$, can be written, with $\mathbf{k} = (0, 0, 1)$,

$$\rho[(\partial u_i/\partial t + u_j\,\partial u_i/\partial x_j)] = F_i - \rho f\epsilon_{ijk}k_j u_k - \partial p/\partial x_i$$
$$+ \mu\,\partial^2 u_i/\partial x_j\,\partial x_j$$

or

$$\rho\, D\mathbf{u}/Dt = \rho\mathbf{F}_b + \mathbf{F}_C - \text{grad } p + \mu\,\nabla^2\mathbf{u} \quad (4 \text{ equations in } \mathbf{u} \text{ and } p)$$

where $\mathbf{F}_C$ = Coriolis "force"; $\mathbf{F}_b$ = body force.

The Inviscid Approximation:

$$\rho[(\partial u_i/\partial t + u_j\,(\partial u_i/\partial x_j)] = \rho F_{bi} - \rho f\epsilon_{ijk}k_j u_k - \partial p/\partial x_i$$
$$\partial\rho/\partial t + \partial(\rho u_k)/\partial x_k = 0$$

Problems

1. Show that the stress tensor is symmetric by taking moments about an axis.

2. Consider the momentum equation for one-dimensional flow along a horizontal streamline, inviscid, and constant density. Obtain the relation governing this flow and apply it to the flow in a converging circular region with radius r. Plot p, u, and r versus x for (a) a linearly decreasing $r(x)$; and (b) for a linearly varying $u(x)$.

3. Derive an expression for the vertical pressure variation in the atmosphere as a function of temperature from the momentum equation for a static fluid plus the ideal gas law. Let $T = T_0 - \gamma z$, where the lapse rate γ is

γ (°K/km)	z range (km)
6.5	0–12
0	12–22
−1	22–32

Plot $T(z)$ and $P(z)$.

4. The flow between infinite parallel plates is said to be parabolic. Determine the velocity distribution from fundamental principles of viscous flow. (*Hint:* One boundary condition is the maximum flow speed, U_m). Assume incompressibility and steady state. Discuss no-slip, flow rate, shear stress at the plates, rotationality, and vorticity.

5. The flow in a circular pipe with steady viscous flow parallel to the axis is called Poiseuille flow. Use the equations in Example 6.2 with $u_\theta = u_r = 0$, and $w =$ the flow along the pipe, to determine the velocity distribution across the radius.

6. A special case for the Navier–Stokes equations that can be solved (reduced to two terms) is the flow near a surface that has small oscillations, $a \sin \omega t$. Consider the flow as a function of height to be $u = f(z)\, e^{-pz}$, p is constant. Obtain an expression for $f(z)$. *Hint:* Show that the motion is described by the diffusion equation.

7. Use assumptions of the Navier–Stokes equations (momentum) to explain why an object falling from a great height reaches a steady terminal speed. (Given that F_{drag} is proportional to U^2).

8. A horizontal blast of air hits the side of a ship and is deflected 90° upward. The effective area of the ship normal to the wind is 5 m by 20 m. Determine the force experienced by the ship if the wind velocity is 30 m/sec.

9. Very slow flow over a sphere is called Stokes flow. The velocity components are

$$u_r = U \cos \theta[1 + \tfrac{1}{2}(R/r)^3 - \tfrac{3}{2}(R/r)]$$

$$u_\theta = U \sin \theta[-1 + \tfrac{1}{4}(R/r)^3 + \tfrac{3}{4}(R/r)]$$

Calculate the viscous stress tensor in cylindrical coordinates.

10. Show that a Newtonian fluid that is incompressible and with constant viscosity obeys the relation

$$\mathbf{\nabla} \cdot \mathbf{\tau} = \mu \, \nabla^2 \mathbf{u}$$

Chapter 7 | Conservation of Energy

The conservation of energy is the third great principle of basic physics to be invoked in the development of the equations for fluid flow. Yet, when we consider energy there is not a unique quantity we can substitute into the general conservation equations as we have done in previous chapters with the density and momentum. In many cases in atomic physics it is necessary to combine the energy principle with the first principle invoked, conservation of mass, to create a useful set of equations. But in the study of geophysical fluid dynamics the conservation of energy generally remains independent. Thus, from this perspective, the first task is to identify the categories of energy that are important to atmospheric dynamics.

Energy exists in many forms. Some of the basic forms we may encounter in atmospheric physics are:

1. *Kinetic* (mass in motion)
2. *Potential* (position in a force field)
3. *Work* (motion in a force field)
4. *Internal* (molecular motion and structure)
5. *Radiant* (involving molecular absorption and radiation)

6. *Chemical* (molecular chemical reactions)
7. *Electromagnetic* (motion in a magnetic field, electrification of particles)

To complement our field equations, the basic energies listed need to be related to the flow properties of *mass, density, pressure,* and *temperature.* First we know that *kinetic* and *potential* energies are forms of *mechanical* energy and are intimately related to mass ($\frac{1}{2}mv^2$ and mg Δz, respectively). Another manifestation of energy, *work,* is usually involved in a situation where a mechanical effort is applied. Work can also be expressed in terms of forces and motion. Next, *internal* energy is affected by a change in any local flow or thermodynamic parameter. In addition, changes in internal energy occur throughout a volume from *radiant* energy when gases in the atmosphere absorb radiation. The radiant or *heat* energy absorbed at the surface is a fundamental source of energy for atmospheric and oceanic motion.

In the nineteenth century, Hermann von Helmholtz expressed the energy principle simply as: *the sum of the kinetic and potential energies is equal to a constant.* However, when heat is absorbed or dissipated, the conservation of energy balance must also include *thermal* energies. Examples of thermal energies are frictional heating, heat conduction, and radiation absorption. Each of these involves the transfer of heat, which affects the internal energy. This transfer is usually quantified with respect to *temperature.* Since temperature is the most basic measurable energy parameter, this chapter relates the internal energy to the fluid temperature. In addition, an internal energy parameter, *enthalpy,* is introduced. It combines the thermodynamic and potential energies.

The last two energy categories mentioned, *chemical* and *electromagnetic,* become important in pollution studies and magnetosphere studies, respectively. They require additional terms to be added to the equations developed in this chapter.

The science of thermodynamics is a huge logical discipline built on fundamental postulates. The basic conservation of energy statement is the first law of thermodynamics. The second law discusses the nature of heat transfer processes. These processes are the thermodynamic means of converting energy from one form to another. And finally, because this subject is centered around energy dealing with heat, an additional postulate is invoked to establish the definition of *temperature.* This parameter indicates the internal energy of a substance.

The fluid dynamics of meteorology requires only a small sample of the rich storehouse of scientific inquiry available in thermodynamics. Still, we need a fairly broad introduction to the terminology and fundamentals of thermodynamics in order to extract the limited, but essential, information that we require.

Thermodynamic processes drive all motion on the planet. The earth's atmosphere–land–water system is a gigantic thermodynamic engine converting solar radiant energy into the winds, ocean currents, and hydrologic cycle. Usually, we isolate various *systems* on particular scales or domains of interest. These can range from the entire planet, for long-term climatological studies, to domains even smaller than an individual convective plume. An example of a system is the land–sea–breeze domain. Here, the unequal heating caused by the reflection/absorption differences of water and earth causes the direction of the local wind to change with the time of day. Other continent–ocean systems govern the larger-scale effects of monsoon dynamics. The land–sea–ice boundaries produce assorted local weather conditions based on differential heating within a much smaller system. There are systems used to define an individual storm—ranging in sizes from hurricanes, to thunderstorms, to a convective cell. And finally, a system that includes the heating variations due to the seasons and latitudinal effects is of basic importance to general circulation modeling.

Despite the thermodynamic origins of the earth's dynamic forcing of its fluid, for short periods thermodynamic processes can be ignored and heating considered simply the antecedent source of the pressure gradient. Still, as general circulation models get more accurate and analyses of storms and climate resolve more detail, more understanding of thermal processes is needed. The thermodynamic underpinnings of, and subtle modifications to, the driving forces become more and more important. Thus, no treatment of atmospheric dynamics is complete without including the energy equation. When the temperature becomes a significant dependent variable this equation is absolutely necessary to close our set of equations. The complete set of equations will include continuity, state, and three momentum equations, and the energy equation.

Thermodynamics deals in particular with end-states of systems when energy changes form. We will investigate a useful application of the energy equation that relates the conditions at one station to another. This will be similar to the $\Sigma \, \rho V A$ (constant) and $\Sigma \, \rho V^2 A$ (constant) relations found in the control volume versions of the conservation of mass and momentum (Chapter 3). However, we first need a field equation for the energy. To do this we need to identify the total energy being conserved and the quantities that serve as the sources and sinks of energies.

One definition of energy (by Planck) is *that which has the ability to produce external effects*. Thus, the first step is to identify the pertinent energy forms that affect the atmospheric fluid flow. Then we need to write a conservation statement for the sum of these energies. This derived total energy may change due to energy sources and/or sinks in a given system. Inside the domain of a single problem, we may observe energy to have several forms, which may or may not be able to convert from one to another.

In general, it is fairly easy to convert *mechanical energies* into *thermal energy*. Many common examples exist. Simply rubbing hands together, applying brakes, or pumping up a tire all convert mechanical to thermal energy. The conversion is reflected by the temperature increase of the system. These common processes involve friction and cannot be reversed. On the other hand, the conversion of thermal energy to mechanical energy is not as common. This process can be accomplished only through the action of changing pressures and requires compressibility of the fluid. Such a process can generally be run either way and is called *reversible*. The storage of the latent heat of evaporation due to the phase change of liquid water to vapor in the atmosphere is an important process of thermal energy conversion. This sink of energy (or source, during the process of condensation) is a significant reversible energy transport mechanism. It must be addressed in the study of cloud physics. It is also a significant contributor to the energetics of many storm systems.

7.1 The First Law of Thermodynamics

The *first law of thermodynamics* states that in a system of constant mass, energy cannot be created or destroyed. Thus any change in energy stored in a material region equals the net energy transfer across the boundaries of that region. Atmospheric fluid flow problems include energy in several forms, and in each problem there may exist sources and sinks of energy. The important forms of possible energy are the kinetic energy of the flow; internal energy related to temperature; and work energy done by pressure forces, internal stress forces and/or body forces. In a material region, net energy change can take place as heat or as work. In an expression of the first law of thermodynamics, we can conveniently separate the energy balance terms into:

$$\underset{\text{energy}}{\text{change in internal}} = \text{Work energy} + \underset{\text{transfer rate}}{\text{Total heat}}$$

$$\delta E/\delta t = \delta W/\delta t + \delta Q/\delta t \qquad (7.1)$$

In the context of our general conservation statement in Section 5.6, Eq. (7.1) gives the change in internal energy E of the material in a given volume as the sum of the work done on the system and the heat added to the system. We will examine each of the three terms in Eq. (7.1) with respect to volume and/or area integrals over a parcel domain.

7.1.1   Total Energy

We define the *total energy* of a flowing parcel to consist of the *specific kinetic energy* $\delta m |u|^2 / 2 / \delta m = \frac{1}{2} |u|^2 \equiv \frac{1}{2} q^2$, and the *specific internal energy* e. The rate of increase of total energy can be separated into the sum of the changes in kinetic and potential energies. Once we have a quantity to be conserved, we can balance its change over the volume by the flux through the surfaces.

The conservation of $E = \rho(e + \frac{1}{2} q^2)$ may be written

$$\frac{\delta E}{\delta t} = \underbrace{\iiint \frac{\partial}{\partial t} \left[\rho \left(e + \frac{1}{2} q^2 \right) \right] dV}_{\text{local time rate of change}} \quad + \quad \underbrace{\iint \rho \left(e + \frac{1}{2} q^2 \right) \mathbf{u} \cdot \mathbf{n} \, dA}_{\text{inflow/outflow}} \qquad (7.2)$$

where $\frac{1}{2} q^2$ is *kinetic energy* and e is the *internal energy* per unit mass. The kinetic energy is associated with the mean macroscopic motion of the aggregate of fluid. The internal energy represents the kinetic energy of the microscopic motions (e.g., molecular vibration and rotation).

7.1.2   Rate of Work Done on a System

When we have a parcel experiencing distortion within a flow field we want to know the work done on that system. *Work* equals the force times the distance, and the *rate of work* is the force times velocity (constant force). Thus we need to collect all the forces affecting the parcel. In Chapter 4 we distinguished the forces as either body forces or surface forces. In addition, the surface forces have been further separated into the normal pressure forces and the internal stress forces. With this in mind, recall that $\boldsymbol{\sigma}\mathbf{n}$ yields the forces on the parcel surface, so that the work rate of these internal stress forces may be written

$$\mathbf{u} \cdot (\boldsymbol{\sigma}\mathbf{n}) = -p\mathbf{u} \cdot \mathbf{n} + \mathbf{u} \cdot \boldsymbol{\tau}\mathbf{n}$$

Hence, we can write an expression for the total work rate on the parcel as

$$\frac{\delta W}{\delta t} = \underbrace{-\iiint \mathbf{F}_b \cdot \mathbf{u} \, dV}_{\substack{\text{work rate} \quad - \\ \text{due to:}}} \underbrace{- \iint p\mathbf{u} \cdot \mathbf{n} \, dA}_{\text{body forces} \quad -} \underbrace{+ \iint \mathbf{u} \cdot \boldsymbol{\tau}\mathbf{n} \, dA}_{\text{pressure forces} \quad +} \qquad (7.3)$$

Note here that in the atmosphere, $\mathbf{F}_b$ is generally the gravitational force per unit volume. The change in *potential energy* will emerge from this term.

It is important to notice that in Eq. (7.1) the work done *by* the parcel equals a loss of energy from the parcel. Flow work is negative in the flow equations. Thus, if we were to consider a finite volume that contained a windmill that was extracting wind energy from the flow, the work done on the windmill would be a sink term in the total energy budget of the flow.

Example 7.1

Consider a volume of air consisting of a box that surrounds a windmill. The windmill extracts work energy to pump water. Sketch in the various work terms that act on this volume of air.

Solution

We must consider the terms in Eq. (7.3), plus the work extracted by the windmill. The gravitational force is normal to the horizontal flow near the surface. Therefore it does no work. The other terms can be sketched as shown in Fig. 7.1.
    The types of work included are

1. Shear work done by the shear stresses in the fluid acting on the boundaries of the control volume.
2. Pressure work done by the fluid pressure acting on the boundaries of the control volume.

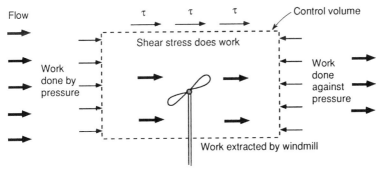

Figure 7.1   The work forces on a control volume. The arbitrary volume includes a windmill in a segment of boundary layer with horizontal flow.

3. Shaft work from a collector inside the control volume and transmitted outside the volume.

===

7.1.3   Rate of Heat Transfer

The change in the internal energy due to heat transfer takes place in atmospheric flows mainly through two mechanisms. The first is the heat change from thermal heat release or absorption from sinks and sources distributed within the volume of the system. These are represented in the single parameter R. The second is a heat change due to *heat conduction* $\mathbf{K}$. This is a vector that shows the direction, in or out, and the magnitude of the heat conduction through the sides of the domain.

$$\delta Q/\delta t = \iiint \rho R \, dV - \iint \mathbf{K} \cdot \mathbf{n} \, dA \tag{7.4}$$

Generally the heat conduction $\mathbf{K}$ is related to the temperature gradient according to the Fourier law for heat conduction,

$$\mathbf{K} = k_h \operatorname{grad} T \tag{7.5}$$

where k_h is the heat conduction coefficient. This coefficient can be established by experiment in a fashion analogous to the viscosity coefficient (Section 1.11).

Once again we have a term where the source or sink of energy is distributed throughout the volume. Atmospheric examples of this include radiative heating and/or latent heat release within the volume. The last term in Eq. (7.4) represents the heat transfer across the surface. It is proportional to a vector component of $\mathbf{K}$ normal to the surface. In this case, from Eq. (7.5), heat flux depends on the normal component of the temperature gradient vector on the boundaries of the domain.

7.2   The Energy Equation

The first law of thermodynamics relates the change in total energy to sink and source terms—work rate and heat transfer. This is similar to the momentum principle relating the change in momentum to sink and source terms—

the forces. The forces also appear in the sink/source terms of the energy equation as they do work. Our goal is to obtain the field equation that describes the energy state at every point in a field. Our method is to write the energy balance for a small volume that will be considered in the limit of an infinitesimal parcel.

In Fig. 7.2, the surface forces are the pressure and the surface stresses: The body force work is done with or against gravitational force. The stress work requires a combination of the components of τn and $\mathbf{u}$. The heat flux depends on the orientation of the temperature gradient relative to each of the surfaces. Finally, there may be radiative-type energy absorption throughout the volume.

With Fig. 7.2 in mind and the results of Section 7.1, we can now write $\delta E/\delta t = \delta Q/\delta t - \delta W/\delta t$ in terms of volume and surface area integrals over our domain, the parcel. We apply the divergence theorem to the conduction, stress, and inflow/outflow surface area terms. Then all of the energy flux terms can be written with respect to volume integrals,

$$\iiint \partial/\partial t\left\{\rho\left(e + \frac{1}{2}q^2\right)\right\} dV + \iiint \mathrm{div}\left\{\rho\left(e + \frac{1}{2}q^2\right)\mathbf{u}\right\} dV$$

$$= \iiint \mathbf{F}_b \cdot \mathbf{u}\, dV + \iiint \mathrm{div}(\sigma\mathbf{u})\, dV + \iiint \rho R\, dV - \iiint \mathrm{div}\,\mathbf{K}\, dV$$

$$(7.6)$$

Now, we let the volume of the parcel approach zero (staying within the continuum of course). We obtain the differential form of the conservation of energy for any point in the field.

In the limit $\delta V \to 0$ of Eq. (7.6),

$$\partial/\partial t\{\rho(e + \tfrac{1}{2}q^2)\} + \mathrm{div}\{\rho(e + \tfrac{1}{2}q^2)\mathbf{u}\} = \mathbf{F}_b \cdot \mathbf{u} + \mathrm{div}(\sigma\mathbf{u}) + \rho R - \mathrm{div}\,\mathbf{K}$$

temporal internal energy + kinetic energy change	+	advective change	=	work of body force	+	work of surface forces	+	source and sinks	−	heat flow

The left side of this equation may be written

$$\rho\{\partial/\partial t(e + \tfrac{1}{2}q^2) + \mathbf{u} \cdot \mathrm{grad}(e + \tfrac{1}{2}q^2)\} + (e + \tfrac{1}{2}q^2)\{\partial\rho/\partial t + \mathrm{div}\,\rho\mathbf{u}\}$$

The second term in { } is identically zero from continuity, leaving

$$\rho\{\partial/\partial t(e + \tfrac{1}{2}q^2) + \mathbf{u} \cdot \mathrm{grad}(e + \tfrac{1}{2}q^2)\} = \rho\{\partial/\partial t + \mathbf{u} \cdot \mathrm{grad}\}(e + \tfrac{1}{2}q^2)$$

The full equation then becomes

$$\rho\{\partial/\partial t + \mathbf{u} \cdot \mathrm{grad}\}(e + \tfrac{1}{2}q^2) = \mathbf{F}_b \cdot \mathbf{u} + \mathrm{div}(\sigma u) + \rho R - \mathrm{div}\,\mathbf{K} \qquad (7.7)$$

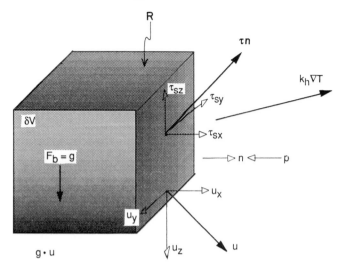

Figure 7.2   A small parcel of fluid with energy change terms. There are forces which must be multiplied by **u** to yield work rates: gravitational **g**, pressure p, and stress $\tau\mathbf{n}$ (shown on one face). There are also radiative transfer energy R and the heat transfer vector $k_h \nabla T$.

Noting that the left side involves the total derivative, we can write

$$\rho D/Dt(e + \tfrac{1}{2}q^2) = \mathbf{F}_b \cdot \mathbf{u} + \mathrm{div}(\boldsymbol{\sigma}\mathbf{u}) + \rho R - \mathrm{div}\ \mathbf{K} \qquad (7.8)$$

In addition, a tensor relation can be employed here,

$$\mathrm{div}(\boldsymbol{\sigma}\mathbf{u}) \equiv \mathbf{u} \cdot (\mathbf{div}\ \boldsymbol{\sigma}) + \boldsymbol{\sigma}\ \mathrm{grad}\ \mathbf{u}$$

to express the stress work in two parts,

$$\rho D/Dt(e + \tfrac{1}{2}q^2) = \mathbf{F}_b \cdot \mathbf{u} + \mathbf{u} \cdot \mathrm{div}\ \boldsymbol{\sigma} + \boldsymbol{\sigma}\ \mathrm{grad}\ \mathbf{u} + \rho R - \mathrm{div}\ \mathbf{K} \qquad (7.9)$$

This is the complete energy equation. The problem of the existence of two terms involving the general stress tensor is still present. However, these terms can be investigated by separating the pressure and stress forces. In the process, we will find that the energy equation involves changes in the mechanical energy (the velocity) and/or thermal energy (the temperature). We will find that the work done on the surface contributes something to both energy equations, part changing the velocity and part changing the internal energy.

7.3   The Mechanical Energy Equation

In the complete energy equation, Eq. (7.9), work energy can go into two places. These are kinetic energy and internal energy. To simplify the analysis of the work done we can construct an equation that involves the change in kinetic energy alone. (Equally, we will isolate the internal energy in the next section.)

Note that in Eq. (7.7) the work effect of the internal surface forces in changing the energy is represented by div($\boldsymbol{\sigma}\mathbf{u}$). When this term is expanded, there are many elements involving pressure, velocity, and surface stress. To make this term more manageable (susceptible to approximation) we separate the pressure and stress work terms.

For the mechanical energy part of Eq. (7.9) we want to concentrate on the kinetic energy. To begin we can obtain an expression for kinetic energy if we multiply the momentum equation by $\mathbf{u}$,

$$\mathbf{u} \cdot \{\rho \, D\mathbf{u}/Dt = \rho\mathbf{F}_b - \text{grad } p + \text{div } \boldsymbol{\tau} = \rho\mathbf{F}_b + \text{div } \boldsymbol{\sigma}\}$$

to get

$$\rho \, D(\tfrac{1}{2}q^2)/Dt = \rho\mathbf{F}_b \cdot \mathbf{u} + \mathbf{u} \cdot \text{div } \boldsymbol{\sigma}$$

$$= \rho\mathbf{F}_b \cdot \mathbf{u} - \mathbf{u} \cdot \text{grad } p + \mathbf{u} \cdot \text{div } \boldsymbol{\tau} \qquad (7.10)$$

This expression involves mechanical energies only (kinetic and work). The first term in Eq. (7.10) indicates that the effect of the body forces work is to cause changes in kinetic energy. The second term on the right is the work done on the parcel by the pressure gradient. When the $\nabla p < 0$ the kinetic energy increases, for there is positive work done. The last term represents the work that must be done against stress gradients as the parcel is advected through a variable stress field. Since atmospheric flows are dominantly horizontal and stress gradients are mainly vertical, this term is generally much less than $\mathbf{u} \cdot \nabla p$.

If we subtract the mechanical energy terms from the complete energy equation, only terms that contribute to thermal energy will be left. We note that there are two contributions from the internal fluid forces to the mechanical energy equation. However, these are not identical to the two terms occurring in the complete energy equation (7.9). The difference is the part of the internal work that goes to change the internal energy.

7.4   The Internal Energy Equation

The internal energy of the fluid gives a measure of the work done on, or by, the fluid. In atmospheric physics, it is generally associated with the temperature. It is of intrinsic interest as one of the "weather" parameters. It also is important in relation to how much moisture, and hence latent heat, can be stored in the air.

The internal energy equation is obtained by subtracting the mechanical energy terms in Eq. (7.10) from the total energy equation (7.9). This leaves

$$\rho \, De/Dt = \boldsymbol{\sigma} \operatorname{grad} \mathbf{u} + \rho \mathbf{R} - \operatorname{div} \mathbf{K}$$

or (7.11)

$$\rho \, De/Dt = -p \operatorname{div} \mathbf{u} + \boldsymbol{\tau} \operatorname{grad} \mathbf{u} + \rho \mathbf{R} - \operatorname{div} \mathbf{K}$$

In this expression, the internal energy e is a scalar—as indeed are all of the terms of this equation. Of special interest is the second term on the right side; this must be a scalar product of two tensors. As such, this term is difficult to evaluate (or even to write out). However, it can be important when the high-frequency turbulent velocity fluctuations acting on the surface stress forces are significant. Since we usually must address turbulence effects by parametrization with respect to the mean, we call this term the mean *dissipation*.

Dissipation, $\Phi \equiv \boldsymbol{\tau} \operatorname{grad} \mathbf{u}$

We note that this term is in the tradition inspired by the generality of the terms in the first law of thermodynamics—that when the energy terms do not balance, one adds a new energy (from a wry comment by Henri Poincare). The dissipation is often "measured" as the quantity left over after all of the other terms have been evaluated.

The other contribution to the internal energy from the internal forces, $-p$ div $\mathbf{u}$, is simply the work of expansion or contraction of the parcel against the normal pressure force. The "direct" thermal energy fluxes are the volumetric sink/source energy term, which is absorbed directly as a change in internal energy, and the heat conduction, which also goes directly to change the internal energy.

When $\nabla \cdot \mathbf{u} = 0$, the internal energy equation involves only three small terms—the dissipation, the sink/source term, and thermal conductivity. Frequently in atmospheric dynamics, all three are very small and the energy equation plays a surprisingly small role. In this case, temperature or thermodynamics does not enter the problem.

7.5   The Enthalpy Equation

The internal energy term e contains effects of the structure and motion of the atoms and molecules. We will consider e to be the specific internal energy for a liquid or gas (single phase). It is primarily a function of temperature. The change in internal energy is parametrized with respect to the quantity of heat per unit mass required to raise the temperature a fixed amount, as discussed in Section 1.10. The pressure work involved when a compressible fluid is considered is accommodated neatly by using the energy term *enthalpy*. For the basic case of dry air the temperature can be related to the internal energy and the pressure under the definition of specific enthalpy,

$$h = e + p/\rho \tag{7.12}$$

The enthalpy provides a relation involving the internal energy and the mechanical properties. The transfer of energy between these two terms is important only for a compressible fluid. When we treat the air as incompressible, we can write

$$e = c_v T = cT \tag{7.13}$$

For a truly incompressible fluid,

$$c_v \equiv \partial e/\partial T]_v = c_p \equiv \partial h/\partial T]_p \tag{7.14}$$

However, air is compressible, and we only approximate it as incompressible for low velocity and small changes in density. When there exist significant thermal changes, new criteria for the incompressible assumptions must be derived. In this text, we treat air as a perfect gas with constant specific heats, so that

$$h = e + p/\rho = e + RT = \int c_p \, dT \approx c_p T \tag{7.15}$$

We can express the rate-of-change equations with respect to the enthalpy by writing

$$Dh/Dt = De/Dt + D(p/\rho)/Dt \tag{7.16}$$

The last term in this equation may be written

$$D(p/\rho)/Dt = (1/\rho) \, Dp/Dt + p \, D(1/\rho)/Dt$$

$$= (1/\rho) \, Dp/Dt - (p/\rho^2) \, D\rho/Dt$$

Then, by using continuity, the last term can be rewritten to get

$$D(p/\rho)/DT = (1/\rho) \, Dp/Dt + (p/\rho) \, \text{div } \mathbf{u} \tag{7.17}$$

At this point we can substitute Eq. (7.17) into (7.16) to get

$$Dh/Dt = De/Dt + (1/\rho) \, Dp/Dt + (p/\rho) \, \text{div } \mathbf{u}$$

or

$$\rho \, Dh/Dt = \rho \, De/Dt + Dp/Dt + p \, \text{div } \mathbf{u} \tag{7.18}$$

Then, using the expression for the total change of internal energy, Eq. (7.11), this may be written

$$\rho \, Dh/Dt = Dp/Dt + p \, \mathrm{div} \, \mathbf{u} + \boldsymbol{\sigma} \, \mathrm{grad} \, \mathbf{u} + \rho R - \mathrm{div} \, \mathbf{K}$$

$$= Dp/Dt + \boldsymbol{\tau} \, \mathrm{grad} \, \mathbf{u} + \rho R - \mathrm{div} \, \mathbf{K} \qquad (7.19)$$

Or, by employing the definition of dissipation, we have

$$\rho \, Dh/Dt = Dp/Dt + \Phi + \rho R - \mathrm{div} \, \mathbf{K} \qquad (7.20)$$

This expression can be combined with the mechanical energy equation (7.10) to form a conservation of enthalpy plus kinetic energy,

$$\rho \, D(h + \tfrac{1}{2}q^2)/Dt = Dp/Dt - \mathbf{u} \cdot \mathrm{grad} \, p + \boldsymbol{\tau} \, \mathrm{grad} \, \mathbf{u} + \mathbf{u} \cdot \mathrm{div} \, \boldsymbol{\tau} + \rho R$$

$$- \mathrm{div} \, \mathbf{K} + \rho \mathbf{F}_b \cdot \mathbf{u}$$

or,

$$\rho \, D(h + q^2/2)/Dt = \partial p/\partial t + \mathrm{div}(\boldsymbol{\tau}\mathbf{u}) + \rho R - \mathrm{div} \, \mathbf{K} + \rho \mathbf{F}_b \cdot \mathbf{u} \qquad (7.21)$$

This equation differs from Eq. (7.9) in the form of the pressure terms on the right side. This difference arises as a consequence of the p/ρ term in the definition of enthalpy, Eq. (7.15). It is inconsequential for incompressible flow, where (7.21) and (7.9) are the same. Thus the enthalpy equation is of use only when the compressibility of the fluid is significant.

7.6   The Moist Atmosphere Energy Equation

One of the important features of the atmospheric flow is the moisture cycle. Precipitation is one of the important "weather" parameters, and the latent heat stored in the moisture is an important factor in weather systems. The evaporation of water requires energy, generally in the form of radiant energy. The process of condensation releases this same energy. In between these two events the dynamics of the atmosphere can shift the latent heat around, dispersing it, concentrating it, or simply transporting it. This takes place on many scales. They range from that of simple convective cumulus production to the dramatic effects in hurricane energetics (where it is a principal factor). It is a factor in the large-scale climatic studies of global heat transport. Thus, it is clearly important to include the latent energy in many calculations. An expedient but effective method is to relate this energy to our basic observable parameter, the temperature.

The latent heat stored in the air will depend directly on how much moisture is in the air. The energy equivalent of the moisture in the air can be related to the mixing ratio,

$$m_x \equiv \text{mass of vapor/mass of dry air} \qquad (7.22)$$
$$= \rho_v/\rho_a$$

The amount of heat required to evaporate a fixed amount of liquid is the *latent heat of vaporization, L.* The amount of heat released due to a phase change from vapor to liquid may be written

$$-L\, Dm_x/Dt$$

This heat is generally released within the volume and can be considered part of ρR, leaving only the radiative body heating rate ρR_r. Inserting these terms into Eq. (7.16) allows it to be written

$$\rho\, D(h + q^2/2 + Lm_x)/Dt = \partial p/\partial t + \text{div}(\boldsymbol{\tau}\mathbf{u}) + \rho R_r - \text{div }\mathbf{K} + \rho\mathbf{F}_b \cdot \mathbf{u} \tag{7.23}$$

Note that we have neglected transport of vapor through the surfaces. If this is important, it must be included as a surface area source term on the right side of Eq. (7.23). Finally, we introduce the geopotential for the gravitational body force,

$$\rho\mathbf{F}_b = \rho\mathbf{g} = -\rho\,\nabla\varphi = -\rho\, d\varphi/dx \tag{7.24}$$

Since φ is not a function of time, and has only the term gz, we can write

$$D\varphi/dx = \partial\varphi/\partial t + u\,\partial\varphi/\partial x + v\,\partial\varphi/\partial y + w\,\partial\varphi/\partial z = w\,d\varphi/dz = \mathbf{F}_b \cdot \mathbf{u}$$

Hence, Eq. (7.23) may be written

$$\rho\, D(h + \varphi + Lm_x + q^2/2)/Dt = \partial p/\partial t + \text{div}(\boldsymbol{\tau}\mathbf{u}) + \rho\mathbf{R}_r - \text{div }\mathbf{K} \tag{7.25}$$

Note that in this equation the geopotential is included in the total derivative on the left side as was shown in Eq. (7.21). The energy quantities in the total derivative occur frequently in atmospheric dynamics and are given names. In this case, with no moisture considerations, the conserved energy term is defined as

$$\text{dry static energy} \equiv e + p/\rho + \varphi = h + \varphi \tag{7.26}$$

When moisture is added, we get the additional term in Eq. 7.25, defining a new conserved quantity,

$$\text{moist static energy} \equiv h + \varphi + L_v m_x \approx C_p T + gz + L_v m_x \tag{7.27}$$

When these terms are substituted in (7.25), we get

$$\rho D(c_p T + gz + Lm_x + q^2/2)/Dt$$
$$= \partial p/\partial t + \text{div}(\boldsymbol{\tau}\mathbf{u}) + \rho R_r - \text{div }\mathbf{K} \tag{7.28}$$

Equation (7.28) provides another relation involving temperature, density, pressure, and velocity in terms of empirical constants. Together with continuity, the momentum equation, and the equation of state, there is a closed set of six equations for six unknowns.

7.7 An Alternative Derivation

As we have done for the conservation of mass and momentum, we can discuss the derivation of the energy equation in terms of a specific small parcel of mass $\rho\, \delta V$ in a Lagrangian frame of reference. We can write the total energy change as we move with the parcel.

The time rate increase = energy supplied + work done
of the total energy   to the parcel   on the parcel

The total energy change (i.e., positive kinetic energy) may be written:

$$D/Dt\{\rho\, \delta V(e + \tfrac{1}{2}q^2)\} = \rho\, \delta V\, D/Dt(e + \tfrac{1}{2}q^2) + (e + \tfrac{1}{2}q^2)\, D(\rho\, \delta V)/Dt$$

0 by continuity

The total energy supply is the sum of the mass times the source or sink energy per unit mass, $\rho\, \delta V R$, plus the heat flux vector. The latter may be written, with the aid of the divergence theorem, as

$$\iint \mathbf{K} \cdot \mathbf{n}\, dA = \iiint \text{div}\, \mathbf{K}\, dV$$

The work done on the parcel includes the body force work plus that done by the internal surface forces. The rate of work done by the external force field is $\rho\, \delta V\, \mathbf{F}_b \cdot \mathbf{u}$.

The work done by the stress can be integrated over the volume using the fact that σ is symmetric, so that we can employ the identity

$$\mathbf{u} \cdot (\sigma\mathbf{n})\, dA = (\sigma\mathbf{u}) \cdot \mathbf{n}\, dA$$

to write the work rate at the surface in a form to which the divergence theorem can be applied,

$$\iint \sigma\mathbf{u} \cdot \mathbf{n}\, dA = \iiint \text{div}(\sigma\mathbf{u})\, dV$$

We then put the above equations together, divide by δV, and consider the limit, $\delta V \to 0$. The resulting differential equation for a field point is

$$\rho(e + \tfrac{1}{2}q^2) = \rho R - \text{div}\, \mathbf{K} + \rho\mathbf{F}_b \cdot \mathbf{u} + \text{div}(\sigma\mathbf{u}) \tag{7.29}$$

Once again, this is the complete energy equation. We can examine the terms in this equation by splitting it into two parts. As in Section 7.3, we get the mechanical energy equation by multiplying the momentum equation by **u**,

$$\rho \, D(\tfrac{1}{2}q^2)/Dt = \rho \mathbf{F}_b \cdot \mathbf{u} + \rho \mathbf{u} \, \text{div} \, \boldsymbol{\sigma} \tag{7.30}$$

We then subtract the mechanical part of the energy, Eq. (7.30), from (7.29) to get the internal energy equation:

$$\rho \, De/Dt = -\text{div} \, \mathbf{K} + \rho R + \boldsymbol{\sigma} \, \text{grad} \, \mathbf{u} \tag{7.31}$$

From this we can see that the body forces are converted into kinetic energy, and the surface forces work has two contributions:

1. Pressure work $[\partial(pu_j)/\partial x_j]$:

    (a) This is the work that goes into compressing the parcel. The force times the deformation changes internal energy through the pressure part of $\boldsymbol{\sigma}$ grad **u**, $[p \, \partial u_j/\partial x_j]$.
    (b) The pressure part of the stress tensor in $\rho \mathbf{u} \, \text{div} \, \boldsymbol{\sigma} = \rho \mathbf{u} \, \text{grad} \, p$ is velocity times force gradient. This changes the kinetic energy [the $u_j \, \partial p/\partial x_j$ term]. This is the energy extracted from the flow. It is equal to the work done on the parcel as it proceeds along the pressure gradient.

2. Viscous work:

    (a) The stress force times deformation changes internal energy through the term
    $$[\tau_{ij} \, \partial u_j/\partial x_i]$$
    (b). The velocity times the force gradient changes kinetic energy through
    $$[u_j \, \partial \tau_{ij}/\partial x_i]$$

The enthalpy equation corresponding to (7.31) is

$$\rho \, Dh/Dt = Dp/Dt + p \, \text{div} \, \mathbf{u} + \boldsymbol{\sigma} \, \text{grad} \, \mathbf{u} + \rho R - \text{div} \, \mathbf{K}$$

or

$$\rho \, Dh/Dt = Dp/Dt + \Phi + \rho R - \text{div} \, \mathbf{K} \tag{7.32}$$

We see that mechanical energy changes are due to the rate of work done by body forces $\mathbf{F}_b \cdot \mathbf{u}$ and stress forces in **u** div $\boldsymbol{\sigma}$. The internal energy changes are related to the stress term through the tensor scalar product, $\boldsymbol{\sigma}$ grad **u**. This is the dissipation, which produces frictional heating. In addition there are changes directly due to heat flux and sinks and sources, R.

By now, we have obtained quite an array of possible energy equations. This reflects the fact that energy has many sources, sinks, and forms. When we are primarily concerned with the dynamics of the flow, the incompressibility assumption decouples the energy equation from the other flow equations.

7.8　Flow along a Streamline

Probably the most productive use of the energy equation for atmospheric flow results from the relation for the change in energy properties along a streamline. When steady-state, inviscid approximations can be made, an integration along a streamline relates the energy at any two points. This can be a great advantage when the flow properties are considered at two sections of a flow with the intervening region bound by streamlines. Then the properties at boundary points, which may be entrance and exits with uniform properties, can be related. The flow in between can be quite complicated.

From the differential form of the general energy equation (7.9), we have an expression for the total change in the internal plus kinetic energy of a parcel as it flows along a streamline,

$$\rho \, D(e + q^2/2)/Dt = \rho \mathbf{u} \cdot \mathbf{F}_b + \text{div} \, (\boldsymbol{\sigma} \mathbf{u}) + \rho R - \text{div} \, \mathbf{K} \qquad (7.33)$$

In the usual atmospheric flow problem, kinetic energy and pressure work are the most significant terms, followed in importance by the gravitational potential energy, the friction work and dissipation (particularly in the boundary layers), and the heat flux terms. (Coriolis force, being a virtual force, does no work. This is also evident from the fact that it is always directed normal to the flow vector.) We will initially consider very simple flows that require many assumptions to obtain the simple equations. However, there are practical applications for even these simple flow situations.

A common way to simplify the problem is to consider cases where variation takes place in one direction only, a one-dimensional problem. The variables are assumed to be uniform in the other directions. An example is a streamline flow, viewed as one-dimensional with speed u in the x-direction, which is arbitrarily aligned with the velocity. (Alternatively, we could consider natural coordinates, where x is the streamwise direction $\mathbf{s}$ and $\mathbf{n}$ is the normal to the streamline (see Section 7.7.1). Consider a flow that is steady-state, inviscid, with density constant, and $\mathbf{K}$ and $R = 0$. The parcel energy changes are due only to body forces and pressure work,

$$\rho u \, d(e + u^2/2)/dx = \rho \mathbf{F}_b \cdot \mathbf{u} - u \, dp/dx \qquad (7.34)$$

These assumptions bring about a great simplification in the term containing the stress tensor. The result is that only the pressure term remains. We can make several observations about the pressure gradient, pressure force, and pressure work in order to understand the last term in Eq. (7.34):

$-\partial p/\partial x$ is the pressure force/unit-volume;
$-1/\rho\ \partial p/\partial x$ is the pressure force/unit-mass;
$-1/\rho\ \partial p/\partial x\ dx = -dp/\rho$ is the work/unit-mass in moving dx;
$-1/\rho\ \partial p/\partial x\ dx/dt = -u/\rho\ \partial p/\partial x$ is the rate of work/unit-mass.

We can also write the last term in Eq. (7.34) as

$$u\ dp/dx = \rho u\ (1/\rho)\ dp/dx = \rho u\ d(p/\rho)/dx$$

Note that when density variation is significant,

$$d(p/\rho)/dx = (1/\rho)\ dp/dx - (p/\rho^2)\ d\rho/dx$$

and the second part of this term is the work done in compressing the parcel. From Section 5.4, this may be written

$$p\ dv_s/dx$$

which is the only term left in the elongation deformation term, div **u**, along the streamline.

We can now write the conservation of energy equation, with φ as a general potential for $\mathbf{F}_b$,

$$\rho u\ d(e + u^2/2)/dx = -\rho u\ d\varphi/dx - u\ dp/dx \tag{7.35}$$

This equation may then be written as

$$\rho u\ (d/dx)[e + u^2/2 + \varphi + p/\rho] = 0 \tag{7.36}$$

Thus, along a streamline (with assumptions of no heat transfer or temperature variations, inviscid, steady-state and nondivergent flow) we have

$$e + q^2/2 + \varphi + p/\rho = h + \varphi + \tfrac{1}{2}q^2 = \text{constant} \tag{7.37}$$

This equation is often called Bernoulli's equation. It has a wide range of applications and is a version of the conservation of energy equation. When the temperature is constant (hence e is a constant) and density is constant, Bernoulli's equation yields quick solutions for the variation of velocity, pressure, and height along a flow stream. It is also called Bernoulli's Law.

Example 7.2

For the leaking tank of Fig. 7.3, calculate the exit velocity of the water as a function of the water depth.

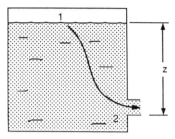

Figure 7.3   Sketch of the flow of fluid through a small exit in a tank.

Solution

Energy losses due to friction are neglected. Bernoulli's equation can be written between two points in the tank. One can be chosen as near the surface at height z. The other is at the exit at height 0. Then,

$$p_1/\rho + V_1^2/2 + z_1 = p_2/\rho + V_2^2/2 + z_2$$

We can assume that $V_1 \ll V_2$. The pressure difference (atmospheric) at 1 and 2 is negligible compared to the other energy terms. Thus,

$$V_2 = (2gz)^{1/2}$$

Thus, the exit velocity decreases as the square root of the height of the surface above the leak.

Example 7.3

Consider the airflow off a bay and into the deep canyons of a city with tall buildings. Using a static pressure measurement at street level of 999 mb, estimate the windspeed on a street at point B in Fig. 7.4. Assume two-dimensional flow at street level; neglect friction. The street width is 20 m. The street is 3 m higher than the bay.

Solution

Assume the change in internal energy is negligible. Then we can write Bernoulli's equation between the Bay and the city street.

$$p_0/\rho + V_0^2/2 + z_0 = p_B/\rho + V_B^2/2 + z_B$$

$$V_B = [2(\{p_0 - p_B\}/\rho + z_0 - z_B) + V_0^2]^{1/2}$$

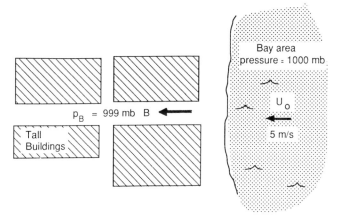

Figure 7.4 Sketch of tall buildings, narrow streets by a bay.

Then,

$$V_B = [2(\{1000 - 999\}) \cdot 10^{-1}/0.00123 + 3) + 5^2]^{1/2}$$

$$= [2(81 + 3) + 25]^{1/2} = 14 \text{ m/sec}$$

The funneling, or Venturi effect, of the narrow streets acts like a converging nozzle to accelerate the flow.

For a particular parcel in a gravitational field with energy transfer, we can write for any two stations that

$$\Delta e + \Delta(q^2/2) + \Delta gz + \Delta(p/\rho) = \text{the gain of energy}$$

$\Delta p/\rho$ is the flow work, on the parcel (at the entrance) and by the parcel (at the exit). (Note that p/ρ is not an intrinsic energy—rather it is a transferred energy.)

Equation (7.37) states that the sum of the internal plus kinetic plus potential (gravitational) plus flow work energy is constant along a streamline. We can also include terms for the heat transfer and rate of work per unit mass by adding terms for their net energy transfer, $(Q - W)$. Then Eq. (7.37) becomes

$$\underbrace{e_1 + p_1/\rho_1 + q_1^2/2 + gz_1}_{\substack{\text{initial energy of} \\ \text{the fluid parcel}}} + \underbrace{(Q - W)}_{\substack{\text{net energy transferred} \\ \text{to the parcel}}} = \underbrace{e_2 + p_2/\rho_2 + q_2^2/2 + gz_2}_{\substack{\text{final energy} \\ \text{of the parcel}}} \tag{7.38}$$

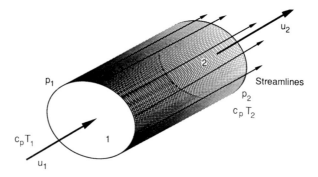

Figure 7.5 A bundle of streamlines, called a streamtube. Along the streamtube there is a change in pressure, temperature, velocity, and height from station 1 to station 2.

This is a practical expression of the general energy equation for the particular flow conditions assumed.

A finite volume can be defined by streamlines as the side walls, with the flow in one end and out the other. A sketch is shown in Fig. 7.5.

Since streamlines have no flow normal to them, a solid surface could be substituted for any surface made up of streamlines. In this case, the energy terms must be multiplied by the mass flow rate to produce the total energy change. However, since the mass flow rate is constant according to continuity, we see that a version of Bernoulli's equation holds for the flow in and out of arbitrary volumes. We will see this expression again for the constant value of I.E. + K.E. + P.E. + work. It has practical value when the "tube" is expanded to be any container with an entry and an outlet. The walls of the container are streamlines. When Eq. (7.38) is multiplied by the mass flow rate and integrated over the flow area, the fluxes can be calculated.

Example 7.4

Energy flux

Consider the energy that is advected past an area enclosed in a streamtube. (See Fig. 7.6) Discuss the energy flux across dA for $\mu = k_h = R = 0$.

Solution

The energy per unit mass advected past dA is

$$[\tfrac{1}{2}q^2 + e + \varphi]\, q \;\cdot \rho\, dA$$

[K.E. + I.E. + P.E.] speed · mass

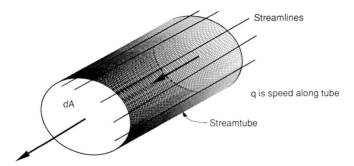

Figure 7.6   Cross section of a streamtube.

Add to this the rate at which normal forces on the surface do work, $pq\,dA$,

$$[\tfrac{1}{2}q^2 + e + \varphi + p/\rho]\,q\rho\,dA$$

In a steady flow field, the energy in the tube is constant and, since the incompressible mass flux $\rho q\,dA$ is also constant, Bernoulli's theorem follows.

$$\tfrac{1}{2}q^2 + e + \varphi + p/\rho = \text{constant}, \qquad \mu = k_h = 0$$

A common special case occurs when there is no change in e or φ.

$$\tfrac{1}{2}q^2 + p/\rho = \text{constant} = P_0/\rho \qquad (7.39)$$

This defines the "total" pressure P_0, obtained when $q = 0$. It is a constant along a streamline and represents the maximum attainable pressure when the kinetic energy is completely converted to pressure. To keep the description in terms of energies, the p/ρ term is sometimes referred to as a potential energy. This can be based on the pressure gradient potential for driving the velocity.

Since Bernoulli's equation is of significant practical value, it is worth looking at from other standpoints. For instance, one version can be obtained from the momentum equation, where the streamline is aligned in the x-direction. Only the mechanical energies will appear.

$$u\,du/dx + (1/\rho)\,dp/dx + \varphi = 0$$

These terms are accelerations/unit mass or, equally, forces. By integrating the force times distance along the streamline, we have

$$\int \text{Force/unit mass} \cdot \mathbf{ds} = \text{work/unit mass}$$

$$\underset{\substack{\delta(\text{K.E.})}}{\Delta u^2/2} + \underset{\substack{+\ \text{pressure} \\ \text{work}}}{\Delta p/\rho} + \underset{\substack{+\ \delta(\text{P.E.}) \\ (\text{work done by gravity})}}{g\ \Delta Z} = \text{constant}$$

where ΔZ is the vertical change as the fluid moves along the arbitrary s-direction. In terms of the total pressure,

$$\underset{\text{dynamic + static + hydrostatic = total (pressures)}}{\tfrac{1}{2}\rho q^2 + p + \rho g z} \;=\; P_0 \tag{7.40}$$

Example 7.5

The arrangement shown in Fig. 7.7 is called a Pitot tube. This instrument for measuring windspeed consists of a slender tube oriented into the wind with ports drilled around the circumference to admit the static pressure to the tube. Another smaller tube is contained within the outer shell. It has an opening facing the wind at the leading edge of the larger tube, thereby measuring the total pressure. Both tubes conduct their pressures to gauges that measure the difference in the two pressures.

Obtain an expression for the freestream velocity in terms of p_m and P_0. Assume the flow is in parallel straight streamlines and the disturbance caused by the tube is small. Also assume that the distance from the freestream flow

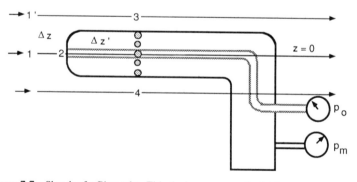

Figure 7.7 Sketch of a Pitot tube. This device measures the total pressure at a stagnation point 2, the static pressure in ports parallel to the stream flow at 3, and can register the Δp on a pressure gauge, $p_o - p_m$.

at point 3 to the wall is small. Then the static pressure at point 3 is measured at the wall intake holes of the tube. The density is given.

Solution

Bernoulli's equation can be written along the streamlines. The center stream-line impacts the tube at the center port, 2. Since the tube is slender, the streamline $1'$–3 slightly above the tube is very nearly undisturbed. We will include the height differential, centerline to freestream stream line, at first.

$$\underset{\text{static}}{p} + \underset{\text{dynamic}}{\rho u^2/2} + \underset{\substack{\text{hydrostatic} \\ \text{pressures}}}{\rho g z} = \underset{\substack{\text{total} \\ \text{pressure}}}{P_0} \text{ (constant)}$$

On streamline 1–2, the flow will come to rest at 2 by symmetry (the stag-nation point), and

$$p_1 + \rho U_1^2/2 = p_2 \equiv P_0 \qquad z \equiv 0$$

On streamline $1'$–3,

$$p_1' + \rho U_1^{2'}/2 + \rho g\, \Delta z = p_3 + \rho U_3^2/2 + \rho g\, \Delta z$$

We assume that

$$U_1' = U_1 = U_3$$

and

$$p_1' = p_3 = p_1 - \rho g\, \Delta z$$

$$p_4 = p_1 + \rho g\, \Delta z$$

Assume the pressure at points 3 and 4 are the same at the wall—the static pressure does not change across the thin boundary layer. This does not vi-olate Bernoulli's law since it is across the streamlines. We will check this in the next section, 7.8.1.

For the *n* (many) holes around the tube, there will be a mean pressure measured in the tube,

$$p_m = p_3 + p_4 + \ldots/n = p_1 \qquad \text{(the static pressure} \\ \text{at the mean height)}$$

We can write Bernoulli's equation for *U* in terms of the pressure difference measured by the Pitot tube and the density,

$$U_1 = [2/\rho(P_0 - p_m)]^{1/2}$$

Thus, a small tube in a flow measures the total pressure at the impact point and the static pressure at ports parallel to the flow. The difference in these pressures is directly proportional to the freestream velocity. This is a very practical flow speed measurement device, and can be found in laboratories and commercial jets.

Example 7.6

A Pitot tube is connected to a wind tunnel as shown in Fig. 7.8. The difference in height of the water manometer is 5 cm. Calculate the wind tunnel air speed.

Solution

We can apply Bernoulli's equation at a point upstream of the Pitot tube and along a streamline that impacts the stagnation point of the tube,

$$p_0/\rho + gz_0 + u^2/2 = p_t/\rho + gz_0$$

The static pressure is measured at the wall. As discussed in Example 7.4, $p_s = p_0$. Thus,

$$u^2/2 = (p_t - p_0)/\rho = (p_t - p_s)/\rho$$

or

$$u = [2(p_t - p_s)/\rho]^{1/2}$$

Substituting for $(p_t - p_s) = \rho_w gd$: ρ (air) $= 1.23$ kg/m^3, and ρ (water) $= 10^3$ kg/m^3,

$$u = [2 \cdot 1000 \cdot 9.8 \cdot 0.05/1.23]^{1/2} = 28.2 \text{ m/sec}$$

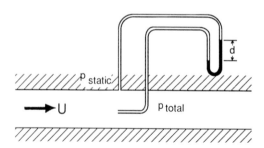

Figure 7.8   Pitot tube connected to a wind tunnel passage where wind flows with velocity U. The difference between total and static pressure results in the fluid-height difference d.

We have assumed the height difference between the stagnation point and the wall was negligible. This difference could easily be calculated in the equation.

7.8.1 *Bernoulli's Equation in Natural Coordinates*

The specialized coordinate system defined by the stream lines is called the *natural coordinate system*. Here the two coordinates are along the flow path and perpendicular to the flow path. Although it is not widely used, it has sufficient advantages in certain flow situations that an example of its use with the Bernoulli relation is informative. One of its liabilities is that the expression for the acceleration, **a**, must include the effects of curvature of the stream lines. The normal lines (n) are perpendicular to the streamlines (s) and point toward the center of curvature of the streamline. The n-lines and s-lines must change direction with the streamlines. At any point, the (s, n) coordinate system defines a plane and the flow is locally two-dimensional.

$$\mathbf{a} = (a_s, a_n) = (\partial V_s/\partial t + V_s\, \partial V_s/\partial s, \quad \partial V_n/\partial t + V_s^2/r) \qquad (7.41)$$

where r is the radius of curvature of the streamline. Note that although $|V_n| = 0$ at all points, if the flow is unsteady, the streamline pattern may be changing with time and we must include $\partial V_n/\partial t$.

The simplifying aspect of these coordinates is in the expression for the velocity, where only one component is not zero,

$$V_s \equiv V, \qquad V_n = 0$$

However, as usual, this simplification is at the expense of another complication—and that is $\partial V_n/\partial x \neq 0$. Since the direction of the coordinates is changing continually, the vector direction of V_n must change, although its magnitude is zero. From Fig. 7.9, we see that $\delta V_n \approx V_s\, \delta\theta \approx V_s\, \delta s/r$. Hence, $\partial V_n/\partial s = V_s/r$.

Let us check to see what form the energy conservation takes in flow along and across the streamlines. To do this we need the inviscid momentum equation in the n-direction. In this case the acceleration equals the sum of the pressure forces plus a component of the gravity force.

$$p\, dA - (p + \partial p/\partial n\; dn)\, dA - (g \cos \theta)(\rho\, dA\; dn)$$

$$= (\rho\, dA\; dn)a_n = (\rho\, dA\; dn)(V^2/r + \partial V_n/\partial t)$$

Thus, the *n-momentum* equation may be written

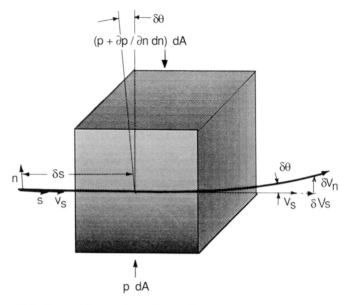

Figure 7.9 A parcel in natural coordinates. The coordinate s is along the streamline and n is directed toward the center of curvature, normal to s. In an increment of time, r moves through an angle $\delta\theta$. The speed V_s may change, and the vector direction of the (zero-length) V_n can change.

$$-\partial p/\partial n \; dn \; dA - \rho g \; \partial z/\partial n \; dA \; dn = \rho(V^2/r + \partial V_n/\partial t) \; dA \; dn$$

since $\cos\theta = \partial z/\partial n$, where the z-direction is the vertical—the direction of **g**.

Or, for steady-state flow,

$$V^2/r + 1/\rho \; \partial p/\partial n + g \; \partial z/\partial n = 0 \qquad (7.42)$$

Similarly, the momentum equation along the steady-state streamline is found to be

$$V \; \partial V/\partial s + 1/\rho \; \partial p/\partial s + g \; \partial z/\partial s = 0 \qquad (7.43)$$

When we integrate (7.43) along the streamline,

$$V^2/2 + \int dp/\rho + \int g \; dz = \text{constant}$$

For constant g and constant ρ,

$$V^2/2 + p/\rho + gz = C \qquad (7.44)$$

This is Bernoulli's equation. This form is consistent with the previous section.

The natural coordinate system allows us to investigate the energy equation when we integrate across the streamline.

In the special case of a straight streamline ($r \to \infty$), Eq. (7.42) becomes

$$\partial p/\partial n + \rho g\, \partial z/\partial n = 0 \qquad (7.45)$$

Then, integrating along the n-direction, we have

$$p + \rho g z = \text{constant} \qquad \text{for} \quad \rho \text{ constant} \qquad (7.46)$$

In this perspective, Bernoulli's equation results with no kinetic energy (no flow). The pressure (static plus hydrostatic) is constant perpendicular to the streamlines.

In the general case with curved streamlines, assume steady flow so that $\partial V_n/\partial t = 0$, and Eq. (7.45) is

$$V^2/r + 1/\rho\, \partial p/\partial n + g\, \partial z/\partial n = 0 \qquad (7.47)$$

If we then integrate in the n-direction, we obtain

$$\int (V^2/r)\, dn + \int (1/\rho)\, \partial p/\partial n\, dn + g \int (\partial z/\partial n)\, dn = \text{constant}$$

or

$$\int (V^2/r)\, dn + \int (1/\rho)\, \partial p/\partial n\, dn + gz = C$$

For constant ρ flow, this becomes

$$\int (V^2/r)\, dn + p/\rho + gz = C \qquad (7.48)$$

However, this is not quite Bernoulli's equation. Apparently we cannot apply Bernoulli's equation normal to any streamline without additional conditions. Since Bernoulli's equation is a powerful integrated form of the energy equation, it is important to learn the conditions required for it to be applicable. We will return to the Bernoulli relation and the circumstances under which it can be used after we have discussed the vorticity.

7.8.2 Compressibility Effects in Bernoulli's Equation

Most of the time we are able to successfully use the incompressible approximation for atmospheric flow. In addition to the simplification in the equations from treating density as a constant, this assumption eliminates the coupling between mechanical and thermal forms of energy. The Bernoulli equation is then simply a mechanical energy equation. However, when the

288   7   **Conservation of Energy**

compressibility of the fluid is significant, the pressure variation must remain an integral.

$$\frac{1}{2} q^2 + gz + \int dp/\rho = \text{constant} \tag{7.49}$$

The integral of this equation along the streamlines in the case where the density is allowed to continuously vary is

$$e_2 - e_1 + (u_2^2 - u_1^2)/2 + g(z_2 - z_1) + \int_1^2 dp/\rho = 0$$

We can get a nice expression that shows the contribution of compressibility to Bernoulli's equation. If we substitute in Eq. (7.36) the identity relating $d\rho$ to dv_s, where specific volume $v_s = 1/\rho$, we get

$$d(p/\rho) = (1/\rho)\, dp - (p/\rho^2)\, d\rho = (1/\rho)\, dp + p\, dv_s$$

Then Eq. (7.36) may be written

$$d(e + u^2/2 + gz) = d(p/\rho) - p\, dv_s \tag{7.50}$$

When this equation is integrated along the streamline, we obtain Bernoulli's equation for compressible flow,

$$e + u^2/2 + gz + p/\rho - \int p\, dv_s = \text{constant} \tag{7.51}$$

The pressure work consists of two parts. The p/ρ term reflects the energy gained from flow along the pressure gradient, and the $p\, dv_s$ term subtracts the work done on the parcel of fluid in the process of compressing it. The last term can be neglected when the $\nabla \cdot \mathbf{u} = 0$ approximation is valid (Section 5.4).

Example 7.7

One way to state that the incompressible version of the energy equation can be used is to examine the ratio,

$$\frac{\text{characteristic mechanical energy}}{\text{characteristic thermal energy}}$$

Consider air flowing at 40 m/sec, at 20°C, with internal energy about 200,000 m^2/sec^2. Is this flow incompressible?

Solution

$$\text{The ratio is } \frac{40^2 \ (\text{m}^2/\text{sec}^2)/2}{200{,}000 \ (\text{m}^2/\text{sec}^2)} = 4 \cdot 10^{-3}$$

When this ratio is very small, the flow may be treated as incompressible in the energy equation. Thus, even this very strong wind is safely assumed to be incompressible for the purposes of using the energy equation.

7.9 Summary

Since our focus is on fluid dynamics, the treatment of thermodynamics has been relatively brief. However, in view of the importance of the energetics approach to all of atmospheric physics, we have introduced some of the phenomenological aspects of energy.

We have derived a conservation of energy relation for the case when heat energy is a factor in our flow. The consequences are that a new variable must be accounted for—either internal energy, enthalpy, or temperature—and a new equation must be invoked. When temperature is to be the new variable, empirical relations relating temperature to internal heat must be defined. This is done with the experimentally determined specific heat constants. The new equation is an expression of the first law of thermodynamics. In it there occur energy terms due to the flow—the pressure work and the internal forces work. Each of these has two contributions, one each to kinetic energy and to internal energy.

We have covered the concept of the first law of thermodynamics that allowed the derivation of a general conservation statement for the energy in a parcel. The general format can be extended to include various energy sinks and sources. Some of these are latent heat release, radiation heating, and frictional dissipation.

The first law deals with the balance of energy terms between two equilibrium states. However, the second law of thermodynamics, dealing with processes and reversibility, has barely been mentioned. This law—that it is impossible to completely convert a given amount of heat into useful mechanical energy—is implied in the concept of reversibility. Only when heat conduction or dissipation is involved is the process generally irreversible. For instance, when two fluids mix to a uniform temperature, energy would be required to restore the fluids to their original conditions. Also, the frictional dissipation is manifest in a temperature rise, which is irreversible.

Our laws of mechanics are generally reversible; that is, the action can be run backward in time, providing a great simplification in the equations. In fact, the most important process in atmospheric dynamics is the reversible adiabatic (with no heat exchange). In this case temperature changes with density and pressure according to the perfect gas law. The most important diabatic effect is probably that of vapor condensation with consequent release of heat within a volume. But fortunately, in most geophysical flow problems, the thermodynamic work has been done before the problem is addressed—in the production of the pressure gradient—and dissipation and heat conduction are small quantities. They can generally be accounted for by adding small corrections to the adiabatic processes.

The integration of the energy equation between two state points along a streamline produced the Bernoulli equation. This is a convenient formula relating conditions at specific points on a control volume. However, the conservation of energy across a streamline is apparently different. We will find that a discussion of vorticity will shed light on this difference.

TYPES OF ENERGY—THERMODYNAMICS

First Law of Thermodynamics

$$\delta E/\delta t = \delta W/\delta t + \delta Q/\delta t$$

Kinetic Energy
Internal Energy
Rate of Heat Transfer

Energy Equation

$$\rho D/Dt(e + \tfrac{1}{2}q^2) = \rho \mathbf{F}_b \cdot \mathbf{u} + \mathbf{u} \cdot \text{div } \boldsymbol{\sigma} + \boldsymbol{\sigma} \text{ grad } \mathbf{u} + \rho R - \text{div } \mathbf{K}$$

Mechanical Energy Equation

$$\rho \, D(\tfrac{1}{2}q^2)/Dt = \rho \mathbf{F}_b \cdot \mathbf{u} - \mathbf{u} \cdot \text{grad } p + \mathbf{u} \cdot \text{div } \boldsymbol{\tau}$$

Thermal Energy Equation

$$\rho \, De/Dt = -p \text{ div } \mathbf{u} + \boldsymbol{\tau} \text{ grad } \mathbf{u} + \rho R - \text{div } \mathbf{K}$$

Enthalpy Equation

$$\rho \, Dh/Dt = Dp/Dt + \Phi + \rho R - \text{div } \mathbf{K}$$

Moisture Effects

Bernoulli's Equation

$$e + q^2/2 + \varphi + p/\rho = h + \varphi + \tfrac{1}{2}q^2 = \text{constant}$$

Bernoulli's Equation along a Streamline

$$\Delta u^2/2 + \Delta p/\rho + g \, \Delta Z = \text{constant}$$

Pitot Tube
Natural Coordinates
Compressibility

Problems

1. Do a force balance on a parcel in a rotating tank of fluid (see the sketch ˙ below). Assume pressure and gravity forces only. Substitute $u_\theta = r\omega$, $a_r = -\rho u_\theta^2/r$, to get a formula for $p(r)$.

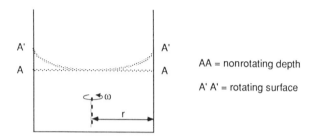

AA = nonrotating depth

A' A' = rotating surface

2. A Pitot tube for measuring air velocity is connected to a differential pressure gauge. If the air temperature is 20°C at standard sea level pressure and the gauge reads a pressure difference of 2 kPa, what is the air velocity?

3. (a) Two conditions of air flow exist at a site with a pressure gauge as shown. There is no flow over A. Over B, horizontal flow velocity is 10 m/sec. Is gauge A reading greater or less than gauge B?

V = 0   V = 10 m/s ⟶

A   B

(b) A spherical probe is used for finding air velocity by measuring the pressure difference between the upstream and downstream points A and B. The pressure coefficients at points A and B are 1.0 and −0.4. The pressure difference $p_A - p_B$ is 5 kPa, and the air density is 1.5 kg/m³. What is the air velocity?

4. You are designing a wind tunnel with a test section cross-sectional area of 15 m² with windspeed of 10 m/sec. Calculate the power needed to operate the wind tunnel. Assume $\rho = 1.2$ kg/m³; exit area is 30 m²; a loss of energy is given as a head loss $= (0.02)(U_T^2/2g)$; negligible energy loss in the entrance and atmospheric pressure at the exit.

5. The cooling system from a power plant discharges heated water into a river. The river is 100 m wide and averages 6 m deep. It flows at an average speed of 1.5 m/sec. If the net amount of heat is 10^4 kW, what is the temperature rise produced in the river?

6. What happens to the energy equation when temperature can be considered constant in a problem?

7. A wind tunnel is constructed with a blower in a rectangular inlet sec-

tion, a converging nozzle, and a test section at the open throat of the nozzle. (See sketch.) Air enters at left and exits through test section at right. The opening is 1 m × 30 cm, and the nozzle exit is 30 cm × 20 cm. The test section velocity is to be 30 m/sec maximum. Assume adiabatic, incompressible flow. What is the minimum power required to drive the tunnel flow?

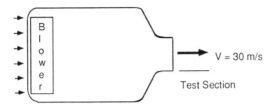

8. Assume natural coordinates (s, n) with (V, V_n). (a) What is the general steady-state inviscid energy equation in the s-direction? (b) What is the general relation in the n-direction? (c) What is (b) when $\nabla \times \mathbf{V} = 0$? (d) What is (a) when the fluid is significantly compressible?

9. Why didn't the Coriolis force appear in Bernoulli's equation?

10. Consider the fluid flow above an open drain. Approximate it as a free vortex located at the origin (see sketch). For $U_1 = 0.7$ m/sec, $U_2 = 0.1$ m/sec, $r_1 = 0.1$ m, and $r_2 = 1.0$ m, use Bernoulli's equation to find the difference in height of the surface at 1 and 2.

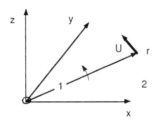

11. Discuss the dynamics of a baseball curve ball in terms of Bernoulli's equations.

Chapter 8 | Vorticity

8.1 Vorticity Characteristics

So far vorticity has entered our discussions only peripherally. However, indications of its importance have persistently arisen throughout the text. In Chapter 1 it was associated with turbulence, even to the point of describing turbulence as "distributed vorticity." In Chapter 2 we introduced the mathematical definition of vorticity, $\nabla \times \mathbf{u}$, and related it to the spin of a parcel at a point. The *vorticity* is thus related to the velocity field through the character of the velocity shear *at a point*. The distinctly different term, *vortex*, was also introduced in that chapter. It was used to describe the various rotating *fields* that we encounter in atmospheric dynamics. There can be confusion between these two terms in atmospheric dynamics because they both play prominent roles. Once again, vorticity is still a point function (just like temperature or pressure) even when the point is a grid point representing a 100-km square of atmosphere or ocean. A vortex is a particular velocity field with fluid rotating around a central point. A sketch of a vortex with several arbitrary points extracted is shown in Fig. 8.1.

In Chapter 4 we found that the velocity gradient field could be conveniently split into two parts. One part, the symmetric tensor is represented by the shearing gradient at a point. It was used extensively in explaining the internal forces in the momentum equation. The second part is characterized by the vorticity vector, which uniquely represents the antisymmetric tensor. Finally, in Chapter 6 we saw that the rules necessary for determining when the Bernoulli equation is applicable would depend on vorticity. Such a basic parameter merits investigation.

The importance of vorticity arises from the fact that it has a governing conservation principle. For inviscid flow, vorticity cannot be created or destroyed. The *vorticity equation* will follow the format of our other field equations: total change in vorticity equals the sum of sinks and sources. It will involve the peculiar arrangement of velocity derivatives that form the definition of vorticity. The vorticity equation will be derived by manipulating the momentum equation. Since we already have obtained a closed set of equations and the new equation is redundant to the momentum equation, the vorticity laws would not appear to be necessary. However, certain phenomena can be more easily understood in terms of vorticity dynamics. The interaction of turbulent elements and cyclone dynamics are two examples of processes which can be better understood by using vorticity concepts. In any event, vorticity provides a different way of looking at the velocity field, one which is particularly important in atmospheric dynamics.

The concept of vorticity is related to the spin of the infinitesimal parcel. In fact, vorticity is a measure of the rotation rate of the principal axes centered at any point in the flow. However, the spin of a fluid parcel is not as simple a definition as that of the solid body, where everything is rigid and all lines rotate together. We can get some insight into the physical linkage between vorticity and rotation (and the velocity gradients) by first examining the two-dimensional flow in two basic flow patterns shown in Fig. 8.2.

In the first case we have a rotating tank where the velocity has come to equilibrium and the fluid rotates as a solid body. In solid-body rotation this can be seen by marking perpendicular axes on a parcel at $t = 0$ on the x-axis some distance from the center of rotation. Then, at later times, both

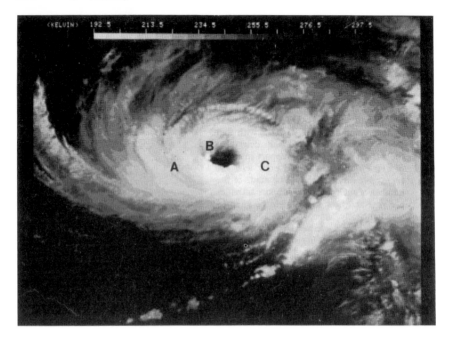

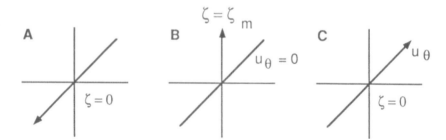

Figure 8.1 A vortex field (a satellite picture of a hurricane) with points A, B, and C selected to illustrate the individual velocity u_θ and vorticity ζ.

axes—in fact all lines in a parcel—will have rotated an angle equal to the rotation angle about the center of rotation. At the y-axis crossing, the axes are rotated 90° in the same sense as the rotation of the entire fluid. In this particular case, the point rotation of the parcel equals the rotation rate of the domain about the center axis. There is no distortion of the cubical parcel. Solid-body rotation is an exceptional case where the rotation of the parcel

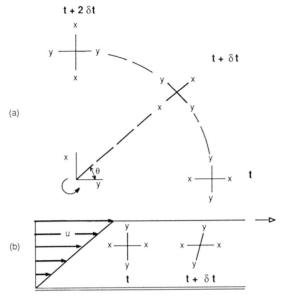

Figure 8.2 (a) Fluid in solid-body rotation in a rotating tank. A vorticity meter is shown at three incremental times. (b) Parallel flow between two plates. Parcel axes are shown at times t and $t + \delta t$.

is the same as the rigid rotation. In general, the fluid parcel is expected to deform under the internal shearing forces present in the flow, as in 8.2b.

8.1.1 One-Dimensional Shear Flow

In the second example, plane parallel shear flow near a boundary, the mean flow of the fluid is moving in straight lines parallel to the surface. However, the flow is decelerating as it approaches the boundary. When considering finite dimensions of the parcel, the y-axis tends to turn due to the incremental velocity in the y-direction, while the x-axis is simply advected along the streamline without turning. So there is rotation at every point (x, y). However, it is not clear what its value should be. One way of defining the rotation of the parcel is to call it the average rotation of any two initially perpendicular lines. Typically, these would be any two axes located in the parcel, either at its center or at a corner. In this case, the vorticity of the parcel is not zero, and the parallel boundary layer flow is rotational. The rotation rate, defined as half the sum of the x and y axes rotation rate, is simply equal to half the y-axis rotation rate. There is distortion of the parcel.

8.1.2   Two-Dimensional Shear Flow

We can envision a shear flow case where the x-axis in the parcel rotates counterclockwise at the same rate that the y-axis rotates clockwise. This will result in a zero mean rotation rate. For this to happen, it is necessary to have a v-component of shear that increases in the x-direction. When this combination of shears exists, there is no rotation by our definition, but considerable distortion does occur.

On the other hand, in example (b) a freestream, uniform, parallel flow beyond the boundary influence will have no distortion, no vorticity, and hence is irrotational. Thus, a rotational region can smoothly adjoin an irrotational one.

In the example of steady flow near a boundary, the no-slip boundary condition is satisfied by the action of the internal stresses. These produce a velocity gradient in the fluid. As the velocity goes from a freestream value to zero at the boundary, the internal stress goes from zero to a maximum value at the boundary. If the fluid flow accelerates as it moves downstream, the velocity and stress gradients will increase. Therefore, the rotation and vorticity will increase. Such a layer is apparently a source of vorticity.

In the case of a rotating tank of fluid that is in solid-body rotation, there are no vertical velocity gradients or viscous terms of evident importance. The freestream flow apparently exists to the boundary. However, to establish such a flow, the action of internal stresses imparting the torque from the rotating bottom of the tank to the fluid would be necessary.

To determine the dynamics of the parcel at a point it is necessary to consider the specific arrangement of shear at each point. We must look at the individual velocity derivatives in each direction. This is the role of the scalar divergence, vector vorticity, and tensor deformation. Previously, we have related the deformation to the strain rate using Stokes' viscosity law in Chapter 6. By this relation we found that the divergence represents a strain-rate component. The nondivergent approximation, lacking this component, permitted great simplifications in the basic equations. In this chapter we will find that the vorticity also represents an important characteristic of fluid flow. Significant simplifications in the equations result when it is zero, particularly in conjunction with zero divergence.

8.1.3   Vorticity of the Parcel

Now we can look at the detailed deformation of the parcel for the case of a parcel with unit depth in a two-dimensional flow. First, we consider the parcel to be a rigid body and look at its reaction to a force **F**.

The parcel shown in Fig. 8.3 will move without deforming. Body forces like gravity can be assumed to act at the center of mass, but surface forces

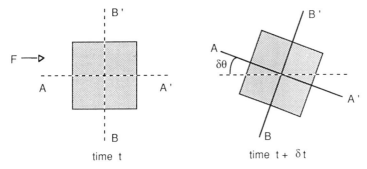

Figure 8.3   The rigid parcel under a force **F**.

and other applied forces are not likely to pass directly through this center. In fact, except for the unusual case of the forces applied exactly at the center of mass, spin is to be expected of all bodies under the action of a force. Thus the parcel will spin with respect to the original coordinate system at a rate $d\theta/dt$. All lines in the parcel will rotate with the same rate.

For a fluid parcel subject to a force, the parcel will deform as shown in Fig. 8.4. Since the fluid parcel is deforming, different lines in a rotating parcel will rotate with different rates. Thus, if we want to assign a unique value to the rotation of the parcel we must consider more than one particular line. We will define the *rotation* of a fluid parcel about any axis as *the average rotation of two perpendicular lines in a plane that is perpendicular to the axis*. The *vorticity* can then be defined at a point as equal to *twice this fluid rotation rate*. The factor of 2 simply eliminates the $\frac{1}{2}$ coefficient that arises in the definition of the mean rotation. We will then be able to show that the vorticity is equal to the curl of the velocity, or $\nabla \times$ **u** (del cross **u**) at that point. Vorticity, (unbold zeta), is always a vector.

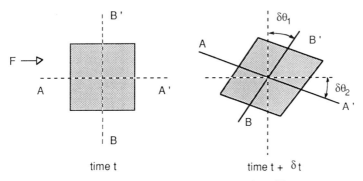

Figure 8.4   The fluid parcel under a force **F**.

On the other hand, if one is approaching vorticity from a strict mathematical direction, sometimes the cross product of **u** is taken as the definition of vorticity. In this case, the equivalence of vorticity to twice the rotation rate will follow. We will find that the vorticity is a naturally occurring parameter in our study of the general deformation of the fluid parcel.

In summary, some of the alternate ways of considering vorticity are

1. Mathematically, $\zeta \equiv \nabla \times$ **u**; the vector part of rot **u**.
2. Physically, $\zeta \equiv 2\ d\theta/dt$.
3. Vorticity is related to the solid-body-like rotation of the parcel.
4. Vorticity is related to the rotation of the principal axes.

Example 8.1

A good feel for vorticity as fluid rotation at a point can be obtained from a discussion of the two-dimensional case. Flow is in the x-y plane and vorticity is in the z-direction only. Consider the parcel at a point shown in Fig. 8.5. Find the vorticity at the point (x, y) in the fluid as the average rotation of the line segments, δx and δy.

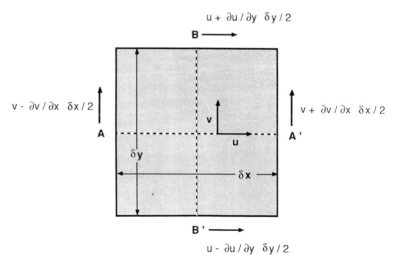

Figure 8.5 The fluid parcel with unit depth in the z-direction centered at $(x, y) = (0, 0)$ in 2-D flow $(u, v, 0)$.

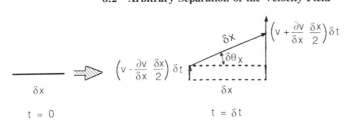

Figure 8.6 Motion of element δx in time δt. The line segment is stretched (or compressed) and turned.

Solution

The rotation rate of a line segment is the angular rate of change, $d\theta/dt$. This rate will depend on the difference between the velocity normal to the line segment at each end of the segment, as shown in Fig. 8.6. Here, the line segment is the incremental side δx, shown at an increment of time after it has been turned by the increment of lateral velocity.

For the small $\delta\theta_x \approx \sin \delta\theta_x \approx$

$$[(v + \partial v/\partial x\ \delta x/2) - (v - \partial v/\partial x\ \delta x/2)]\ \delta t/\delta x$$
$$= \partial v/\partial x\ \delta x/\delta x\ \delta t = \partial v/\partial x\ \delta t.$$

Hence,

$$\delta\theta_x/\delta t = \partial v/\partial x$$

Similarly, the rotation of the δy side is $\delta\theta_y/\delta t = -\partial u/\partial y$.

The average rotation of these two orthogonal lines is defined as the angular rotation rate of the parcel.

$$\delta\theta/\delta t = \tfrac{1}{2}(\partial v/\partial x - \partial u/\partial y)$$

This is half the vorticity, $\zeta = \partial v/\partial x - \partial u/\partial y$ at point (x, y). In this case, $\delta\theta/\delta t$ is half the scalar value of the single component of the vorticity vector in the z-direction, perpendicular to the x-y plane of the velocity shear.

8.2   Arbitrary Separation of the Velocity Field

The local rate of expansion associated with the divergence of a velocity field, and the local vorticity associated with the curl of the velocity field are both independent of the coordinate system. In Section 4.2 we determined that the relative motion of a fluid at a point can be viewed as a combination of a pure strain with zero rotation and pure rotation with angular velocity, $d\theta/dt$. The pure strain part was further separated into an isotropic expansion or contraction associated with the divergence at a point, plus a straining distortion without volume change. The strain portion of the relative motion at a point was used in the stress to rate-of-strain formulation of Section 6.3. We are now ready to investigate the second part of the deformation tensor, the antisymmetric part, rot **u**. We know that this tensor has only three independent components and an associated vector, which is the vorticity $\zeta = \nabla \times \mathbf{u}$.

Consider a general velocity field **u** with

$$\nabla \cdot \mathbf{u} = D \quad \text{and} \quad \nabla \times \mathbf{u} = \zeta$$

where D equals the local rate of expansion per unit volume and ζ is the local vorticity. The **u** can be separated into three parts: a divergent, irrotational $\mathbf{u}^S$; a nondivergent, rotational part $\mathbf{u}^a$; plus any uniform translation **V**. [from Eq. (4.7)]

$$\mathbf{u} = \mathbf{u}^S \quad + \quad \mathbf{u}^a \quad + \quad \mathbf{V}$$

where   (8.1)

$$\nabla \cdot \mathbf{u}^S = D \qquad \nabla \cdot \mathbf{u}^a = 0 \qquad \nabla \cdot \mathbf{V} = 0$$
$$\nabla \times \mathbf{u}^S = 0 \qquad \nabla \times \mathbf{u}^a = \zeta \qquad \nabla \times \mathbf{V} = 0$$

8.2.1 Velocity Potential and Stream Function

We will now use mathematical characteristics of the curl and the divergence to define two potential fields from which the velocity field can be determined. First, since curl grad(any scalar) $\equiv 0$, the condition of irrotationality, $\nabla \times \mathbf{u}_S = 0$, implies that the vector $\mathbf{u}^S$ can be written as the gradient of a scalar. We define this scalar to be φ, the *velocity potential*, with $\mathbf{u}_S = \text{grad } \varphi$.

In a similar fashion, the incompressibility condition on $\mathbf{u}^a$, $\nabla \cdot \mathbf{u}^a = 0$, implies that this velocity component can be written as the curl of some vector, since div curl(any vector) $= 0$. We define this vector to be the vector velocity potential **B**, with $\mathbf{u}^a = \nabla \times \mathbf{B}$. The three-dimensional vector potential **B** is seldom used. However, it is of great value for two-dimensional flow, where it has one component, $\mathbf{B} = (0, 0, \psi)$. The scalar ψ is defined as the *stream function*. This term is used because $\psi = $ constant lines are streamlines. The stream function serves as a scalar potential for the velocity.

This partly physical description of the separation of the velocity field has its counterpart in the general principle of vector analysis. Any general vector function of position can be written as the sum of two vectors of form $\nabla\varphi$

and $\nabla \times \mathbf{B}$, where only $\nabla\varphi$ has nonzero divergence (hence, $\nabla \cdot \nabla \times \mathbf{B} = 0$) with curl $\nabla\varphi = 0$; and only $\nabla \times \mathbf{B}$ has nonzero curl (curl grad $\varphi \equiv 0$) with $\nabla \cdot \mathbf{B} = 0$.

8.2.2 Relation between Velocity, Divergence, and Vorticity

We can get information on how a parcel in a flow field interacts with other parcels by considering how the velocity at a given point affects the velocity at any other point in the field. This is best done by considering the separate contributions of the two parts of the velocity, $\mathbf{u}^S$ and $\mathbf{u}^a$.

First, we will consider the symmetric part of the relative velocity field. From the definitions of $\mathbf{u}^S$,

$$\mathbf{u}^S = \nabla\varphi^S; \qquad \nabla^2\varphi^S = \nabla \cdot \mathbf{u}^S \equiv D \tag{8.2}$$

This Poisson-type equation has solutions:

$$\varphi^S(\mathbf{x}) = -1/(4\pi) \iiint D'/r \, dV(\mathbf{x}') \tag{8.3}$$

where r is the distance from $\mathbf{x}$ to $\mathbf{x}'$. The primed divergence D' is evaluated at $\mathbf{x}'$ for each point of the domain, and V is the volume of fluid, sketched in Fig. 8.7. At any point in the volume we can calculate the velocity, which is a result of the incremental divergence at every other point in the volume, using the potential

$$\nabla\varphi^S = \mathbf{u}^S(\mathbf{x}) = 1/(4\pi) \iiint (\mathbf{r}/r^3) D' \, dV(x') \tag{8.4}$$

The velocity $\mathbf{u}^S$ at any point $\mathbf{x}$ in the domain is the sum of contributions from all of the volume elements $\delta V(\mathbf{x}')$. The incremental velocity from each δV is shown in Fig. 8.7.

$$\delta\mathbf{u}^S(\mathbf{x}) = (\mathbf{r}/r) D' \, \delta V(\mathbf{x}')/(4\pi r^2) \tag{8.5}$$

Equation 8.5 gives the increment of irrotational velocity distribution corresponding to a volume flux $D' \, \delta V(\mathbf{x}')$ at $\mathbf{x}'$ through a surface of radius r from point $\mathbf{x}'$. (In other words, each volume element δV acts like a source in an otherwise expansion-free fluid.) The local elemental source strength at point P' is $D'(\mathbf{x}') \, \delta V'(\mathbf{x}')$. D' is equal to the rate of expansion of this volume element according to

$$\frac{d(\delta V)/dt}{\delta V} = \nabla \cdot \mathbf{u} = D'$$

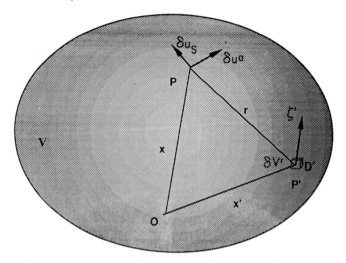

Figure 8.7   An arbitrary volume V and element δV at point P. The axes origin is 0, and the point P' ranges over the volume at distance **r** from P.

In a way similar to the above derivation of the incremental effects of the local expansion rates integrated to give the irrotational component of velocity at point P, we can express the antisymmetric component of relative velocity $\mathbf{u}^a$, in terms of the incremental vorticity at all other points in the domain. We can write $\mathbf{u}^a$ as the curl of a vector potential **B**.

$$\mathbf{u}^a = \nabla \times \mathbf{B} \text{ (vector potential)} \tag{8.6}$$

We can write the vorticity, $\zeta = \nabla \times \mathbf{u}^a$, as

$$\zeta = \nabla \times (\nabla \times \mathbf{B}) = \nabla (\nabla \cdot \mathbf{B}) - \nabla^2 \mathbf{B} \tag{8.7}$$

employing a vector identity.

Since $\nabla \cdot \mathbf{B} = 0$ (which requires that $\zeta \cdot \mathbf{n} = 0$ at the boundaries), we are left with

$$\zeta = -\nabla^2 \mathbf{B}$$

Once again, this is a Poisson-type equation with solutions

$$\mathbf{B}(\mathbf{x}) = (1/4\pi) \iiint \zeta'/r \, dV(\mathbf{x}') \tag{8.8}$$

and the total rotational (or solenoidal) contribution to the velocity field is

$$\mathbf{u}^a(\mathbf{x}) = -(1/4\pi) \iiint r \times (\zeta'/r^3) \, dV(\mathbf{x}') \tag{8.9}$$

We can consider $\mathbf{u}^a(\mathbf{x})$ as the sum of the contributions from each of the $\delta V'(\mathbf{x}')$ making up the volume V,

$$\delta \mathbf{u}^a = -[(\mathbf{r} \times \boldsymbol{\zeta}')/(4\pi r^3)]\, \delta V'(\mathbf{x}') \tag{8.10}$$

In this case, $\delta \mathbf{u}^a$ must be perpendicular to the vector $\mathbf{r}$ between $\mathbf{x}$ and $\mathbf{x}'$.

Thus, at any point $P(\mathbf{x})$ in the volume, there is an incremental velocity consisting of $\delta \mathbf{u}^S + \delta \mathbf{u}^a$ due to the divergence and vorticity at every other point in the volume, $P'(\mathbf{x}')$. The total velocity at P due to the distributed divergence and vorticity over the volume is given by the integrals in Eqs. (8.4) and (8.9).

Example 8.2

Consider a two-dimensional volume with unit depth. There is a singularity at point P' with source strength $D'\,\delta V' = 1000$ cm³/sec and vorticity strength $\zeta'\,\delta V' = 2000$ cm³/sec. Sketch the velocity increments at a point $P(x, y)$ due to the source and vorticity at point P' at distance $\mathbf{r}$. Calculate the speed at $r = 5$ cm and 20 cm.

Solution

We place our coordinates centered at P. (See Fig. 8.8) The singularities are at some point $\mathbf{r}$ away. Hence we know there will be an induced velocity along $\mathbf{r}$ due to the source (or sink) and an induced velocity perpendicular to $\mathbf{r}$ due to the vorticity.

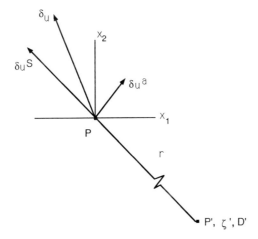

Figure 8.8 The components of $\delta\mathbf{u}$ at P due to D' and ζ' at P'.

If the flow is nondivergent and irrotational with the exception of the singularities at point P', then the flow at $P(x, y)$ will be

$$\delta \mathbf{u} = \delta \mathbf{u}^S + \delta \mathbf{u}^a$$

$$= (D' \mathbf{i}_r + \zeta' \mathbf{i}_\theta)\, \delta V'/(4\pi r^2)$$

Hence at $r = 5$ cm,

$$\delta \mathbf{u}(5\text{ cm}) = (1000\mathbf{i}_r + 2000\mathbf{i}_\theta)/4\pi \cdot 25$$

$$= (3.2\mathbf{i}_r + 6.4\mathbf{i}_\theta)\text{ cm/sec}$$

and

$$|\delta u| = 7.16\text{ cm/sec}$$

At $r = 20$ cm,

$$\delta \mathbf{u}(20\text{ cm}) = (1000\mathbf{i}_r + 2000\mathbf{i}_\theta)/4\pi \cdot 400$$

$$= (0.2\mathbf{i}_r + 0.4\mathbf{i}_\theta)\text{ cm/sec}$$

and

$$|\delta u| = 0.45\text{ cm/sec}$$

In both cases, the velocity drops off in accordance with the inverse square law.

The parallel development of the two components of the velocity can be seen in the recapitulation of the equations,[1]

$\mathbf{u}^S$	$\mathbf{u}^a$
$\mathbf{u}^S = \nabla \varphi^S$	$\mathbf{u}^a = \nabla \times \mathbf{B}$
$\nabla \cdot \mathbf{u}^S = \nabla^2 \varphi_S \equiv D$	$\nabla \times \mathbf{u}^a = -\nabla^2 \mathbf{B} = \zeta,\quad \nabla \cdot \mathbf{B} = 0$
$\varphi^S = -1/4\pi \iiint D'/r\, dV(x')$	$\mathbf{B} = (1/4\pi) \iiint \zeta'/r\, dV(x')$
$\mathbf{u}^S = \nabla \varphi^S = 1/4\pi \iiint \mathbf{r}/r^3 D'\, dV(x')$	$\mathbf{u}^a = \nabla \times \mathbf{B} =$
	$\quad -1/4\pi \iiint \mathbf{r} \times \zeta/r^3\, dV(x')$
$\delta\mathbf{u}^S(x) = 1/4\pi r^2\, \mathbf{r}/r\, D'\, \delta V(x')$	$\delta\mathbf{u}^a = -(\mathbf{r} \times \zeta')/(4\pi r^3)\, \delta V(x')$

---

[1] Note that there is an analogy between the development in this section and that for electromagnetic theory. In that case a relation between electrical current and magnetic field is obtained. Called the Biot–Savart law, it results when current replaces ζ and magnetic field replaces $\mathbf{u}$.

8.3   Kinematics of Vorticity

We have determined that the mathematical definition of vorticity, $\nabla \times \mathbf{u}$, is related to the turning rate of our parcel of fluid. It is twice the angular velocity. This is why a very small paddle wheel placed in a flow can be a measure of the vorticity at that point. The vorticity is twice the rotation rate of the wheel. We assume that it is measuring twice the rotation rate of the small parcel at this location (i.e., it is a point measurement). While vorticity is associated with the skew-symmetric portion of the deformation, it can also be obtained from the complete deformation rate tensor by multiplying by ϵ_{ijk}. Since this tensor is skew-symmetric, its effect is to select out the three independent terms of the rotation tensor.

$$\epsilon_{ijk}\, \partial u_k / \partial x_j = \epsilon_{ijk}(e_{kj} + \cap_{kj}) = \epsilon_{ijk}\cap_{kj} = \zeta_i \qquad (8.11)$$

Since

$$\text{div curl } \mathbf{u} = \nabla \cdot \zeta \equiv 0 \qquad (8.12)$$

vorticity is a solenoidal vector. Thus, we can apply many of the ideas developed for a nondivergent velocity field to a vorticity field.

Example 8.3

Assume a flow field is observed to satisfy $\mathbf{u} = A x \mathbf{i} + B y \mathbf{j}$.
(a) What are the restrictions on A and B for continuity?
(b) Is the flow irrotational?

Solution

(a) To satisfy continuity $\partial u / \partial x + \partial v / \partial y = 0$, $\partial u / \partial x = A$ and $\partial v / \partial y = B$. Thus, $A + B = 0$ and $A = -B$.
(b) Check to see if $\nabla \times \mathbf{u} = 0$.

$$\partial u / \partial y = 0; \; \partial v / \partial x = 0. \text{ Thus, } \partial v / \partial x - \partial u / \partial y = 0$$

and the flow is irrotational.

8.3.1   The Vorticity Filament

In analogy to the definition of a streamline, consider a line such that its tangent is parallel to the local vorticity for all points. This line is defined as a vorticity filament. (It is often called a vortex line, but this is likely to cause confusion with the line vortex, to be defined in Section 8.4.4.)

In the case shown in Fig. 8.9, only ζ_2 is not zero and the vorticity filament is parallel to x_2.

8.4 The Vorticity Tube

The vorticity filament hasn't the great practical usage of its counterpart, the streamline. However, if we pursue the analogous definition to a bundle of streamlines forming a streamtube, we obtain the vorticity tube, which does have practical application. Thus, we consider a reducible closed curve C as a singly connected region (e.g., no cylinder inside). From the surface confined by C we can construct a three-dimensional volume called the vorticity tube. We define the sides of the tube as made up of all vorticity filaments that pass through C_1 and terminate at another closed curve, C_2. This is defined as a *vorticity tube*, shown in Fig. 8.10. (Although the term vortex tube is in general use, we are reserving the term vortex for the fixed and free vortex fields).

Closed curves such as the C in Fig. 8.10 define a set of vorticity filaments that pass through them at some specific time. They can be thought of as marking a collection of the fluid material. Such lines are sometimes called material lines. They always pass through the same fluid parcels. The history of the material lines is of specific interest in atmospheric dynamics. The lines contain the same fluid for all time as they move and distort with the flow.

Just as $\nabla \cdot \mathbf{u} = 0$ requires that streamtubes cannot end in the fluid, $\nabla \cdot \zeta = 0$ requires vorticity tubes to not end in the fluid. They may form closed loops, extend to infinity, or terminate at a surface where $\zeta \to 0$.

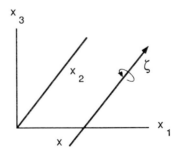

Figure 8.9 A vorticity filament for the case $\zeta = \zeta_2$ at point $(x, 0, 0)$.

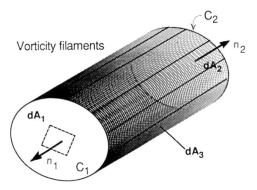

Figure 8.10 The vorticity tube.

According to the divergence theorem, the vorticity flux across the surface area of the tube, S, equals the divergence of the vorticity within the bounded volume. In the case of our vorticity tube with caps A and A' on the ends, we can write the integral over the entire surface of the vorticity tube as

$$\iint \boldsymbol{\zeta} \cdot \mathbf{n}\, dA = \underbrace{\iint \boldsymbol{\zeta} \cdot \mathbf{n}_2\, dA_2}_{\text{end}} - \underbrace{\iint \boldsymbol{\zeta} \cdot \mathbf{n}_1\, dA_1}_{\text{end}} + \underbrace{\iint \boldsymbol{\zeta} \cdot \mathbf{n}_3\, dA_3}_{\text{sides}}$$

Now, this integral can be related to the volume integral of div $\boldsymbol{\zeta}$ using the divergence theorem,

$$\iint \boldsymbol{\zeta} \cdot \mathbf{n}\, dA = \iiint \text{div } \boldsymbol{\zeta}\, dV = 0 \qquad (\text{since div curl } u = 0)$$

and $\boldsymbol{\zeta}$ is normal to $\mathbf{n}_3$; hence $\boldsymbol{\zeta} \cdot \mathbf{n}_3 \equiv 0$, and

$$\iint \boldsymbol{\zeta} \cdot \mathbf{n}_1\, dA_1 = \iint \boldsymbol{\zeta} \cdot \mathbf{n}_2\, dA_2 = \text{constant} \qquad (8.13)$$

Thus, the scalar quantity,

$$\iint \boldsymbol{\zeta} \cdot \mathbf{n}\, dA$$

is constant along the tube, even though the cross-sectional shape and area may change. It is defined as the *strength of the vorticity tube*. We expect this quantity to be useful, as it is a conserved property. However, the vorticity flux is not a physically tangible property to most of us. Fortunately, it can be related to a more palatable parameter, circulation.

8.4.1   Circulation

The *circulation* associated with a closed curve S is defined as

$$\Gamma \equiv \oint \mathbf{u} \cdot \mathbf{dS} \qquad (8.14)$$

Γ is positive for counterclockwise circulation (the right-hand rule is applied). Consider the circulation in the x-y plane of the small element of Fig. 8.11. We can write the circulation around the element as,

$$\delta\Gamma = [v + (\partial v/\partial x)\,\delta x - v]\,\delta y + [u - (u + (\partial u/\partial y)\,\delta y)]\,\delta x$$

$$= (\partial v/\partial x - \partial u/\partial y)\,\delta x\,\delta y$$

or

$$\delta\Gamma/\delta A = \partial v/\partial x - \partial u/\partial y = \zeta^z$$

The vorticity is equal to the circulation per unit area.

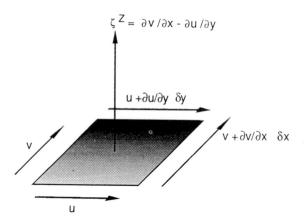

Figure 8.11   A small element with $\delta\Gamma = \mathbf{u} \cdot \delta s$.

Example 8.4

By examining the circulation of a small circular parcel, show that the strength of a small circular vorticity tube is equal to the circulation around the vorticity tube. Relate it to the vorticity of the parcel viewed as a point in the field.

Solution

Consider the small element δA shown in Fig. 8.12, with incremental circulation $\delta\Gamma = \mathbf{u} \cdot \delta\mathbf{s}$. From the definition of the circulation (u_Θ is constant),

$$\delta\Gamma \equiv \lim_{\delta r \to 0} \oint \mathbf{u} \cdot \mathbf{dS} = \oint u_\theta \, dS = u_\theta(2\pi \, \delta r)$$

Hence, we can write for the incremental area of the parcel,

$$u_\theta = \delta\Gamma/(2\pi \, \delta r) \qquad \text{or} \quad \delta\Gamma = 2\pi \, \delta r \, u_\theta \qquad (8.15)$$

We know that in the limit of a very small parcel, the rotation can be viewed as solid-body rotation. It is associated with the vorticity at the point of the parcel. From solid body rotational velocity, $u_\theta = \delta r \, d\theta/dt$, we get

$$u_\theta \, \delta S = (\delta r \, d\theta/dt)(2\pi \, \delta r) = 2 \, (d\theta/dt) \, \pi \, \delta r^2$$

$$= 2 \, (d\theta/dt) \, \delta A = \zeta \delta A$$

or

$$\delta A = u_\theta \, 2\pi \, \delta r/\zeta \qquad (8.16)$$

Therefore, from Eqs. (8.15) and (8.16), in the limit for a small parcel we obtain

$$\lim_{\delta A \to 0} \frac{\delta\Gamma}{\delta A} = \frac{2\pi \, \delta r \, u_\theta}{2\pi \, \delta r \, u_\theta/\zeta} = \zeta \qquad (8.17)$$

or vorticity equals the circulation per unit area at a point.

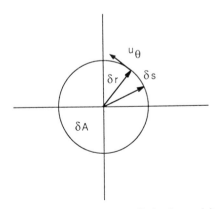

Figure 8.12   A cross section of a small circular vorticity tube.

Of course this is identical to the result for a square. In the limit of $\delta A \rightarrow$ 0, the shape of the parcel does not matter.

The circulation around an element can be used to relate the circulation around a finite curve to the enclosed vorticity. Consider the fluid divided into a mesh as shown in Fig. 8.13.

The total circulation around S can be approximated as the sum of the circulation around all contained mesh units with incremental circulation $\delta\Gamma_i$.

$$\Gamma = \sum_i \delta\Gamma_i$$

The adjacent sides of individual mesh units have equal and opposite $\delta\mathbf{u} \cdot \delta\mathbf{s}$ contributions to the total circulation at all interior points. There is a non-canceled contribution only from the sides of units that adjoin the bounding surface. Therefore the sum of all mesh circulations in A will be equal to the sum of the boundary contributions.

8.4.1.a   Stokes' Theorem Relating Circulation to Vorticity

We can place the relationship between the vorticity and the circulation on a formal level by using Stokes's theorem. This theorem relates the line

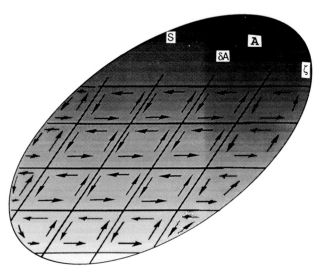

Figure 8.13   A curve S enclosing an area A with circulation.

integral of a vector around a finite curve to the areal integral of the curl of the vector over the area contained within the curve. We have seen that the vorticity is the circulation per unit area. Since the circulation is the line integral of the velocity around the curve, and the vorticity is the curl of the velocity at a point, Stokes's theorem will relate these two quantities for a finite region. Note that translational velocities give no net contribution to the circulation.

Stokes's theorem states that the line integral of $\mathbf{u} \cdot d\mathbf{s}$ around the perimeter is equal to the integral of the curl $\mathbf{u}$ over the enclosed area,

$$\oint \mathbf{u} \cdot d\mathbf{S} = \iint \text{curl } \mathbf{u} \cdot \mathbf{n} \, dA \equiv \iint \zeta \cdot \mathbf{n} \, dA = \Gamma \qquad (8.18)$$

Therefore, in addition to the definition of Γ as a line integral of the tangential velocity around a closed curve, we can think of circulation as the flux of vorticity across the surface enclosed by the curve. We can also consider Γ as the strength of the vorticity tube defined by the curve S. The circulation is an important concept because it can be determined from a measured velocity field, and it is conserved for the vorticity tube.

8.4.1.b Vorticity of a Vorticity Tube Segment

In atmospheric dynamics we frequently deal with a thin two-dimensional fluid layer with a finite vorticity tube extending from top to bottom. We can combine the conservation of circulation with that of mass to derive a practical rule for the behavior of such a vorticity tube segment.

First, consider part of a vorticity tube where

$$\delta\Gamma = \zeta \, \delta A = \text{constant} \qquad (8.19)$$

Vorticity tubes are buried in the fluid and move with the fluid, enclosing the same material at all times. Hence, a segment of a vorticity tube with length dL (Fig. 8.10) defines a material portion of the fluid. In homogeneous flow the contained mass is preserved, so that

$$\delta M = \rho \, \delta A \, \delta L = \text{constant}$$

Then, since the circulation $\delta\Gamma = \zeta \, \delta A$ is also constant,

$$\delta M / \delta\Gamma = (\rho/\zeta) \, \delta L = \text{constant} \qquad (8.20)$$

Thus, in constant-density flow ζ varies inversely with δL, the length of a vorticity tube segment, as pointed out by Helmholz in 1868. With constant density, if we stretch the tube (increase δL), then ζ must increase. This is an increase in circulation per unit area, and therefore the u_θ will increase. This characteristic behavior will appear when we consider the variation of vorticity in the fluid as it is described by the vorticity transport equation.

In atmospheric applications, the flow is often treated as two-dimensional horizontal. The absolute or total vorticity, $\zeta_T \equiv \zeta + f$, can be expressed for constant density by Helmholz's law as

$$\frac{\zeta_T}{L} = \text{constant} \equiv potential\ vorticity \tag{8.21}$$

The scale length L is taken as the effective height of the "parcel" of air. For a uniform region, Stokes's theorem allows us to apply Eq. (8.21) to a large uniform mass of air. The fact that ζ_T has two independent parts, where f is the vertical component of the earth's vorticity, means that Eq. 8.21 can be satisfied in a variety of ways.

Example 8.5

Consider the flow of a vortex in water flowing through a channel with a hump in the bottom as shown in Fig. 8.14. What is the behavior of the vorticity tube segment shown?

Solution

We have

$$\zeta_1/L_1 = \zeta_2/L_2 = \zeta_3/L_3 = \zeta_4/L_4 = \zeta_5/L_5$$

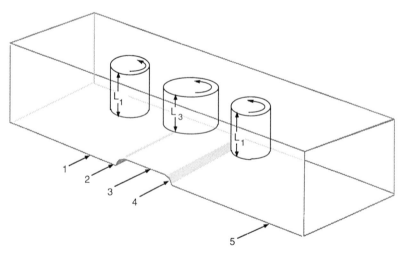

Figure 8.14 A sketch of a vertical vortex tube segment experiencing shortening due to bottom topography.

Also,

$$\Gamma = \zeta_1 A_1 = \zeta_2 A_2 = \zeta_3 A_3 = \zeta_4 A_4 = \zeta_5 A_5$$

Since

$$L_1 > L_2 > L_3 < L_4 < L_5 = L_1$$

$$\zeta_1 > \zeta_2 > \zeta_3 < \zeta_4 < \zeta_5 = \zeta_1$$

The vorticity decreases over the bump, then increases as it returns to the low part of the channel. Similarly, the area must increase over the bump and then decrease when the length grows to L_1.

Example 8.6

Consider the effects of an atmospheric flow of a uniform, steady passage of an airmass over a mountain range under the constraint ζ_T/L = constant for (a) westerly flow; (b) easterly flow. In each case, briefly describe what is happening to L and vorticity at the stations shown in Fig. 8.15 and 8.16.

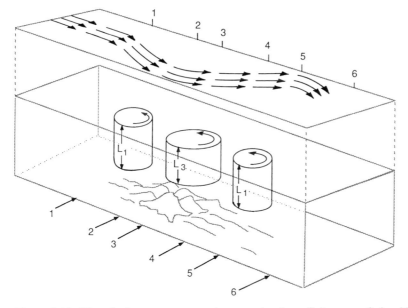

Figure 8.15 Westerly flow over a mountain range showing a finite mass of air with vorticity, $\zeta + f$.

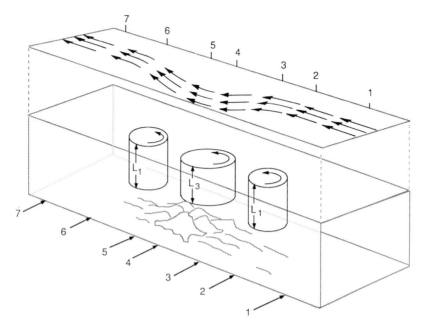

Figure 8.16   Easterly flow over a mountain range of air mass with vorticity $\zeta + f$.

Solution

We have two contributions to the vorticity of the air parcel. One is from the planetary vorticity and one from the relative vorticity. Thus, a change in L can be balanced by changing either or both of the local vorticity or the Coriolis component (the latter by changing the latitude).

(a) For a uniform westerly flow, the flow must move as shown in the cross-sections at the top of each figure for the reasons given at each station.

Station descriptions for Fig. 8.15 are as follows.

1. Assume $\zeta_{\text{local}} = 0$ so that $\zeta_T = f$. Then $f/L_1 \equiv$ potential vorticity.

2. L is decreasing; according to Eq. (8.21), $\zeta_2 + f_2$ must decrease, approaching either anticyclonic curvature or southward movement. The local vorticity is increased by turning south so that both mechanisms act in consort.

3. L is maximum and $(\zeta_3 + f_3)/L_3$. However, there is continued southward movement. A new equilibrium $(\zeta_4 + f_4)/L_3$ is sought, but ζ must reverse curvature to cyclonic, at f_4. At 3, f is still decreasing, and an overshoot occurs northwards.

4. ζ now has a positive cyclonic contribution, f is minimum, and $(\zeta_4 + f_4)/L_4$. Now L begins to increase.

5. With both northward movement and anticyclonic tendency increasing total vorticity, an equilibrium is easily maintained. ζ will decrease to zero at some point where f_5/L_5 is the equilibrium state again. However, f is increasing.

6. An overshoot, an anticyclonic ridge to balance the increased planetary vorticity, occurs here, and so on.

Because equilibrium is not reached until trajectories with excess f contributions are attained, a series of waves exists downstream.

(b) For a uniform easterly flow, the flow pattern will be different because the two contributions do not have the same sign in the initial turn.

Station descriptions for Fig. 8.16 are as follows.

1. Uniform flow with $\zeta/L = f_1/L$.

2. As L decreases, if the $\zeta + f$ decrease is taken care of by a ζ change to anticyclonic, there is a turn northward, which implies an f increase and a more negative ζ. This eventually would act to turn the flow around—not a solution for uniform flow over a mountain range.

A possible solution which is always in equilibrium occurs if f decreases initially. Then a cyclonic turning lets a ζ increase balance the f decrease, and flow can turn in equilibrium even before experiencing a change in L. As L decreases and f decreases, less positive contribution from ζ is needed to balance $(\zeta + f)/L_2$.

3. Equilibrium is obtained at f_3/L_3 with no contribution from ζ; however, f is still decreasing.

4. There is an anticyclonic maximum at an f_3 minimum: $(\zeta_3 + f_3)/L_3$.

5. Both ζ and f increase through a point where $f_5/L_5 = f_1/L_1$.

6. ζ will reach a small anticyclonic maximum while $f_6 < f_1 = f_7$.

7. With no other gains or losses, $f_7/L_7 = f_1/L_1$.

Here, the early start in adjustment of ζ_T with change in f leads to a symmetric solution.

Notice the important difference at each station 5. In both cases, L is increasing; hence $\zeta + f$ must increase and f is increasing. However, in (a) ζ is decreasing and in (b) ζ is increasing. The net effect is to restore uniform flow quickly in case (b), but to allow waves in case (a).

8.4.2 The Line Vortex

A concentrated line vortex is a mathematical concept with significant practical applications. It is defined as the limit of a vorticity tube for a small area while the circulation is kept constant.

$$\lim_{\delta A \to 0} \text{vorticity tube} = \text{line vortex;} \quad \Gamma \text{ constant}$$

The line vortex is thus a line singularity—or, in two dimensions, a point singularity—with circulation Γ. The circulation is the characteristic feature of the line vortex and is called its strength. From Eq. (8.16), we can determine the flow around an infinite straight-line vortex in a field with no other sources of vorticity. The flow will have tangential velocity inversely proportional to the radial distance from the point of vorticity,

$$u_\theta = r\, d\theta/dt = \Gamma/(2\pi r) \qquad (8.22)$$

This can be seen from the definition of circulation as the line integral around a circle of radius r, since the total contained circulation is that of the line vortex. The vorticity is the circulation per unit area, $\Gamma/(\pi r^2)$.

The line vortex can be viewed as a bundle of vorticity filaments as shown in Fig. 8.17. It has a vorticity distribution that produces a circulation Γ and associated velocity

$$u_\theta = r\, d\theta(r)/dt = r\zeta(r)/2 = (r/2)\,[\Gamma/\pi r^2] = \Gamma/(2\pi r) \qquad (8.23)$$

Since the velocity is of form $u_\theta \propto C/r$, this is a free vortex. The velocity field around the line vortex is the irrotational *free-vortex* field.

The distinction between the vorticity filament and the line vortex is significant—they are two different things. The *vorticity filament* is a line of fixed vorticity with a point concept of the velocity field in the immediate vicinity as solid-body-like rotation. Thus, $ds = r\, d\theta$ and $u_\theta = ds/dt$, or

$$u_\theta = r\, d\theta/dt = r\zeta/2 = Cr$$

The local vorticity at a point (of a parcel) associated with vorticity ζ is a *forced vortex* velocity field.

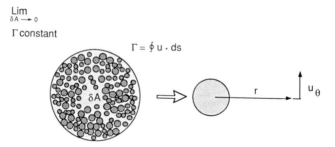

Figure 8.17   The line vortex with associated velocity u_θ.

The line vortex is a useful concept when the flow field of interest is large compared to the concentrated area of rotation. This model provides us with a good approximation even if the line vortex is curved in space, provided that its radius of curvature $\mathbf{R}$ satisfies $r/\mathbf{R} \ll 1$. It yields a good first approximation for the winds in the vicinity of tornados, whirlwinds, whirlpools, contrails, coherent vortices, cyclones, and hurricanes. This singularity in an otherwise irrotational field supplies a crucial ingredient to the potential flow theory discussed in Chapter 9.

8.5 Vorticity Transport Equation

One of the reasons that vorticity is such an important parameter in atmospheric flows is its quality of persistence. When a fluid mass acquires vorticity, this "arrangement" of the velocity field is quite stable and tends to last a relatively long time compared to other flow patterns on the same scale. In the atmosphere, cyclones and hurricanes last for days, containing the most important aspects of the daily weather pattern. There are also vortex motions on scales of PBL-size vortices embedded in a typical PBL, or tornados spawned on the thunderstorm scale; and on down to the three-dimensional vorticity characterizing the boundary-layer turbulence. Also, in the ocean there are vortex eddies with scales in 10–100 km and lifetimes from days to years. To investigate the role of vorticity, we start with the definition of vorticity as a characteristic of the relative motion at a point of the velocity field. We then construct an equation for vorticity from our equation of motion.

Recall that for an antisymmetric tensor there is an associated vector with elements equal to the three independent elements of the tensor. For the rotational part of the velocity deformation tensor, rot $\mathbf{u}$, that vector is the vorticity vector. Thus the action of the rotation tensor on any vector $\mathbf{v}$ can also be obtained by the cross product of ζ with that vector,

$$(\text{rot } \mathbf{u})\mathbf{v} = \zeta \times \mathbf{v} \quad \text{with} \quad \zeta \equiv \nabla \times \mathbf{u} \quad (8.24)$$

Since our basic equation of motion, the momentum equation, involves the velocity as the dependent variable, by taking the cross product of all terms we will obtain an equation involving vorticity. We will develop this equation using symbolic notation and vector identities, starting with the momentum equation.

$$\rho \, D\mathbf{u}/Dt = \rho \left\{ \partial \mathbf{u}/\partial t + (\mathbf{u} \cdot \text{grad})\mathbf{u} \right\}$$

$$= -\text{grad } p + \mu \text{ div grad } \mathbf{u} - \rho f \mathbf{k} \times \mathbf{u}$$

where we have used the alternate expressions,

$$(\text{grad } \mathbf{u})\mathbf{u} = (\mathbf{u} \cdot \text{grad})\mathbf{u}$$

$$\mathbf{F}_c = f\epsilon_{ijk}k_j u_k = f\mathbf{k} \times \mathbf{u}, \qquad \mathbf{k} \equiv (0,\,0,\,1)$$

Recall that div grad $\mathbf{u} \equiv \nabla^2\mathbf{u}$ in rectangular coordinates only.

Divide through by ρ to minimize the occurrence of ρ. The advective term and the viscous term can be broken into parts. This is advantageous since curl $\mathbf{u} \equiv \zeta$ occurs and gradients of scalars result. The latter will vanish when the curl of the equation is taken. We will need several of the tensor relations from Chapter 4, starting with

$$(\mathbf{a} \cdot \text{grad})\mathbf{a} = \text{grad } a^2/2 + (\text{curl } \mathbf{a}) \times \mathbf{a}$$

and

$$\text{div grad } \mathbf{a} = \text{grad div } \mathbf{a} - \text{curl curl } \mathbf{a} \tag{8.25}$$

Thus, the momentum equation may be written

$$\partial\mathbf{u}/\partial t + \text{grad } u^2/2 + (\text{curl } \mathbf{u}) \times \mathbf{u} = -1/\rho \text{ grad } p$$

$$+ \nu \text{ grad}(\text{div } \mathbf{u}) - \nu \text{ curl curl } \mathbf{u} - f\mathbf{k} \times \mathbf{u}$$

Substituting vorticity for curl $\mathbf{u}$,

$$\partial\mathbf{u}/\partial t + \text{grad } u^2/2 + \zeta \times \mathbf{u} + 1/\rho \text{ grad } p$$

$$- \nu \text{ grad}(\text{div } \mathbf{u}) + \nu \text{ curl } \zeta + f\mathbf{k} \times \mathbf{u} = 0 \tag{8.26}$$

Take the curl of this equation,

$$\text{curl } \partial\mathbf{u}/\partial t + \text{curl}(\text{grad } u^2/2) + \text{curl}(\zeta \times \mathbf{u}) + \text{curl}(1/\rho \text{ grad } p)$$

$$- \text{curl}[\nu \text{ grad}(\text{div } \mathbf{u})] + \text{curl}(\nu \text{ curl } \zeta) + \text{curl}(f\mathbf{k} \times \mathbf{u}) = 0 \tag{8.27}$$

Now use the identities:

$$\text{curl grad } \varphi = 0$$

(to eliminate the second and fifth terms);

$$\text{curl}(\mathbf{a} \times \mathbf{b}) = \text{div}\{(\mathbf{a};\,\mathbf{b}) - (\mathbf{b};\,\mathbf{a})\}$$

and

$$\text{div}(\mathbf{a};\,\mathbf{b}) = (\mathbf{b} \cdot \nabla)\mathbf{a} + (\text{div } \mathbf{b})\mathbf{a}$$

[noting that $(\mathbf{b} \cdot \nabla)\mathbf{a} \equiv (\nabla\mathbf{a})\mathbf{b}$]; to get

$$\nabla \times (\zeta \times \mathbf{u}) = (\mathbf{u} \cdot \nabla)\zeta + (\nabla \cdot \mathbf{u})\zeta - (\zeta \cdot \nabla)\mathbf{u} - (\nabla \cdot \zeta)\mathbf{u}$$

and

$$\nabla \times (f\mathbf{k} \times \mathbf{u}) = (\mathbf{u} \cdot \nabla)f\mathbf{k} + (\nabla \cdot \mathbf{u})f\mathbf{k} - (f\mathbf{k} \cdot \nabla)u - (\nabla \cdot f\mathbf{k})u$$

The last terms in each of these expansions equals zero (div curl $\mathbf{u}$ and $\partial f/\partial z = 0$). Substituting these two identities into Eq. (8.27),

$$\partial \zeta/\partial t + (\mathbf{u} \cdot \nabla)\zeta + (\nabla \cdot \mathbf{u})\zeta - (\zeta \cdot \nabla)\mathbf{u} + \nabla \times (1/\rho \text{ grad } p)$$

$$+ \nu\nabla \times \nabla \times \zeta + (\mathbf{u} \cdot \nabla)f\mathbf{k} + (\nabla \cdot \mathbf{u})f\mathbf{k} - (f\mathbf{k} \cdot \nabla)\mathbf{u} = 0 \quad (8.28)$$

Use Eq. (8.25) to change the viscous term,

$$\nu \text{ curl curl } \zeta = -\nu \text{ div grad } \zeta$$

Several terms can be combined to form the *total vorticity* term,

$$\partial \zeta/\partial t + (\mathbf{u} \cdot \nabla)(\zeta + f\mathbf{k}) + (\nabla \cdot \mathbf{u})(\zeta + f\mathbf{k}) - [(\zeta + f\mathbf{k}) \cdot \nabla]\mathbf{u}$$

$$- \nu \nabla^2\zeta + \nabla \times (1/\rho) \text{ grad } p = 0 \quad (8.29)$$

When density variations can be neglected, the last term can be dropped. This is a good approximation in the atmosphere if ρ is a function of pressure only. In most cases, we can assume $w \ll u, v$ and negligible horizontal change in density.

Thus, with $\nabla \times (1/\rho) = 0$ and substituting

$$[(\zeta + f\mathbf{k}) \cdot \nabla]\mathbf{u} = (\nabla\mathbf{u})(\zeta + f\mathbf{k}), \quad (8.30)$$

$$\frac{\partial \zeta}{\partial t} + (\mathbf{u} \cdot \nabla)(\zeta + f\mathbf{k}) + (\text{div } \mathbf{u})(\zeta + f\mathbf{k}) - (\nabla\mathbf{u})(\zeta + f\mathbf{k}) - \nu \nabla^2\zeta = 0$$

This is the basic vorticity equation for many practical applications. It is an equation for the rate of change of angular momentum of a parcel, accounting for the rotating frame of reference. Note that we have picked up three terms from the advective portion of the total velocity derivative. This is a consequence of a vector ζ (which depends on $\mathbf{u}$ gradients) being advected by $\mathbf{u}$. In addition to the local plus advective change making up the total derivative, there are three terms that change the vorticity. These are (1) an area change preserving circulation and thereby changing vorticity due to the divergence of $\mathbf{u}$; (2) distortion effects due to the velocity gradient tensor operating on the vorticity vector; and (3) the tangential viscous stress forces imparting rotation by diffusion, or generating vorticity at a boundary.

In atmospheric synoptic scale problems, the domain extends over a large enough scale (500–1000s km) that the change in Coriolis parameter must be included. The significant vorticity term is $\zeta + f\mathbf{k} \equiv \zeta_T$, defined as the *total vorticity*. Since $\partial f\mathbf{k}/\partial t = 0$, Eq. (8.30) may be written

$$\partial(\zeta + f\mathbf{k})/\partial t + (\mathbf{u} \cdot \text{grad})[\zeta + f\mathbf{k}]$$

$$= [(\zeta + f\mathbf{k}) \cdot \text{grad}]\mathbf{u} - (\zeta + f\mathbf{k})(\text{div } \mathbf{u}) + \nu \nabla^2\zeta \quad (8.31)$$

This form of the vorticity equation has widespread use in synoptic meteorology and oceanography.

We can write the *vorticity equation* compactly.

$$D\zeta_T/Dt = -\zeta_T \text{ div } \mathbf{u} + (\zeta_T \cdot \nabla)\mathbf{u} + \nu \nabla^2 \zeta$$

Rate of	=	Divergence	+	Distortion of parcel,	+	Diffusion	
change of		area change		change radius gyration,		between	(8.32)
vorticity		(horizontal)		twisting, tilting		parcels.	

The vorticity equation is somewhat similar to the momentum equation. The total change in vorticity depends on several sink/source terms, and the viscous term is identical. The complete equation is difficult to solve—it is a nonlinear vector equation with a tensor operating on the velocity vector in the distortion term. Consequently, approximations are generally made to simplify the terms.

8.5.1 The Advection Term

Since many atmospheric flows are approximately horizontal and two-dimensional, the vertical component is the only significant component of vorticity in these cases. Also, the Coriolis force varies with y (latitude) only. Then, the vorticity advection term $(\mathbf{u} \cdot \text{grad}) \zeta_T$ is greatly simplified. Thus,

$$(\mathbf{u} \cdot \nabla)\zeta_T = (u \, \partial/\partial x + v \, \partial/\partial y + w \, \partial/\partial z)(0, 0, \zeta^z) + v \, \partial f/\partial y$$

$$= (u \, \partial/\partial x + v \, \partial/\partial y + w \, \partial/\partial z)\zeta^z + v \, \partial f/\partial y \qquad (8.33)$$

Equation (8.32) becomes a scalar equation, with $\zeta^z = \zeta$, $\zeta_T = \zeta_T^z = \zeta + f$,

$$\partial\zeta/\partial t + (\mathbf{u} \cdot \nabla) \zeta + v \, \partial f/\partial y = \zeta_T \, \partial w/\partial z - \zeta_T \nabla_H \cdot \mathbf{u} + v \, \partial^2\zeta_T/\partial z^2$$

In addition, the viscous diffusion can generally be neglected for large-scale flow. We then get

$$\partial\zeta/\partial t + (\mathbf{u} \cdot \nabla) \zeta + v \, \partial f/\partial y = \zeta_T \, \partial w/\partial z - \zeta_T \nabla_H \cdot \mathbf{u} \qquad (8.34)$$

Equation (8.32) has a different approximate version for mesoscale or planetary boundary layer scales, where the variation in Coriolis parameter is negligible ($f \ll \zeta$), and frequently div $\mathbf{u} \approx 0$, but the viscous term is important, so that

$$D\zeta/Dt = (\zeta \cdot \text{grad}) \mathbf{u} + v \nabla^2\zeta \qquad (8.35)$$

Vorticity change	=	Distortion	+	Diffusion

Example 8.7

Derive Eq. (8.32) from the component momentum equations. To simplify this illustration of the large number of terms that arise when dealing with the individual components, leave out most of the terms in the first few equations, carrying only sufficient terms to illustrate the procedure.

Solution

Subscripts denote differentiation, superscripts denote components.
  Continuity:

$$u_x + v_y + w_z = 0$$

We can generate vorticity component terms like $\zeta^z = v_x - u_y$ by forming these terms out of the component momentum equations.

$$u_t + \ldots + wu_z + \ldots - \nu(u_{xx} + u_{yy} + u_{zz}) = 0$$

$$v_t + \ldots + wv_z + \ldots - \nu(v_{xx} + v_{yy} + v_{zz}) = 0$$

Then, cross differentiate the *u*-momentum equation with respect to *y* and the *v*-momentum equation with respect to *x*.

$$u_{yt} + \ldots + w_y u_z + wu_{zy} + \ldots - \nu(u_{xxy} + u_{yyy} + u_{zzy}) = 0$$

$$v_{xt} + \ldots + w_x v_z + wv_{zx} + \ldots - \nu(v_{xxx} + v_{yyx} + v_{zzx}) = 0$$

Subtract the second equation from the first (to form vorticity terms):

$$(v_x - u_y)_t + \ldots + w(v_x - u_y)_z + w_x v_z - w_y u_z - w_z(v_x - u_y) + \ldots$$

$$= \nu\{(\ldots + (v_x - u_y)_{zz}\}$$ (8.36)

Here, continuity has been used to write

$$u_x u_y - v_x v_y + u_y v_y - u_x v_x = -(u_x + v_y)(v_x - u_y) = -w_z(v_x - u_y)$$

Now, collecting all of the terms for $v_x - u_y$, the *z*-component of ζ in Eq. (8.36), we get

$$\zeta^z_t + u\zeta^z_x + v\zeta^z_y + w\zeta^z_z = -w_x v_z + w_y u_z + w_z \zeta^z + \ldots + \nu(\zeta^z_{zz} + \zeta^z_{yy} + \zeta^z_{zz})$$

The rest of the terms can be combined to form components of ζ to get

$$\zeta^z_t + u\zeta^z_x + v\zeta^z_y + w\zeta^z_z = w_x \zeta^x + w_y \zeta^y + w_z \zeta^z + \nu \nabla^2 \zeta^z$$

or

$$D\zeta^z/Dt = (\zeta \cdot \mathrm{grad})\, w + \nu\, \nabla^2 \zeta^z \qquad (8.37)$$

This is the z-component equation for ζ. If next we calculate $\partial/\partial y$ (w-momentum equation) $-\ \partial/\partial z$ (v-momentum equation) we would get the x-component of the ζ equation. Another similar calculation would yield the y-component equation.

Adding all three components gives

$$D\zeta/Dt = (\zeta \cdot \mathrm{grad})\mathbf{u} + \nu\, \nabla^2 \zeta$$

or

$$D\zeta_i/Dt = \zeta_j\, \partial u_i/\partial x_j + \nu\, \partial^2 \zeta_i/\partial x_j\, \partial x_j \qquad (8.38)$$

These are the vorticity transport equations. We have written the first term on the right side in the symbolic form as a scalar $(\zeta \cdot \nabla)$ times a vector $\mathbf{u}$. Note this may also be written as the vector ζ times the velocity gradient tensor,

$$(\mathrm{grad}\ \mathbf{u})\ \zeta = \begin{pmatrix} u_x & u_y & u_z \\ v_x & v_y & v_z \\ w_x & w_y & w_z \end{pmatrix} (\zeta^x,\ \zeta^y,\ \zeta^z)$$

$$= [\ldots,\ \ldots,\ w_x\zeta^x + w_y\zeta^y + w_z\zeta^z]$$

The z-component of this term is

$$w_x(w_y - v_z) + w_y(u_z - w_x) + w_z(v_x - u_y)$$

$$= w_x w_y - v_z w_x + w_y u_z - w_x w_y + w_z v_x - w_z u_y$$

$$= -w_x v_z + w_y u_z + w_z v_x - w_z u_y$$

as found above.

The order of writing the advective term is clearly important, since

$$(\zeta \cdot \nabla)\mathbf{u} = (\nabla \mathbf{u})\zeta \neq \zeta(\nabla \mathbf{u})$$

Example 8.8

In the large-scale mean atmospheric flow at mid-latitudes, there are long (3000–10000 km) waves embedded in the flow, called Rossby waves. As the flow proceeds along the wavy path, it experiences alternating curvature.

(See Fig. 8.18.) This results in alternating relative vorticity. The waves need not be stationary and can move laterally along a zonal line. Use Eq. (8.34) to discuss their movement.

Solution

We look at the terms in Eq. (8.34),

$$\partial\zeta/\partial t + u\,\partial\zeta/\partial x + v(\partial\zeta/\partial y + \partial f/\partial y) + w\,\partial\zeta/\partial z = \zeta_T\,\partial w/\partial z - \zeta_T\,\nabla_H \cdot \mathbf{u}$$

and neglect the underlined terms to get

$$\partial\zeta/\partial t + u\,\partial\zeta/\partial x + v\,\partial f/\partial y = -\zeta_T\,\nabla_H \cdot \mathbf{u} \qquad (8.39)$$

Now, apply this equation in the mid-troposphere where $\nabla_H \cdot \mathbf{u} \approx 0$, (called the level of non-divergence, roughly 8 km for Rossby waves). This leaves

$$\partial\zeta/\partial t + u\,\partial\zeta/\partial x + v\,\partial f/\partial y = 0 \qquad (8.40)$$

We can use this equation to analyze the flow in Fig. 8.18. First, consider f as constant.

At point A, u and $\partial\zeta/\partial x$ are positive; hence $u\,\partial\zeta/\partial x > 0$. The only way to keep $D\zeta/Dt = 0$ is to have $\partial\zeta/\partial t < 0$. This can be done by the whole wave system moving to the east, bringing a local decrease of relative vorticity to A.

At point B, $u > 0$ but $\partial\zeta/\partial x < 0$, and $u\,\partial\zeta/\partial x < 0$. With constant f this would also require eastward movement of the wave system to keep vorticity conserved at B.

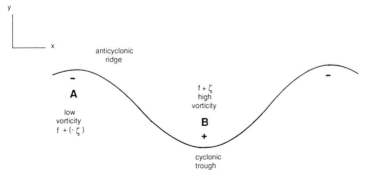

Figure 8.18 Rossby waves in a mean westerly flow.

When the advection of relative vorticity is dominant, the wave system tends to move eastward. The greater the value of u, the faster the waves must move.

Next we consider the effect when the planetary vorticity is not constant and look at $v \, \partial f/\partial y$.

At point A, v is negative, and $v \, \partial f/\partial y < 0$. This requires an increase in local vorticity, which can be accomplished with westward movement of the wave system. At point B, $v \, \partial f/\partial y > 0$, and the balancing decrease in local vorticity also requires westward movement of the wave.

Since the relative vorticity effect and the planetary vorticity effect are opposite, there is likely an equilibrium at some point, depending on wave characteristics (u, v, $\partial \zeta/\partial x$) and latitude ($\partial f/\partial y$). Based on the requirement that $D\zeta_T/Dt = 0$, we can state:

1. At a given mid-latitude, a certain wavelength of a Rossby wave will remain stationary.
2. Shorter waves and larger u velocities will move eastward ($\lambda \approx 3000$ km).
3. Longer waves can move westward ($\lambda \approx 10{,}000$ km).

Finally, add the divergence effect, $-(\zeta + f) \, \nabla_H \cdot \mathbf{u}$, using Eq. (8.39).

f is positive and generally greater than ζ, so that ζ_T is increased by convergence and decreased by divergence. The more vorticity present initially, the greater is this effect. The convergence in cyclones generates more intense cyclones, whereas divergence in anticyclones generally weakens the anti-cyclone (provided $f > |\zeta|$).

The vorticity transport equation is not a convenient expression to be used for determining velocity—it would be very difficult to integrate to get velocities. Still, vorticity is an important entity in its own right, and many phenomena can be qualitatively understood in terms of vorticity. It is important to atmospheric and oceanic flow in the study of turbulence and in large-scale motion. In fact, Eq. (8.34), together with a thermodynamic equation, are said to form the basic equations of dynamic meteorology.

The pressure gradient and gravity forces do not appear in the vorticity equation. Under our assumptions, they do not change the vorticity of a parcel. This is because vorticity is associated with the solid-body-like rotation of the parcel, and requires a torque to impart spin. Surface forces can cause rotation, but body forces pass through the center of mass and do not create angular momentum. However, when density gradients are included, the center of mass will not coincide with the geometric center, and rotation can result. This condition is important in certain atmospheric and oceanic flows where density variations can create vorticity in stratified conditions. The $\nabla \times \nabla p/\rho$ term will then add vorticity due to density-pressure gradient interactions whenever there is significant horizontal density gradient.

8.5.2   The Distortion Terms

The $(\zeta \cdot \text{grad})\ \mathbf{u}$ term in Eq. (8.35), which is called the distortion term, represents generation (or destruction) of ζ by a stretching or turning of the vorticity filaments. This term can be illustrated for an element δs on a vorticity filament (Fig. 8.19), where

$$\mathbf{s} = (s, n); \qquad u \equiv (u_s, u_n)$$

The inviscid, nondivergent, vorticity equations in these coordinates are:

$$D\zeta_s/Dt = \zeta_s\, \partial u_s/\partial s + \zeta_n\, \partial u_s/\partial n \approx \zeta_s\, \partial u_s/\partial s$$

and

$$D\zeta_n/Dt = \zeta_s\, \partial u_n/\partial s + \zeta_n\, \partial u_n/\partial n \approx \zeta_s\, \partial u_n/\partial s \qquad (8.41)$$

The $\delta\mathbf{u}$ component parallel to $\mathbf{ds}$, δu_s, causes stretching (or contraction) of the vorticity filament segment $\mathbf{ds}$. The $\delta\mathbf{u}$ component perpendicular to $\mathbf{ds}$, δu_N, causes twisting of $\mathbf{ds}$ (although $|\zeta_n| = 0$ at a point, the vector ζ is changing along the vorticity filament, since it is turning in space). This behavior of the vorticity segment is similar to that of a material line in the velocity field. There is a stretching component and a rigid turning of the line segment.

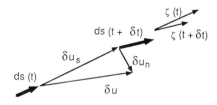

Figure 8.19   An element of a vorticity filament.

8.5.2.a   Two-Dimensional Vorticity Transport Equation

Equation (8.32) is related to the conservation of *angular momentum*. It does not show conservation of vorticity. Vorticity can be changed (created) by the stretching and turning terms.

However, in two-dimensional flow, ζ is perpendicular to **u** and the plane of flow. Thus there is no variation in the ζ direction, and therefore no turning or stretching. This can also be seen by looking at Eq. (8.32) in index notation, with $\nabla \cdot \mathbf{u} = 0$, and writing ζ as $\nabla \times \mathbf{u}$,

$$D\zeta_i/Dt = [\partial u_i/\partial x_j][\epsilon_{jlm} \, \partial u_m/\partial x_l] + \nu \, \partial^2 \zeta_i/\partial x_j \, \partial x_j \qquad (8.42)$$

In two-dimensional flow, in the second term above, i, j, l, and m, can be 1 or 2 only, so that ϵ_{jlm} must always equal zero, leaving

$$D\zeta_i/Dt = \nu \, \nabla^2 \zeta_i \qquad (8.43)$$

Thus, there is no change in the vorticity due to distortion caused by velocity gradients. In *two-dimensional inviscid* flow, *vorticity* is conserved.

8.5.3   The Viscous Term

The viscous term in Eq. (8.32) is a diffusion of vorticity term. It takes the same form as the incompressible diffusion of momentum term, where the total change in momentum, $D\mathbf{u}/Dt$, is proportional to $\nu \, \nabla^2 \mathbf{u}$. Also, in two-dimensional flow the vorticity transport equation is the same as the temperature diffusion equation,

$$DT/Dt = K/\rho \, c_p \, \nabla^2 T \qquad (8.44)$$

In two-dimensional flow, viscosity is the only way to impart or eliminate vorticity, since the other terms are zero, as shown in the previous section [Eq. (8.43)]. Just as in the momentum equation, this term is important in the region of a boundary layer. At the wall, vorticity can be generated through this viscous term, and then diffused into the flow.

When turbulence is thought of as a distribution of vorticity elements, the time history of the turbulent vorticity elements is related to the diffusion by viscosity according to Eq. (8.32). The vorticity elements will interact, stretching and twisting, becoming smaller until the dissipation term takes over. We will discuss this process after developing the perturbation equations.

8.6   Vorticity Characteristics

Vorticity is a vector field associated with the dynamics of the flow field. Specifically, it concerns the velocity derivatives rather than the velocity. When the flow is incompressible with uniform density, there are similarities between the vorticity and velocity descriptions. Note that if grad ρ is not zero, gravity can change $D\zeta/Dt$.

Vorticity	Velocity
$\nabla \cdot \zeta = 0$	$\nabla \cdot \mathbf{u} = 0$
Vorticity filaments are tangent to z.	Streamlines are tangent to $\mathbf{u}$.
Vorticity filaments cannot end in fluid or at a boundary.	Streamlines cannot end in fluid or at a boundary.
Vorticity flux = strength of the vorticity tube, and is conserved along tube.	Mass flux = mass flow rate. Mass is conserved along tube.
Behavior of a vorticity filament is described by	Behavior of a material element is described by
$D\zeta_i/Dt = (\partial u_i/\partial x_j)\zeta_j$	$D\,\delta x_i/Dt = (\partial u_i/\partial x_j)\,\delta x_j = \delta u_i$
Vorticity is conserved in	Mass is conserved.
2-D, inviscid flow, $\nabla \cdot \mathbf{u} = 0$.	
Vorticity equation, $\nabla \cdot \mathbf{u} = 0$:	Velocity (momentum/u.mass):
$D\zeta/Dt = (\zeta \cdot \nabla)\,u + \nu\,\nabla^2\zeta$	$D\mathbf{u}/Dt = \mathbf{F}_b - \nabla p/\rho + \nu\,\nabla^2\mathbf{u}$

The equations in vorticity or velocity can each have advantages in different circumstances. Both quantities have viscous dissipation . . . they decrease by molecular diffusion in proportion to u. Since u is transported by molecules, so too is $\nabla \times \mathbf{u}$ and, therefore, ζ.

The sources of momentum, or velocity change, are $\mathbf{F}_b$ and $\nabla p/\rho$. These forces are independent of the magnitude of velocity. In contrast, the vorticity transport equation (for smaller scales) has no ∇p terms. p has been eliminated in the derivation of the vorticity equation by cross-differentiation. However, there is a new term $(\zeta \cdot \nabla)\,\mathbf{u}$, which provides rules for vorticity distribution. Since vorticity is a factor in this term, it states that vorticity will increase or decrease in relation to how much vorticity is already present. When we write out the parts of the total derivative, we see that the vorticity is a factor in all of the terms. The action of each of these terms is:

$$\partial \zeta/\partial t = -(\mathbf{u} \cdot \nabla)\,\zeta + (\zeta \cdot \nabla)\,\mathbf{u} + \nu\,\nabla^2\zeta \qquad (8.45)$$

local	=	advective	+	twisting	+	molecular
change		term for		stretching		diffusion
		nonuniform ζ		terms (3-D)		of ζ

8.6.1 Helmholtz's Laws

Although the concept of vorticity was first introduced in the 1750s (as $\nabla \times \mathbf{u}$), it was not used productively until Helmholtz stated the following laws in 1858 for inviscid fluid flow.

1. A fluid element without rotation originally cannot gain rotation.
2. Fluid parcels remain part of the same vorticity tube for all time.
3. Vorticity tubes must be closed or end on boundaries.

Helmholtz's law can be stated in an expression more pertinent to atmospheric applications: In inviscid flow, a vorticity tube moves with the fluid, and its strength remains constant when there is uniform density or when density varies with pressure only.

A related law in terms of the circulation was derived by Kelvin, starting with the identity,

$$\frac{D}{Dt} \oint \mathbf{u} \cdot \mathbf{ds} \equiv \oint \frac{D\mathbf{u}}{Dt} \cdot \mathbf{ds}. \qquad (8.46)$$

This equation is most useful when $D\mathbf{u}/Dt$ is the gradient of a scalar, in which case the right side is zero. This is true of inviscid homogeneous flow where

$$D\mathbf{u}/Dt = \mathbf{F}_b + \nabla(p/\rho) = -\nabla[\Phi + p/\rho]. \qquad (8.47)$$

The result,

$$\frac{D\Gamma}{Dt} = \frac{D}{Dt} \oint \mathbf{u} \cdot \mathbf{ds} = 0, \qquad (8.48)$$

is *Kelvin's circulation theorem:* the Γ associated with the line integral around a material curve is constant. This is an expression of the persistence of vorticity in an inviscid fluid once it has been created. Or, if the initial vorticity is zero, it remains zero.

There is an important atmospheric application of the relation (Eq. 8.46) between the circulation and the circular integral of the total velocity derivative. In the atmosphere, the Coriolis component of the circulation, $2\Omega \cdot d\mathbf{A}$, is added to the local circulation Γ. Here, $d\mathbf{A}$ is the area enclosed by the line integral around s when s is in a vertical plane. In addition, consideration of the general case when density is not necessarily a function of pressure alone leaves a nonzero line integral on the right side of Eq. 8.46. Hence,

$$\frac{D(\Gamma + 2\Omega \cdot d\mathbf{A})}{Dt} = -\oint \frac{dp}{\rho} \qquad (8.49)$$

This is called the *Bjerknes circulation theorem*. It differs from Kelvin's theorem by the addition of the planetary circulation term and by allowing for the horizontal variation of density.

Example 8.9

There are many cases where the differential heating of the surface resulting from different heat absorption characteristics of the surface leads to a local atmospheric circulation. This flow can be investigated by applying Eq. (8.49) to the region of a vertical slice of area that spans a land–sea boundary and has a limited height. There are constant pressure and density lines, as shown in Fig. 8.20. Assume pressure lines are essentially horizontal. Constant density lines will tilt when daytime heating decreases ρ over land, and vice versa for nighttime cooling. Use an average ρ and a constant ΔP for approximations of the line integral.

Solution

Since "total circulation" $\Gamma + f \cdot A$ is calculated around S by the line integral of dp/ρ, we can write

$$\Delta\Gamma_T = -\oint dp/\rho \approx \Delta p[1/\rho_\ell - 1/\rho_s] \tag{8.50}$$

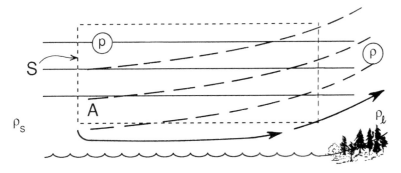

Figure 8.20   The land–sea breeze vertical profile (ρ_s = density over the sea; ρ_ℓ = density over land). Constant-pressure lines are horizontal. Constant-density lines curve upward over land. $\rho_\ell > \rho_s$ at the same height. The perimeter curve S encloses area A.

The constant-pressure lines are nearly horizontal. During daytime, faster heating of the land means $\rho_M < \rho_S$. This means the circulation from Eq. (8.50) is positive—counterclockwise. Hence the flow is from sea to land at the surface. The situation is reversed at night.

8.7 Application of Momentum, Energy, and Vorticity Equations to Bernoulli's Law

In Section 7.8 we introduced Bernoulli's law as a special case of the energy equation along a streamtube. We can now elaborate on this useful law and the conditions when it can be applied.

First, we look at the dynamic version of Bernoulli's equation. This is obtained from the momentum equation for constant-temperature, nondivergent flow,

$$\partial u/\partial t + (\mathbf{u} \cdot \mathrm{grad})\mathbf{u} = \mathbf{F} - \nabla p/\rho + v\,\nabla^2\mathbf{u}$$

As was done in the derivation of the vorticity equation, we use the vector identity,

$$\mathbf{u} \times \nabla \times \mathbf{u} = \tfrac{1}{2}\nabla(\mathbf{u} \cdot \mathbf{u}) - (\mathbf{u} \cdot \nabla)\,\mathbf{u} = -(\nabla \times \mathbf{u}) \times \mathbf{u}$$

Placing this in the momentum equation, we have

$$\partial\mathbf{u}/\partial t + \mathrm{curl}\ \mathbf{u} \times \mathbf{u} + \mathrm{grad}\ \tfrac{1}{2}q^2 = -\mathrm{grad}\ \Phi - \mathrm{grad}(p/\rho) + v\,\nabla^2\mathbf{u}$$

or

$$\partial\mathbf{u}/\partial t + \zeta \times \mathbf{u} = -\nabla(\Phi + p/\rho + \tfrac{1}{2}q^2) + v\,\nabla^2\,\mathbf{u} \qquad (8.51)$$

The first term on the right side is the sum of the kinetic energy, gravitational potential energy, and the flow work. We have defined this term as $H = \Phi + p/\rho + \tfrac{1}{2}q^2$, where q is the speed and Φ is the gravitational geopotential.

For steady state, inviscid flow, Eq. (8.51) becomes

$$\nabla H = -\zeta \times \mathbf{u} \qquad (8.52)$$

Since the gradient of H is normal to u and ζ, H is constant along any streamline or vorticity filament. In addition, it is zero everywhere if $\zeta = \nabla \times \mathbf{u} = 0$, which is the condition of irrotationality.

Thus, for steady-state, inviscid, irrotational flow, Bernoulli's law holds for the entire flow, along or across streamlines. When the flow is inviscid, and initially irrotational, it will always remain irrotational according to the vorticity transport equation. Bernoulli's law can be used at any two points of the flow.

Equation (8.52) is simply the mechanical energy equation. If temperature variation is a factor, the complete energy equation can be used to define a general H_0 (which includes internal energy, $H_0 = e + \Phi + p/\rho + q^2/2$). In this instance, Eq. (8.52) is obtained only as a special case when the flow is steady, nonconducting, inviscid, and isentropic.

8.7.1 Bernoulli's Equation in Streamline Coordinates

We can now continue the investigation of Bernoulli's law in streamline co-ordinates that was begun in section 7.8.1. There, we were left with an extra term, $\int V^2/r\, dn$, in our integration across the streamlines. An expression for this term can be obtained by considering the circulation around a parcel. The two-dimensional projection of the parcel is shown in natural coordinates in Fig. 8.21.

In the natural coordinate system, subscript n designates the normal component to the streamline and is not an index. The velocity V is directed along the streamline s. Note that although $\mathbf{V}_n$ is indicated in Fig. 8.21, its magnitude is zero at each point. However, because the nature of the natural coordinate system is to constantly change direction, we must keep track of the rate of change of the vector $\mathbf{V}_n$. At an incremental distance away from a point (e.g., from the left corner of the parcel), the new direction of the

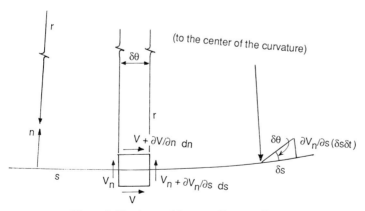

Figure 8.21 A parcel in streamline coordinates.

velocity on a curved streamline has an incremental component of velocity in the n-direction. This component is changing at the rate $\partial V_n/\partial s \ \delta s$.

We obtain the Bernoulli equation for each direction. Using

$$\tan \delta\theta \approx \delta\theta = [(\partial V_n/\partial s) \ \delta s \ \delta t]/\delta s = (\partial V_n/\partial s) \ \delta t$$

and

$$\tan \delta\theta \approx \delta\theta = \delta s/r$$

Hence,

$$\partial V_n/\partial s = (1/r) \ \delta s/\delta t = V/r$$

and the circulation around the parcel is

$$\delta\Gamma = V \ \delta s + (\partial V_n/\partial s \ \delta s)\delta n - (V + \partial V/\partial n \ \delta n) \ \delta s$$

$$= (\partial V_n/\partial s - \partial V/\partial n) \ \delta s \ \delta n$$

$$= 2 \ (\delta\theta_z/\delta t) \ \delta A$$

Or, in the limit $\delta s \to 0$,

$$\delta\Gamma = \zeta \ \delta A$$

Now, if the flow is irrotational,

$$\zeta = \partial V_n/\partial s - \partial V/\partial n = 0$$

and

$$\partial V/\partial n = \partial V_n/\partial s = V/r \tag{8.53}$$

hence,

$$V \ \partial V/\partial n = V^2/r \tag{8.54}$$

Returning to the n-component of the momentum equation,

$$\partial V_n/\partial t + V_s \ \partial V_n/\partial s + (1/\rho) \ \partial p/\partial n + g_n = 0 \tag{8.55}$$

where, from the geometry shown in Fig. 8.22,

$$g_n = -g \cos \delta\theta = -g \ \delta z/\delta n$$

$$g_s = -g \sin \delta\theta = -g \ \delta z/\delta s$$

For steady irrotational flow we can substitute

$$\partial V_n/\partial s = \partial V/\partial n \qquad \text{and} \qquad V^2/r = V \ \partial V/\partial n$$

to get

$$V \ \partial V/\partial n + (1/\rho) \ \partial p/\partial n + g \ \partial z/\partial n = 0 \tag{8.56}$$

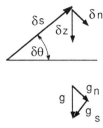

Figure 8.22 The gravity force in streamline coordinates.

We can now integrate in the *n*-direction to get Bernoulli's equation.

$$\tfrac{1}{2}V^2 + p/\rho + gz = \text{constant} \tag{8.57}$$

In summary, to have Bernoulli's equation be valid across a streamline we have assumed the flow is

1. Inviscid
2. Steady-state
3. Constant ρ
4. Irrotational

Example 8.10

Use natural coordinates to discuss the nature of the vortex flow velocity fields and Bernoulli's equation for (a) irrotational flow; (b) rotational flow.

Solution

(a) Irrotationality implies that

$$\partial V_n/\partial s = \partial V/\partial n = -\partial V/\partial r = V/r$$

thus,

$$dV/V = -dr/r$$

and

$$\ln V = -\ln r + \ln C$$

or

$$\ln(Vr) = \ln C$$

hence,

$$Vr = C$$

We have seen that $V = C/r$ is the velocity distribution of the free vortex. Thus a free vortex is irrotational (except at $r = 0$).

Bernoulli's equation along the streamline is obtained from

$$\partial V/\partial t + V \, \partial V/\partial s = -(1/\rho) \, \partial p/\partial s - g_s \qquad (8.58)$$

For steady state and $g_s = g \, \partial z/\partial s$,

$$\partial/\partial s[V^2/2 + p/\rho + gz] = 0$$

Hence, $H = V^2/2 + p/\rho + gz$ is constant along a streamline.

For the n-component of δV we found that Eq. (8.57) yielded Bernoulli's relation for any two points in the domain of irrotational flow.

(b) In the case of a rotational field,

$$V = (d\theta/dt) \, r \qquad \text{or} \quad d\theta/dt = V/r$$

and

$$\partial V_n/\partial s = V/r$$

The vorticity is

$$\zeta = 2 \, d\theta/dt = 2V/r = \partial V_n/\partial s - \partial V/\partial n = V/r - \partial V/\partial n$$

Therefore,

$$-\partial V/\partial n = V/r = \partial V_n/\partial s$$

Substitute this derivation of the first term in the steady-state version of Eq. (8.40),

$$V \, \partial V_n/\partial s + (1/\rho) \, \partial p/\partial n + g \, \partial z/\partial n = 0$$

to get

$$-V \, \partial V/\partial n + (1/\rho) \, \partial p/\partial n + g \, \partial z/\partial n = 0$$

When this is integrated for constant ρ,

$$-V^2/2 + p/\rho + gz = \text{constant} \qquad (8.59)$$

Thus when considering a direction across streamlines in rotational flow, there is an important difference between this relation and the Bernoulli law along a streamline—the sign of the first term. This result can be used in cases of a forced vortex. For instance, the surface contour in a rotating tank full of liquid can now be found using Eq. (8.59). However, the main point here is to illustrate that incorrect answers result when Bernoulli's equation is used under the wrong conditions.

===

Example 8.11

In the atmosphere, an air mass frequently follows a curved path. Use Eq. (8.55) to obtain a relation for $V(r, f)$. Consider irrotational flow, uniform flow, and curved flow.

Solution

From Eq. (8.55),

$$\partial V_n/\partial t + V_s \, \partial V_n/\partial s + (1/\rho) \, \partial p/\partial n + g_n = 0$$

From geometry, $\partial V_n/\partial s = V/r$,

$$\partial V_n/\partial t + V^2/r + (1/\rho) \, \partial p/\partial n + g \, \partial z/\partial n + fV = 0$$

Under the irrotational condition, $\partial V_n/\partial s = \partial V/\partial n = V/r$,

$$V^2/r = V \, \partial V/\partial n$$

Thus,

$$V^2/r + (1/\rho) \, \partial p/\partial n + g \, \partial z/\partial n + fV = 0$$

and

$$V \, \partial V/\partial s + (1/\rho) \, \partial p/\partial s + g \, \delta z/\delta s = 0$$

For uniform, straight-line flow, $R \to \infty$,

$$(1/\rho) \, \partial p/\partial n = -fV$$
$$\partial p/\partial s = \rho V \, \partial V/\partial s$$

and for steady-state, $\partial p/\partial n$ only,

$$V_g \equiv -1/(\rho f) \, \partial p/\partial n$$

the geostrophic wind.

For a finite R,

$$(1/\rho) \, \partial p/\partial n = -fV - V^2/r; \qquad \partial p/\partial s = 0$$

and isobars are parallel to streamlines.

Finally, the equation for this case is

$$f(V - V_g) = -V^2/r$$

or

$$V^2/r + fV - fV_g = 0$$

Hence, solving for V,

$$V = [-f \pm (f^2 + 4fV_g/r)^{1/2}]/(2/r)$$

$$= (fr/2)[-1 \pm (1 + 4V_g/(rf))^{1/2}]$$

Only certain combinations of the sign of r and V_G correspond to physically realizable flows with V positive and real. This is the gradient wind, which depends on the pressure gradient, Coriolis force, and radius of curvature (centrifugal force).

Summary

DEFINITIONS

1. $\zeta \equiv \nabla \times \mathbf{u}$
2. *Solid-body-like rotation,* $\zeta \equiv 2\, d\theta/dt$
3. *Circulation:* $\Gamma \equiv \oint \mathbf{u} \cdot d\mathbf{S}$
4. *Vorticity filament*
5. *Vorticity tube*
6. *Line vortex*

ASSOCIATED VELOCITY

$$\mathbf{u} = \mathbf{u}^S + \mathbf{u}^a + \mathbf{V}$$

$$\nabla \cdot \mathbf{u}^S = D \qquad \nabla \cdot \mathbf{u}^a = 0$$

$$\nabla \times \mathbf{u}^S = 0 \qquad \nabla \times \mathbf{u}^a = \zeta$$

$$\mathbf{u}^S = \nabla\varphi \qquad u^A = \nabla \times \mathbf{B}$$

VORTICITY TRANSPORT EQUATION

$\nabla \times$ (momentum equation) + ($\nabla \cdot \mathbf{u} = 0$ and identities)

yields

$$D\zeta_i/Dt = \zeta_j\, \partial u_i/\partial x_j + \nu\, \partial^2 \zeta_i/\partial x_j\, \partial x_j$$

or

$$D\zeta/Dt = (\zeta \cdot \nabla) \cdot \mathbf{u} + \nu\, \nabla^2 \zeta$$

Total change in vorticity is from:

Stretching (increase dL)
Turning (changes components; zero in two-dimensions)
Diffusion by viscosity

Notes (for homogeneous or ρ a function of p only)

1. Viscosity is necessary to *create* or *destroy* vorticity.
2. Vorticity is present at *boundaries* where no-slip requires viscosity.
3. Vorticity *persists* in inviscid flow.
4. Vorticity is a *characteristic* of turbulent flow.

Problems

1. Consider the flow given by the velocity potential.

$$\varphi = x^2 - 2xy$$

Is this flow irrotational? ($u_i = \delta\phi/\delta x_i$)

2. Given: $\mathbf{u} = (2xy + z,\ x^2y + z,\ x + y + z)$, is this flow irrotational?

3. Consider the flow given by

$$u = (3x + y^2 + 2z)t,$$

$$v = (2xy - 2zy)t,$$

$$w = (x^2 - y^2 + 2y - z^2 - 2)t.$$

Is the flow rotational?

4. Consider a tornado with concentrated vorticity at the center. If you measure the wind velocity tangential to the circular flow (u_θ) at 2 km to be 20 m/sec, what is the circulation? what is the vortex strength? What is the wind speed at 500 m from the center with this approximation?

5. A uniform synoptic scale flow of westerlies encounters a mountainous region, and the flow appears to turn to the south. What equation would you use to analyze this?

6. In an inviscid fluid flow, how can the vorticity change? What about in an inviscid, two-dimensional flow?

7. Suppose a ship has a rotating cylinder of height H, radius R mounted vertically midship. This gives rise to a lateral force and has been suggested as a method of propulsion. You could calculate the force created by the flow if you knew the pressure distribution; the pressures if you knew the velocities near the surface (assuming potential flow). Given the ψ, calculate the force for $U = 10$ m/sec, $R = 3$ m, $H = 30$ m ($u = \delta\psi/\delta y$, $v = -\delta\psi/\delta x$).

$$\psi = Ur \sin\theta - R^2 U/r \sin\theta - RU \ln(r/R)$$

which represents a free-stream flow + a source + a line vortex at $r = 0$. This produces a rotating cylinder in the freestream.

8. If we think of turbulence as an ensemble of vortex elements in a homogeneous fluid, and suppose that the trend is to smaller and smaller turbulent elements, what is the overall trend in vorticity, neglecting viscosity?

9. Using a conservation statement for a vortex tube segment (Helmholz's law), say what happens to a cyclone as it moves from over the high Rocky Mountain plateau to over the low plains.

10. The Burgers' vortex is described in cylindrical coordinates with

$$u_r = -ar, \qquad u_z = az,$$

$$u_\theta = \Gamma/(2\pi r)[1 - \exp(-ar^2/2v)]$$

What is the vorticity field for this flow? What are each of the terms in the vorticity equation?

Chapter 9 | Potential Flow

9.1 Introduction

The Navier–Stokes equations, when they include the continuity, energy, and state equations, are a complete set. This set may total from four to six equations, depending on the importance of the temperature and density variations. However, in general, this set of equations is unsolvable by analytic means. Fortunately, in many important problems with practical import, thermal effects are negligible. In addition, we do not need to include all of the forces we allowed for when we developed the momentum equations. Thus, in specific circumstances, the flow equations become amenable to analytic solutions. The circumstances occur under conditions of incompressibility and irrotationality. The elegant mathematical framework that is then applicable yields solutions that apply in what is called *ideal flow*. Despite the restrictive assumptions, these solutions have a surprisingly wide range of practical applications. However, they do fail in many circumstances where the required conditions are not met.

By now, we have a good familiarity with the mathematical terms in the equations involving divergence, viscosity, and rotationality. Yet, without extensive practical experience, we cannot know whether actual flows behave as though they were nondivergent, inviscid, or irrotational. One way to gain an idea of the behavior of a particular flow problem is to obtain the equations under these approximations. Then it is possible to solve for the predicted flow and to test the solutions against observations. This has been done for the potential flow solutions. The elegant mathematical solutions presented in this chapter have been found to have far-reaching practical applications.

POTENTIAL FLOW ASSUMPTIONS

Incompressible Approximation

$$\nabla \cdot \mathbf{u} = 0$$

$$\nabla \times (\nabla \varphi) = 0$$

$$\rightarrow \mathbf{u} = \nabla \varphi$$

Irrotational Approximation and Inviscid Approximation

$$\nabla \times \mathbf{u} = 0$$

$$\nabla \cdot (\nabla \times \mathbf{B}) = 0$$

$$\rightarrow \mathbf{u} = \nabla \times \mathbf{B}$$

Two-Dimensional Flow

$$\partial u/\partial x + \partial v/\partial y = 0$$

$$u \equiv \partial \psi/\partial y, \quad v \equiv -\partial \psi/\partial x$$

Ideal Flow Theory (also often called Potential Flow)

$$\nabla^2 \varphi = 0 \qquad \nabla^2 \psi = 0$$

The following material outlines the potential flow approximation procedure based on the work of previous chapters. First, we will summarize the assumptions and the resulting governing equations. Then we will spend the remainder of the chapter exploring the solutions to these equations.

9.2 The Velocity Potential

We have discussed the existence of a potential function several times in this text. The velocity potential will be a function specified at each point in the field, from which the velocity field can be obtained by differentiation. The existence of the velocity scalar potential function is implied purely by the mathematical identity, curl grad $\equiv 0$. From the definition of irrotational flow, $\zeta \equiv$ curl $\mathbf{u} = 0$. Therefore in irrotational flow there will exist a scalar φ such that

$$\mathbf{u} = \text{grad } \varphi = (\partial\varphi/\partial x, \ \partial\varphi/\partial y, \ \partial\varphi/\partial z) \tag{9.1}$$

Here, φ is called the *scalar velocity potential*.

If the scalar potential field is known, then the velocity vector field is obtained at all points by differentiation. This allows us to combine the component velocity equations (in u, v, and w) into a single equation in the scalar φ. Now, the velocity potential always exists when there is no vorticity. So, due to the close connection between viscous effects and rotation in the fluid, an assumption that the flow is inviscid is in general equal to assuming irrotationality. When the flow is frictionless and irrotational, the velocity potential is a very useful device. This is especially true when we can also assume that the flow is incompressible, so that

$$\text{div } \mathbf{u} = \text{div grad } \varphi = \nabla^2\varphi = 0 \tag{9.2}$$

The velocity potential then satisfies Laplace's equation. Therefore, any harmonic function is a possible velocity potential. This category of velocity potentials encompasses a large body of mathematical solutions. The manipulation of these solutions is called *ideal flow theory,* or sometimes simply *potential flow theory.* Strictly, the distinction between potential flow and ideal flow is compressibility. A potential flow exists for compressible flow, even though the flow is not nondivergent. However, this category is not as important as the case with nondivergence. Hence, ideal flow is often referred to as potential flow.

Example 9.1

Calculate the velocity for the given φ. Is this a valid incompressible flow? Is it rotational?

$$\varphi = \frac{C}{2}(x_1^2 - x_2^2)$$

Solution

Given the velocity potential field $\varphi(x_1, x_2)$ we can calculate the velocity field $U(x_1, x_2)$. The velocity is grad φ,

$$U = (\partial\varphi/\partial x_1, \partial\varphi/\partial x_2, \partial\varphi/\partial x_3) = (C/2)(2x_1, -2x_2, 0)$$

Check this velocity field for incompressibility by calculating $\nabla \cdot \mathbf{u}$.

$$\partial u_1/\partial x_1 + \partial u_2/\partial x_2 + \partial u_3/\partial x_3 = C - C = 0$$

Thus, this φ distribution is a valid incompressible flow.

It is clear that $\nabla \times \mathbf{u} = 0$, since $\mathbf{u} = \nabla\varphi$ and $\nabla \times \nabla\varphi \equiv 0$. However, it is instructive to check. Since this is a two-dimensional flow, there is only one component of vorticity,

$$\zeta_3 = \partial u_1/\partial x_2 - \partial u_2/\partial x_1 = 0 - 0 = 0$$

Thus, $\nabla \times \zeta = 0$ and this $\varphi(x, y)$ represents an irrotational flow.

9.3 The Stream Function for Two-Dimensional Flow

There exists another potential function for determining the velocity, called the stream function. This is a function that also yields the velocity field under differentiation and which exists independent of φ. In Section 8.2, we found a definition for the existence of a vector potential $\mathbf{B}$ in a way similar to that of finding the velocity potential. The requirement for the existence of this function comes from the definition of incompressible flow, div $\mathbf{u} = 0$. Then, from the mathematical identity, div curl $\equiv 0$, there exists a vector $\mathbf{B}$ such that $\mathbf{u} = \nabla \times \mathbf{B}$. $\mathbf{B}$ is called the *vector potential*.

There is an advantage in being able to represent a velocity field in terms of a scalar field. It is not evident how the vector potential denotes any advantage over the velocity vector in representing the velocity field. However, when the flow is two-dimensional, the vector potential has only one component, the scalar ψ. This is the called the stream function. We have already discussed the streamlines from a physical point of view, in Section 1.8. The specification of the stream function ψ is equivalent to specifying the streamlines in the field. Since streamlines are a powerful graphical representation of the flow field, ψ is an important field variable in two-dimensional fluid mechanics. In two-dimensional flow we ignore the vector potential **B** and specify the single nonzero scalar component of $\mathbf{B} = (0, 0, \psi)$ as the stream function. It is common to reserve the term *potential* for φ.

Frequently, with no knowledge of **B**, the *stream function* is simply defined from the two-dimensional nondivergence condition (the incompressible continuity equation),

$$\operatorname{div} \mathbf{u} = \partial u/\partial x + \partial v/\partial y = 0$$

One can then define a function such that

$$\partial \psi/\partial x = -v, \quad \text{and} \quad \partial \psi/\partial y = u \tag{9.3}$$

This allows continuity to be identically satisfied,

$$\psi_{xy} - \psi_{yx} \equiv 0$$

When, in addition, the flow is irrotational, we have

$$\partial v/\partial x - \partial u/\partial y = 0$$

and therefore

$$\partial^2 \psi/\partial x^2 + \partial^2 \psi/\partial y^2 = \nabla^2 \psi = 0 \tag{9.4}$$

Equations (9.2) and (9.4) are forms of *Laplace's equation.* Any function that satisfies this equation can be used to define an incompressible, irrotational velocity field. This function can be used to specify either the velocity potential or the stream function. When either the velocity potential or the stream function are given, there will always be another solution of Laplace's equation that defines the concurrent stream function or velocity potential. In other words, if we have one set of ψ and φ lines, which represent a particular velocity field, we automatically have another velocity field by exchanging the designations of ψ and φ.

From the definition of ψ in Eq. (9.3), the constant ψ lines are everywhere tangent to the velocity. This is the definition of a *streamline;* hence the term *stream function.* From the definition of the scalar potential in Eq. (9.1), the velocity is in the direction of maximum change of the velocity potential (the gradient of φ). Thus the velocity vector is normal to constant φ lines. Since they are everywhere perpendicular to each other, constant ψ and φ lines form an orthogonal set. We can check this by looking at the slopes.

1. On lines of constant ψ,

$$d\psi = (\partial\psi/\partial x)dx + (\partial\psi/\partial y)\, dy = 0$$

and

$$dy/dx = \frac{-\partial\psi/\partial x}{\partial\psi/\partial y} = -(-v/u) = v/u \qquad (9.5)$$

2. On lines of constant ψ,

$$d\varphi = (\partial\varphi/\partial x)\, dx + (\partial\varphi/\partial y)\, dy = 0$$

and

$$dy/dx = \equiv \frac{\partial\varphi/\partial x}{\partial\varphi/\partial y} = -u/v \qquad (9.6)$$

The slopes of constant ψ lines are negative reciprocals of constant φ lines, defining their orthogonality.

Therefore, in two-dimensional irrotational flow, a reciprocal solution exists for all velocity potentials and stream functions. The network of lines of constant φ and ψ form the basis of a wide range of flow solutions called potential flows.

Example 9.2

Consider the velocity potential given by

$$\varphi = \frac{C}{2}(x_1^2 - x_2^2) \qquad (9.7)$$

Calculate the velocity and equation for the potential lines.

Solution

We can obtain the velocity directly from the potential function,

$$(u_1, u_2) = (\partial\varphi/\partial x_1, \, \partial\varphi/\partial x_2) = (Cx_1, \, -Cx_2)$$

There are two independent equations in ψ from the definition of the stream function,

$$u = Cx_1 = \partial\psi/\partial x_2$$

$$v = -Cx_2 = -\partial\psi/\partial x_1$$

Thus, on integration,

$$\psi = Cx_1x_2 + f_1(x_1) + A$$

and also,

$$\psi = Cx_1x_2 + f_2(x_2) + A$$

Equating these two expressions for ψ shows that $f_1(x_1) = f_2(x_2)$ plus a constant, and therefore both must equal zero. Thus,

$$\psi = Cx_1x_2 + A$$

In this case, the streamlines and potential lines are orthogonal sets of hyperbolas.

We can summarize this section in a few lines:

$$\nabla \times \mathbf{u} = 0 \qquad \text{implies} \qquad \nabla \times (\text{grad } \psi) \equiv 0$$

$$\rightarrow \mathbf{u} = \text{grad } \varphi, \tag{9.8}$$

and

$$\nabla \cdot \mathbf{u} = 0 \qquad \text{implies} \qquad \nabla \cdot (\text{curl } \mathbf{B}) \equiv 0$$

$$\rightarrow \mathbf{u} = \text{curl } \mathbf{B} \tag{9.9}$$

where φ is a scalar potential for the velocity and $\mathbf{B}$ is a vector potential. Now, when we retain for a moment the full vector potential, $\mathbf{B} = (\psi_1, \psi_2, \psi_3)$, and use the fact that $\nabla \cdot \mathbf{B} = 0$, we have

$$\mathbf{u} = (\partial\varphi/\partial x, \ \partial\varphi/\partial y, \ \partial\varphi/\partial z)$$

$$\mathbf{u} = (\partial\psi_3/\partial y - \partial\psi_2/\partial z, \ \partial\psi_1/\partial z - \partial\psi_3/\partial x, \ \partial\psi_2/\partial x - \partial\psi_1/\partial y) \tag{9.10}$$

We will use only the two-dimensional case, where $\psi_1 = \psi_2 = 0$ and $\psi_3 \equiv \psi$, so that

$$\mathbf{u} = (\partial\psi_3/\partial y, \ -\partial\psi_3/\partial x) \tag{9.11}$$

9.3.1   Uniform Flow

When there is a purely translational flow $\mathbf{V}$ such that it is uniform and constant, then this is a potential flow field since $\nabla \cdot \mathbf{V} = \nabla \times \mathbf{V} = 0$. The stream functions are straight, parallel lines in the flow direction, while the potential lines are perpendicular to the flow.

Example 9.3

Get the stream function and potential function for the uniform flow in the x_1 direction as shown in Fig. 9.1.

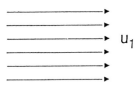

Figure 9.1 Uniform flow in the x_1 direction.

Solution

Start with the velocity,

$$(u_1, 0) = (\partial\psi/\partial x_2, -\partial\psi/\partial x_1)$$

Integrate to get

$$\psi = u_1 x_2 + C_1$$

The stream function lines are horizontal and parallel to the flow. The $x_2 = 0$ line can be designated as the $\psi = 0$ line.

The potential function can be obtained from its relation to u_1,

$$\partial\varphi/\partial x_1 = u_1$$

Hence,

$$\varphi = u_1 x_1 + C_2$$

The $\varphi = 0$ line can be chosen as the x_1 axis.

When the flow is at an angle θ to the coordinates, it is often easier to write these potentials in polar coordinates,

$$(\varphi, \psi) = (U_r \sin\theta, U_r \cos\theta)$$

Since $\partial u_1/\partial x_1 = \partial u_1/\partial x_2 \equiv 0$, the flow is nondivergent and irrotational.

9.3.2   Ideal Fluid Flow Vorticity Equation

When we have irrotational flow we can write a single scalar equation in ψ replacing the vector momentum equations. Then, if the flow is also non-divergent in two dimensions, we can use the definition of the stream function to write the two-dimensional vorticity transport equation in terms of ψ.

$$\partial \nabla^2\psi/\partial t + \partial\psi/\partial y \, \partial \nabla^2\psi/\partial x - \partial\psi/\partial x \, \partial \nabla^2\psi/\partial y = \nu \nabla^4\psi \qquad (9.12)$$

This is one equation in the scalar stream function ψ. It is fourth order, with viscous terms as highest order. When it is solved for ψ, the velocity and vorticity fields follow immediately by differentiation.

Finally, we note that a flow can satisfy continuity but not irrotationality. Thus, the stream function can exist for a rotational, viscous flow. Similarly, a velocity potential can exist for compressible flow which is irrotational but not nondivergent. However, these cases do not constitute ideal fluid flows.

9.4   Potential Motion with Circulation Compared to Rotational Flow in a Free Vortex

We have discussed rotation from several different perspectives. In potential flow the flow fields are irrotational; hence, the parcel at any point has no solid-body-like rotation. However, the rotating velocity field—the free vortex—is an important potential flow field. It is associated with a concentrated "point" of vorticity—the line vortex. The parameter connecting the vorticity to the finite flow field is the circulation. This section deals with these relations.

An irrotational flow can be defined in several equivalent ways:

1. With respect to vector analysis, where curl $\mathbf{u} = 0$.
2. In engineering fluid dynamics, as zero rotation of the fluid parcel.
3. In synoptic meteorology, as zero circulation per unit area.

A simple rectilinear *line vortex* (Fig. 9.2) can be used to illustrate the relation between these definitions. Consider the potential flow defined by $\varphi = C\theta$, a two-dimensional problem independent of z. The lines where θ is constant are rays from the origin of a constant angle. The stream function lines, which must be orthogonal to the potential lines, are circles.

The symmetry of this circular flow suggests that the equations will be simplest if written in cylindrical coordinates.

In cylindrical coordinates, we can obtain the velocity field from the given potential,

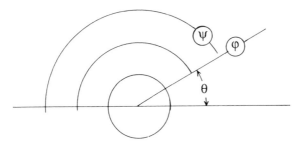

Figure 9.2   The potential representation of a two-dimensional circular vortex.

$$\mathbf{u} = \nabla\varphi = (\partial\varphi/\partial r,\ 1/r\ \partial\varphi/\partial\theta,\ \partial\varphi/\partial z)$$

$$= (0,\ C/r,\ 0)$$

$$= (u_r,\ u_\theta,\ u_z) \equiv (0,\ U,\ 0) \tag{9.13}$$

This is the velocity field in a free vortex. First, we need to check for potential flow validity.

$$\nabla\cdot\mathbf{u} = 1/r\ \partial(ru_r)/\partial r + 1/r\ \partial u_\theta/\partial\theta + \partial u_z/\partial z = 0$$

$$= \partial u_r/\partial r + u_r/r + 1/r\ \partial u_\theta/\partial\theta + \partial u_z/\partial z = 0$$

Then, substitute $\mathbf{u} = \nabla\varphi$ to get

$$\nabla^2\varphi = 1/r\ \partial/\partial r(r\ \partial\varphi/\partial r) + 1/r^2\ \partial^2\varphi/\partial\theta^2 + \partial^2\varphi/\partial z^2$$

$$= \partial^2\varphi/\partial r^2 + 1/r\ \partial\varphi/\partial r + 1/r^2\ \partial^2\varphi/\partial\theta^2 + \partial^2\varphi/\partial z^2$$

Since $\partial\varphi/\partial r = 0$, $\partial\varphi/\partial z = 0$, and $\partial\varphi/\partial\theta = C$, $\partial^2\varphi/\partial\theta^2 = 0$, this potential ($\varphi = C\theta$) satisfies the criterion for a potential function that $\nabla^2\varphi = 0$. The only exception is at the singular point $r = 0$, where u grows unbounded. The origin must be excluded from our defined potential field.

This potential is a multivalued function as $\varphi(0)$ over-lies $\varphi(0 + 2\pi)$. We can restrict φ to $0 \le \theta < 2\pi$, and as long as the velocity is continuous, we do not need to worry about the discontinuity in φ.

9.4.1   Singularities

Consider the areas in the domain where such a vortex exists at $(0,0)$, shown in Fig. 9.3. The flow field is irrotational everywhere (except at the singular point at the origin, where $r = 0$, u and $\zeta \le \infty$).

The circulation around any area is related to the vorticity inside the circuit by Stokes's law (Section 8.4.1). Thus the value of the circulation for the

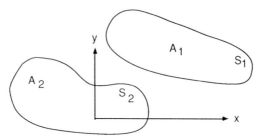

Figure 9.3   Two domains around a vortex at (0,0).

closed curves of Fig. 9.3 depends on whether or not the origin is inside the area. For circuit 1,

$$\Gamma_1 = \oint \mathbf{u} \cdot \mathbf{ds}_1 = \varphi(\theta_o) - \varphi(\theta_o) = 0$$

For any circuit containing the singularity at the origin,

$$\Gamma_2 = \oint \mathbf{u} \cdot \mathbf{ds}_2 = \varphi(\theta_o + 2\pi) - \varphi(\theta_o) = 2\pi C \qquad (9.14)$$

For any region that includes the origin, the circulation is contributed entirely by the point vorticity. The element of vorticity is constant, concentrated at $r = 0$.

Often, one treats a finite region of vorticity as a point compared to a much larger domain. In this case, the circulation around this region will depend on the contour integral and the area enclosed, so that

$$\Gamma = \int\int \zeta \, dA \qquad (9.15)$$

When the vorticity is constant in the region, $\Gamma = \zeta A$. The flow associated (or "produced") by the concentration of vorticity at $r = 0$ is irrotational everywhere except at $r \rightarrow 0$.

There is some looseness about the connection between the singularity and the velocity field surrounding it. Often, the velocity is said to have been produced by the singularity. This is true when we look at the physical nature of the singularities. In this case, the velocity is often forced by a concentrated source near the center of the free-vortex flow. Tornados and cyclones are examples. Here, the singularity is a physical manifestation of the two-dimensional flow approximation breaking down at the central point. The vertical flow near the center serves as a sink or source of mass, where $\nabla \cdot \mathbf{u}$

$\neq 0$. For instance, in a laboratory rotating tank experiment, a drain at the center furnishes the singularity.

For a cyclone, the vertical flow in the eye of the storm produces the singular value of $\nabla \cdot \mathbf{u} = D$ at the central point. The vorticity is the inevitable result of imperfect alinement of forces in the initial flow and the ensuing balance between centrifugal and pressure gradient forces. In the atmosphere, the planetary vorticity can furnish the small perturbation in the initial flow that will establish the vorticity flow direction. However, from the perspective of the two-dimensional flow definitions, a perturbation in the mean velocity field may be the initiator of the flow which then results in the singularity. We shall simply refer to the velocity field as that which is associated with the point singularity.

9.4.2 Rotation and the Vortex

Rotation has been defined as the rotating portion of the complete deformation of the fluid, given by $\partial u_i / \partial x_j$. The rotation information is found to be contained in the antisymmetric tensor, rot $\mathbf{u}$, while the vorticity is the vector associated with the three independent components of rot $\mathbf{u}$. The general mathematical relation between the area rotation (vorticity) and the line integral of the velocity around the area (the circulation), was expressed through Stokes' theorem. This relates the line integral of a vector around a perimeter S to the integral over the contained area of the normal component of the curl of the vector. We can therefore write

$$\iint \text{rot } \mathbf{u} \, dA = \iint \text{curl } \mathbf{u} \cdot \mathbf{n} \, dA = \iint \zeta \cdot \mathbf{dA} = \oint \mathbf{u} \cdot \mathbf{ds} = \Gamma \quad (9.16)$$

The free vortex is associated with the flow around a line vortex. The vorticity and rotation are zero for all elements that do not include the origin. The circulation around any areas that do not contain any vorticity is also zero. We will look closer at the meaning of irrotationality for the parcel that is rotating around the vortex center, since the flow field around a line vortex is a practical approximation to some real atmospheric flows.

Although we envision vorticity or rotation as the rigid-body rotation of the fluid element, we must be aware that even at the continuum scale the parcel has deformation. The fluid parcel cannot support shearing stress, hence cannot be rigid. Our continuum approximation, "small enough that properties are uniform across the parcel," will not apply to the angular rotation, which depends on the orientation of each rotating line. We can check the effect of rotation on a parcel in the flow by examining the distortion of individual elements (e.g., the sides of the parcel) as they rotate around the vortex center. The rotation depends on the component velocity gradients that will be the point values for the velocity field as the parcel size approaches zero.

Consider elements marked by axes **r**, θ, as in Fig. 9.4.

In the rigid-body-like part of the rotation, both line elements rotate counterclockwise. Since $u_\theta = C/r$, and $\Gamma = 2\pi C$ for line element A of Fig. 9.4, we have

$$\delta\theta = (u_\theta/r)\,\delta t = (C/r^2)\,\delta t = \Gamma/(2\pi r^2)\,\delta t \qquad (9.17)$$

However, the finite dimension of the B line element aligned along **r** means that there exists an increment of velocity between the inner and outer ends. This will turn the line element clockwise through an angle

$$\delta\theta' \approx \tan\delta\theta' = (\partial u_\theta/\partial r)\,\delta r\,\delta t/\delta r$$

$$= -\Gamma/(2\pi r^2)\,\delta t \qquad (9.18)$$

This equation gives the part of the rotation due to the relative motion discussed in connection with the parcel deformation in Section 4.2. It was also discussed in relation to the definition of vorticity in Section 8.1. As long as we define the rotation of a parcel with sides A and B as the average of the two orthogonal vector rotations, it is $\frac{1}{2}(\delta\theta + \delta\theta') = 0$. The parcel is distorting but has zero vorticity.

We have previously noted that there is always a coordinate system such that the relative motion consists of pure distortion plus solid-body rotation. We can sometimes locate the principal axes by considering two sets of perpendicular lines in the parcel moving in a line vortex field. The cross in Fig. 9.5 is found to have zero net rotation as it rotates around the z axis. The solid lines show distortion due to the different rotation of the lines, which were arbitrarily oriented in the x and y directions. However, these lines rotate an equal amount in opposite directions, so that by our definition of rotation rate, the average rotation of two perpendicular lines, rotation =

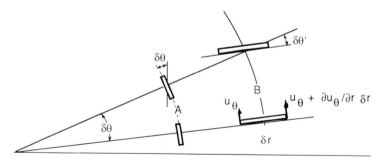

Figure 9.4   Elements of a rotating parcel in a free vortex.

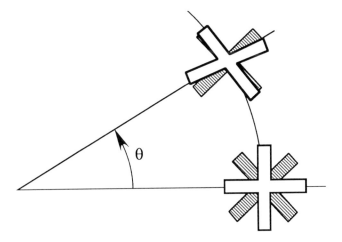

Figure 9.5 Perpendicular line elements in a parcel rotating about the z-axis in free vortex flow.

0. If we had chosen the axes which were at 45° from the solid lines originally, the two effects would be balanced for both axes. Thus, the crosshatched lines remain perpendicular and do not rotate. These are the principal axes. This illustrates why we can say that the vorticity represents the rotation rate of the principal axes, with no need for averaging.

9.5 Potential Flow (of an Ideal Fluid)

The assumptions of an incompressible, irrotational, inviscid flow reduce the equations of motion to Laplace's equation in the potential function. Therefore, any harmonic function represents a possible potential flow. Some of these mathematical solutions may have practical applications if their flow fields approximate an actual flow. We will discuss an example to illustrate this point.

Consider the velocity potential (discussed in Example 9.2),

$$\varphi = \frac{C}{2}(x^2 - y^2) \qquad (9.19)$$

If we assume ideal flow, then all of the information necessary to construct the flow field is contained in Eq. (9.19).

The constant φ lines are first plotted, as in Fig. 9.6. Then the velocity can be calculated from

$$\mathbf{u} = \nabla\varphi = C(x, -y) \qquad (9.20)$$

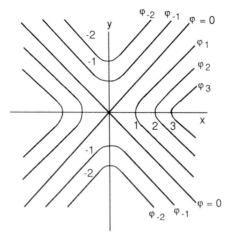

Figure 9.6   Velocity potential for $\varphi = \frac{1}{2}C(x^2 - y^2)$.

and plotted as in Fig. 9.7. The velocity vectors are perpendicular to the constant potential lines.

Now we should check to see if the flow field corresponding to this potential field represents an ideal flow.

The check for incompressible flow is

$$\boldsymbol{\nabla} \cdot \mathbf{u} = \partial u/\partial x + \partial v/\partial y = C - C = 0$$

The check for irrotationality is

$$\boldsymbol{\nabla} \times \mathbf{u} = \zeta^z = \partial u/\partial y - \partial v/\partial x = 0 - 0 = 0$$

(This is a formality, since $\mathbf{u} = \boldsymbol{\nabla}\varphi$ guarantees $\boldsymbol{\nabla} \times \mathbf{u} = 0$.)

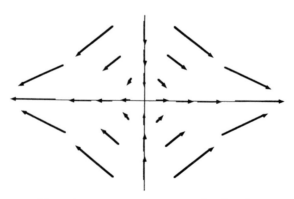

Figure 9.7   Velocity field for $\varphi = \frac{1}{2}C(x^2 - y^2)$.

We can also check to see that Laplace's equation is satisfied.

$$\nabla^2 \varphi = \partial^2 \varphi / \partial x^2 + \partial^2 \varphi / \partial y^2 = C - C = 0: \qquad \text{Harmonic } \varphi$$

(Again, a formality, since $\nabla \cdot \mathbf{u}$ and $\nabla \times \mathbf{u} = 0$ guarantee $\nabla^2 \varphi = 0$.)

We can now obtain equations for the stream function. Use the definition of the stream function as

$$(u, v) = (\partial \psi / \partial y, -\partial \psi / \partial x) = C(x, -y); \qquad \nabla^2 \psi = 0$$

Thus,

$$\partial \psi / \partial y = Cx$$

can be integrated to obtain

$$\psi = Cxy + A + f_1(x) \tag{9.21}$$

and

$$\partial \psi / \partial x = Cy$$

can be integrated to obtain

$$\psi = Cxy + A' + f_2(y) \tag{9.22}$$

Subtracting Eq. (9.22) from Eq. (9.21), we get

$$f_1(x) - f_2(y) + 0 = 0; \qquad \text{hence}, f_1(x) = f_2(y) = 0$$

Note that the constants of integration are arbitrary and we can set

$$\psi(0, 0) = A = A' = 0$$

We therefore have $\psi = Cxy$ and dy/dx ($\psi = $ constant) $= -y/x$.

It is apparent from the figures and discussion that the constant ψ lines are normal to the constant φ lines. This could be shown formally by checking that the slopes of each are negative reciprocals, as was done in Section 9.3. Also note that the values at $x, y = 0, 0$ are $u, v = 0, 0$. This is a stagnation point of the flow.

The stream functions of Fig. 9.8 are rectangular hyperbolas, as are the constant-potential function lines of Fig. 9.6. Flow is parallel to the constant-stream function lines and normal to constant-potential lines at all points.

We can visualize general real-flow situations that may be approximated by the flow represented in Fig. 9.8. Since by definition, there is no flow across a streamline—a constant ψ line—one can substitute a solid surface for any constant ψ line.

Since the streamline value is arbitrary, A can be set such that $\psi = 0$ is a significant streamline. In other words, the coordinate axes x_1 and x_2 can

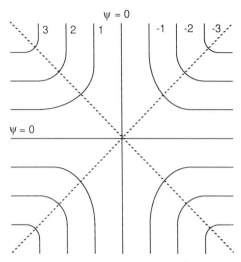

Figure 9.8   Stream functions for $\varphi = \frac{1}{2}C(x^2 - y^2)$.

be represented by $A = 0$. Thus the flow in any quadrant models the flow in a corner (see Fig. 9.9.)

If a line such as $\psi = -1$ is considered solid, the flow might be that through a turning vane, or the cup of an anemometer. For $\psi \neq 0$ the flow approximates that over a family of curves. It might model the flow in a particular shape of valley or an upslope wind, as in Fig. 9.10.

When the x_2 axis is taken as a solid surface, we have a model of a jet impinging on a wall, as in Fig. 9.11.

Finally, the entire flow can model the flow of two opposing jets directed together at the origin, as in Fig. 9.12.

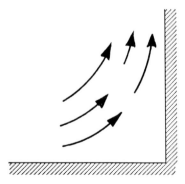

Figure 9.9   Model of flow in a corner.

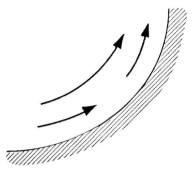

Figure 9.10 Model of flow in a curve.

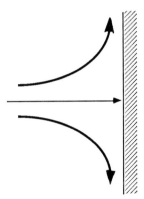

Figure 9.11 Model of flow into a solid surface.

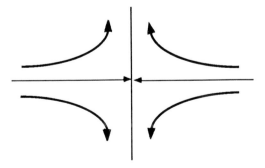

Figure 9.12 Model of flow of two jets opposing.

The procedure of developing flows from potential flow solutions is often called the *method of singularities*.

9.6 The Method of Singularities

Frequently the entire domain of flow in a problem involves regions wherein different forces govern the flow. The most frequent example is the flow above a surface. Near the surface, we have stated that the viscous forces must be accounted for in a region called the boundary layer. However, sufficiently far from the surface, the viscous forces may be small compared to the other forces determining the flow. Although the inside flow (in the boundary layer) may be nondivergent, it is not likely to be irrotational, since flows where viscosity plays an important role are also rotational. However, the outside flow can quite possibly be irrotational in addition to nondivergent. Thus we may have a potential-flow solution for the region outside the influence of the boundary and a boundary-layer solution next to the surface. The two solutions may touch, overlap, or require another solution for a region in between.

When the problem involves a complex domain wherein two or more governing sets of approximate equations are appropriate, the procedure for obtaining solutions is

1. Isolate the regions where the different regime flow is likely to exist (hopefully, but not necessarily, inviscid, nondivergent, and irrotational). Approximate the equations appropriate to each region.
2. Solve the simplified equations in each domain.
3. Attempt to match the solutions at the boundaries.

This is not always going to work. The separate regions may not be evident, the simplified equations may not be solvable, and step three can depend critically on steps one and two. However, in the case of plane potential flow, where the basic flow is uniform except for singular points, the steps are very easy. The regions of step one are points where mass is either created or destroyed, and/or the vorticity is infinite. The combinations of a very few singularities provide a wide variety of solutions. These include source/sinks and vortices. When we add these singularities to a uniform field potential, all potential flow solutions with practical applications can be generated. The equations governing the flow are Laplace's, $\nabla^2 \varphi = 0$, $\nabla^2 \psi = 0$. They are linear, and superposition of solutions is allowed. The domain of the singularity is simply a point, which is removed or hidden from the applied domain of the flow solution.

Thus, besides the uniform flow (which is two-dimensional, parallel flow), there are only two basic singularities to learn. They can be thought of as lines normal to the plane of flow.

9.6.1   The Line Source/Sink

Since our basic potential field is nondivergent, the only allowed occurrence of a source or sink is as a singular point in the flow field. Let the singularity be at (x, y) or $(r,\theta) = (0, 0)$. The cylindrical symmetry of the flow about point singularities in a two-dimensional field suggests the advantageous use of cylindrical coordinates. The flow emerging from a source at the origin is shown in Fig. 9.13.

The flow rate is given by

$$Q = \oint u_r ds = \int_0^{2\pi} u_r r\, d\theta \equiv \beta \tag{9.23}$$

where β is the source strength,

$$\beta = 2\pi r u_r, \qquad u_r = \beta/(2\pi r), \qquad u_\theta = 0 \tag{9.24}$$

This velocity field can be checked for nondivergence,

$$\nabla \cdot \mathbf{u} = (1/r)\, \partial(ru_r)/\partial r + (1/r)\, \partial u_\theta/\partial_\theta = (1/r)\, \partial(\beta/2\pi)/\partial r = 0$$

except at $r = 0$, which is the location of the singularity. Checking this velocity field for irrotationality, we find

$$\nabla \times \mathbf{u} = \zeta^z = (1/r)[\partial(ru_\theta)/\partial r] - (1/r)\, \partial u_r/\partial\theta = 0$$

except perhaps when $r = 0$.

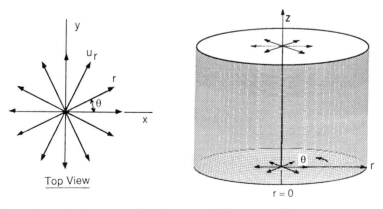

Figure 9.13   The line source. Velocity u_r is radially outward. There is no change along the z-direction.

We can obtain the velocity potential and the stream function for the source by integration of the equations for the velocity in terms of these functions:

$$u_r = \partial\varphi/\partial r = (1/r)\,\partial\psi/\partial\theta = \beta/(2\pi r) \tag{9.25}$$

Thus

$$\varphi = (\beta/2\pi)\ln r + \text{constant} \tag{9.26}$$

and

$$\psi = (\beta/2\pi)\theta + \text{constant} \tag{9.27}$$

The constants can be set to zero. A source will have positive β, and the case of a sink is obtained when the strength is negative. The sign of the vortex is positive for right-hand-rule rotation.

9.6.2   The Line Vortex

The line vortex is a vorticity tube with radius approaching zero. The tube is assumed to be normal to the plane of flow—in the z-direction of the (x, y) or (r, θ) plane, as shown in Fig. 9.14.

We have the circulation from the line integral around any circle centered at $(r, \theta) = (0, 0)$,

$$\Gamma \equiv \oint u_\theta\,ds = u_\theta\,2\pi r \tag{9.28}$$

where

$$u_\theta = \Gamma/(2\pi r) = (1/r)\,\partial\varphi/\partial\theta = -\partial\psi/\partial r$$

This can be integrated to yield φ and ψ.

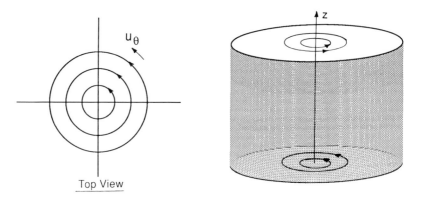

Figure 9.14   The line vortex. Velocity u_θ is tangential with no change in the z-direction.

We could also obtain φ and ψ from our results in Section 9.6.1. The constant φ and ψ lines in the solution for a source are a set of orthogonal lines, made up of circles and rays emanating from $(0, 0)$. One can interchange φ and ψ to get another set of orthogonal φ and ψ,

$$\varphi = (\Gamma/2\pi)\,\theta$$

and

$$\psi = -(\Gamma/2\pi)\ln r \qquad (9.29)$$

These expressions are the same as would be obtained from integration of the field around a line vortex. In this case, the streamlines are circles at constant r and the equipotential lines are rays. We can now check for ideal flow by evaluating the divergence and curl of $\mathbf{u}$.

$$\nabla \cdot \mathbf{u} = 0 \quad \text{and} \quad \nabla \times \mathbf{u} = 0, \quad (\text{except at } r = 0)$$

It is also valuable to note that for any closed line segment including the origin,

$$\Gamma = \oint \mathbf{u} \cdot \mathbf{ds} = \iint \nabla \times \mathbf{u}\, dA$$

and

$$\beta = \oint \mathbf{u} \cdot \mathbf{n}\, ds = \iint \nabla \cdot \mathbf{u}\, dA \qquad (9.30)$$

The circulation and source strength are not zero due to the vorticity and a source at $r = 0$.

9.6.3 Superposition of Singularities

9.6.3.a Source and Vortex

We can create potentials representing different flows by adding singularities and simple parallel flows (with corresponding simple potentials). For the singularities, we will find that the source and the vortex are all that are needed to create many practical flows (each may be plus or minus). These singularities can each stand alone (in static fluid), have a simple parallel flow field superimposed (a potential flow), or be used in superposition of singularities. We will find that simple combinations of sinks/sources, vortices, and a uniform flow yield many flow patterns that are suggestive of actual geophysical flows. One case that immediately arises is the superposition of a sink and a vortex.

We create the velocity potential field by simply adding the potentials for a sink and a vortex,

$$\varphi = \varphi_{sink} + \varphi_{vortex} = -(\beta/2\pi) \ln r + (\Gamma/2\pi)\theta \qquad (9.31)$$

$$\psi = \psi_S + \psi_V = -(\beta/2\pi)\theta - (\Gamma/2\pi) \ln r \qquad (9.32)$$

The resulting flow pattern is shown in the stream function plot of Fig. 9.15. The velocity making up the pattern shown in Fig. 9.15 is

$$\mathbf{u} = [\partial\varphi/\partial r, \partial\varphi/(r\partial\theta)] = [-\beta/(2\pi r), \Gamma/(2\pi r)] \qquad (9.33)$$

The converging vortex is a model of the inward spiraling flow within a cyclone or a whirlpool. The singularity at the center, where fluid would be piling up in two dimensions, is accounted for in the atmosphere by a small region of upward flow. Here, the two-dimensional plane flow assumption is no longer valid. This effect is clear from looking at the streamlines.

One can calculate the equation for a streamline from Eq. (9.33).

$$r_{streamline} = \exp\{2\pi\psi/\Gamma\}\exp\{-\beta\theta/\Gamma\} \qquad (9.34)$$

The singular nature of the origin is evident from the streamline pattern shown in Fig. 9.15.

It is interesting that this simple combination of singularities gives such a successful model for flows from spiral galaxies to the bathtub drain. One might expect the sink alone to model the latter flow, for instance. However, observations indicate that the sink plus vortex is the stable flow pattern. Evidently the smallest perturbation of the flow into a sink is sufficient to produce the vortex component. This results in the strong attractor solution for this flow situation. For the bathtub drain, this might be a plus or minus rotation. For the atmosphere, the Coriolis force usually provides the determining direction.

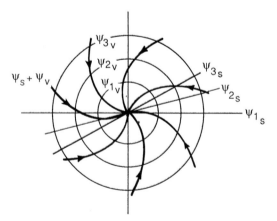

Figure 9.15   The stream functions for a sink (ψ_S) plus a vortex (ψ_v).

Example 9.4

Consider modeling the flow outside the core of a tornado using a free vortex. The velocity is determined to be tangential at 30 m/sec at a distance of 500 m from the center of the tornado. What is the circulation of the tornado? What can you say about the velocity field?

Solution

For the free vortex,

$$u_\theta = C/r$$
$$C = u_\theta \cdot r = 30 \cdot 500 = 15{,}000 \text{ m}^2/\text{sec}$$

and the circulation is

$$\Gamma = \oint u_\theta \cdot dS = 30 \cdot 2\pi \cdot 500 = 94{,}250 \text{ m}^2/\text{sec}$$

We can write

$$u_\theta = 15{,}000/r$$

for the velocity in the free vortex. However, as $r \to 0$, this is not a good model. In fact, we know that across the center of the tornado, the velocity must switch directions. So we might expect $u = 0$ at the center. This will require some additional concept to complete the field.

The vortex provides an excellent model for many observed flows from hurricanes to turbulent elements. It is therefore worthwhile to look further at potential flows that involve the vortex, and at the simplest of interactions between more than one vortex.

9.6.3.b   A Source in a Uniform Flow

If we superimpose a source and a uniform flow we obtain streamlines as sketched in Fig. 9.16. Note that one streamline, called the stagnation streamline, splits into two as it encounters the effects of the source.

The *uniform flow* can be represented by potential and stream functions that can be written (with flow at an angle α to the x–y coordinates).

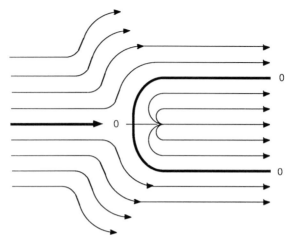

Figure 9.16 Stream functions for a source in a uniform flow. The $\psi = 0$ line bifurcates to form the outline of a blunt body.

$$\varphi = U_\infty(x \cos \alpha + y \sin \alpha) = U_\infty r \cos \theta$$

$$\psi = U_\infty(y \cos \alpha - x \sin \alpha) = U_\infty r \sin \theta \qquad (9.35)$$

In Fig. 9.16, $\alpha = 0$ and the total stream function and potential function are

$$\psi = \psi_{\text{source}} + \psi_{\text{uniform}} = (\beta/2\pi)\,\theta + U_\infty r \sin \theta$$

$$= (\beta/2\pi) \tan^{-1}(y/x) + U_\infty y \qquad (9.36)$$

$$\varphi = \varphi_{\text{source}} + \varphi_{\text{uniform}} = (\beta/2\pi) \ln r + U_\infty r \cos \theta$$

$$= (\beta/4\pi) \ln(x^2 + y^2) + U_\infty x \qquad (9.37)$$

By symmetry, there is a stagnation point at the point where the flow divides to pass around the source. This is the point where the outward velocity from the source equals the freestream velocity. The velocity is

$$u_r = (1/r)\,\partial\psi/\partial\theta = \beta/(2\pi r) + U_\infty \cos \theta$$
$$u_\theta = -U_\infty \sin \theta \qquad (9.38)$$

This velocity is zero on the $\theta = \pi$ line at $r = \beta/(2\pi U_\infty)$. At this point the stagnation point divides (it cannot end). Therefore, setting $\psi = \psi_0$ at the stagnation point,

$$\psi_0 = \beta/2$$

The ψ_o streamline is given by

$$\psi_o = \beta/2 = \beta/2\pi\theta + U_\infty r \sin \theta \qquad (9.39)$$

and

$$r = \beta(\pi - \theta)/(2\pi U_\infty \sin \theta) \qquad (9.40)$$

9.6.3.c   Flow over a Cliff

Since any streamline can be replaced by a solid body, the top half of the flow over ψ_o looks like the flow over a bluff body. It has been used as a model for the sea-breeze flow encountering a cliff at the water's edge.

The velocity at any point (r, θ) can be obtained from Eq. (9.25),

$$U^2 = u_r^2 + u_\theta^2 = (1/r \, \partial\psi/\partial\theta)^2 + (\partial\psi/\partial r)^2$$

$$= U_\infty^2[1 + r_o^2/r^2 + 2r_o/r \cos \theta] \qquad (9.41)$$

where $r_o \equiv \beta/(2\pi U_\infty)$.

It is possible to smooth the model of the cliff somewhat by adjusting the parameter r_o, which depends on the strength of the source.

9.6.3.d   Source Plus Sink to Produce the Doublet

A useful new singularity called a *doublet* can be generated by placing a source and a sink very close together, at $x = \pm\epsilon$, as shown in Fig. 9.17.

Here we can write the stream function as

$$\psi = \psi_{SO} + \psi_{SI}$$
$$= (\beta/2\pi)[\tan^{-1} y/(x + \epsilon) - \tan^{-1} y/(x + \epsilon)] \qquad (9.42)$$

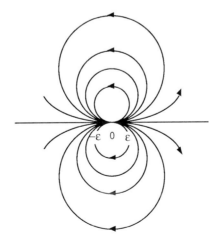

Figure 9.17   Stream functions for a closely spaced source plus sink at $x = \pm\epsilon$.

A *doublet* is obtained by letting $\epsilon \to 0$ *while keeping* $\beta\epsilon$ *constant* $\equiv Q_o$, the *strength of the doublet*. Hence this singularity is a superposition of an infinite source and infinite sink. When the terms in Eq. (9.31) are expanded about the origin and the limit is taken, $\epsilon \to 0$ while $\beta\epsilon = Q_o$, we get

$$\psi = -(Q_o/2\pi)[y/(x^2 + y^2)] = -(Q_o/2\pi)\sin\theta/r \qquad (9.43)$$

$$\varphi = -(Q_o/2\pi)[x/(x^2 + y^2)] = -(Q_o/2\pi)\cos\theta/r \qquad (9.44)$$

The valuable use of this singularity arises when it is placed in a uniform flow, which results in the flow pattern around a cylinder, shown in Fig. 9.18.

The stream function here is

$$\psi = U_\infty y - (Q_o/2\pi)[y/(x^2 + y^2)] \qquad (9.45)$$

The streamline $\psi = 0$ is

$$0 = U_\infty y - (Q_o/2\pi)[y/r^2]$$

or

$$r = [Q_o/(2\pi U_\infty)]^{1/2} = R, \qquad \text{a constant}$$

This is a circle. The flow inside the circle is from source to sink with the singularity contained within. However, the flow outside accurately represents the flow past a cylinder. Note that when $y = 0$, $\psi = 0$ for $-\infty \le r \le \infty$, so that the zero streamline divides at the leading and trailing edge of the cylinder. The flow solution also represents the flow over a half-circular cylinder sitting on the ground (the airflow over a Quonset hut, for instance).

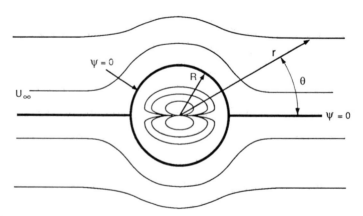

Figure 9.18   Flow of a uniform stream past a doublet. The $\psi = 0$ streamline bifurcates and rejoins to form a circle.

Example 9.5

Calculate the pressure distribution around a cylinder in a freestream flow, U_∞.

Solution

From the stream function for a doublet plus a uniform flow [Eqs. (9.34) and (9.42)],

$$\psi = U_\infty r \sin\theta - (Q_0/2\pi r) \sin\theta$$

and the velocity is

$$(u_r, u_\theta) = [(1/r)\, \partial\psi/\partial\theta, \ -\partial\psi/\partial r]$$

$$= [U_\infty \cos\theta - Q_0/(2\pi r^2) \cos\theta,$$

$$-U_\infty \sin\theta - Q_0/(2\pi r^2) \sin\theta]$$

$$= \{U_\infty \cos\theta[1 - Q_0/(2\pi U_\infty r^2)],$$

$$-U_\infty \sin\theta[1 + Q_0/(2\pi U_\infty r^2)]\}$$

Since

$$Q_0/(2\pi U_\infty) = R$$

the velocities on the surface of the cylinder are

$$[0, \ -2U_\infty \sin\theta]$$

There is a maximum velocity, clockwise, at $\theta = \pi/2$ equal to $-2U_\infty$.

The pressure distribution is determined by Bernoulli's equation for this inviscid, irrotational flow,

$$p + \rho u_\theta^2/2 = p_\infty + \rho U_\infty^2/2$$

If this were the flow over a building, we could assume that the p_∞ was approximately atmospheric, the U_∞ was given, and the pressure depended on θ.

If a vortex was superimposed at the cylinder center, the flow would be similar except the stagnation point would be moved toward the top or bottom (depending on the sign of the vortex). This would have the effect of causing a net force in the lateral direction. This can be seen from Bernoulli's law and from the fact that the velocities are now decreased on the stagnation-point side. Finally, in aerodynamics, the cylinder can be transformed into an airfoil shape, along with the streamlines, velocities, and pressure distribution.

9.7   The Idealized Vortex

When measurements are taken through a cyclone or a hurricane, the observed velocity field is as shown in Fig. 9.19.

This velocity distribution could be approximated with ideal flow theory using the measured values of velocity at different points to calculate values for the circulation and angular rotation. The velocity field at $r > r_c$ could then be determined from Eq. (9.33). However, the core region is not insignificant, and we would like to have a model that yields winds for $r < r_c$.

The irrotationality of the flow in the field surrounding a vortex core suggests the idea of separating the flow field into a potential flow and a local idealized vortex singularity that produces the rotational effects in the rest of the field. The processes that create divergence and/or rotation are swept

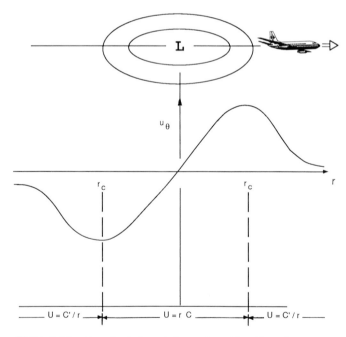

Figure 9.19   Aircraft windspeed observations taken through a vortex.

into the singularity and simply parametrized with a "strength" of the singularity. Thus, one can replace a vortex with an approximation consisting of an inner vortex core with constant rotation plus a potential flow outside of the core associated with the vortex strength, Γ. The value of Γ can be determined by the strength of the inner core rotation. It will be

$$\oint \mathbf{u} \cdot \mathbf{ds} = 2\pi r u_\theta \tag{9.46}$$

at $r \geq r_c$, the radius of $u_\theta \equiv U$ is maximum. Beyond that point, the velocity will decrease in inverse proportion to the radius in the manner of a free vortex.

Inside the core is the forced vortex. In the atmosphere this will be a region where three-dimensional effects are important and the flow is not irrotational. The velocity of a forced vortex is

$$u_\theta \equiv u = r \, d\theta/dt \tag{9.47}$$

Outside the core, there is a region that is two-dimensional, irrotational, and nondivergent. This is a potential flow region where

$$u = \Gamma/(2\pi r) \tag{9.48}$$

At the radius of the core, $r = r_c$, and we have

$$\Gamma = 2\pi r_c U = 2\pi r_c^2 \, d\theta/dt = 2A_c \, d\theta/dt = \zeta A_c$$

Therefore, outside the core,

$$u = (r_c/r)U = d\theta/dt \, (r_c^2/r) = \zeta r_c^2/(2r) \tag{9.49}$$

The rotation rate and the velocity are shown in Figure 9.20. The actual rotation rate and velocity observed in a hurricane are also indicated.

Example 9.6

Calculate and plot the streamlines, circulation, velocity, vorticity, and pressure variation for a tornado modeled as an ideal vortex. Assume the maximum winds are known to be 80 m/sec at 100 m from the center.

Solution

We have seen that the free vortex models the outer flow with

$$u_\theta = C/r$$

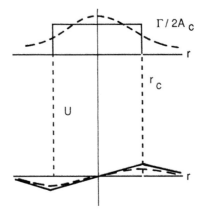

Figure 9.20   Rotation and velocity in an ideal vortex (———) and that generally observed (———).

We now have another solution, the forced vortex for the flow inside the core of the tornado.

$$u_\theta = C_2 r = \omega r$$

We use the forced vortex out to the point of maximum velocity. The free vortex must then match velocity at the radius of the core, R.

$$C/R = \omega R$$

Hence,

$$C = \omega R^2$$

The velocity inside the core is

$$u_\theta = \omega r = (80/100)r = 0.80\,r \text{ m/sec}$$

Outside the core,

$$u_\theta = \omega R^2/r = 0.8 \cdot 100^2/r = 8000/r \text{ m/sec}$$

The circulation in the free vortex is

$$\Gamma = \oint u_\theta \cdot dS = 80 \cdot 2\pi \cdot 100 = 50{,}265 \text{ m}^2/\text{sec}$$

The circulation in the forced vortex is then

$$\Gamma = 2\pi u_\theta = 2\pi\omega r^2$$

It increases as r^2 to the maximum at $2\pi\omega R^2$ and is constant for $r \geq R$.

The free vortex is irrotational, and $\zeta = 0$ for $r \geq R$. The vorticity inside the core is

$$\zeta = \zeta_z = \partial u_\theta / \partial r + (u_\theta / r) = \partial (\omega r) / \partial r + \omega r / r = 2\omega$$

This is the constant vorticity of the forced vortex.

The velocity potential for the free vortex is

$$\varphi = (\Gamma / 2\pi)\, \theta$$

These are radial lines emanating from the origin in the core. However, they do not apply in the forced vortex region, since this is a rotational domain and no velocity potential is defined.

The flow can be approximately incompressible in both regions, and the streamfunctions can be found from u_θ. Inside,

$$\partial \psi / \partial r = u_\theta = \omega r$$

Integrating yields

$$\psi = \tfrac{1}{2}\omega R^2 + C_1$$

Outside the core, we have found that

$$\psi = (\Gamma / 2\pi) \ln[r/R] + C_2$$

If we set the $r = R$ line as the zero stream function, then both constants are zero. Hence,

$$\psi = \begin{cases} \tfrac{1}{2}\omega(r^2 - R^2), & 0 \leq r \leq R \\ (\Gamma / 2\pi) \ln[r/R], & R \leq r \leq \infty \end{cases}$$

The pressure field can be calculated from Bernoulli's relation in the irrotational outer region.

$$p/\rho + u_\theta^2/2 = p_o/\rho = \text{constant}$$

Hence,

$$p = p_o - \rho u_\theta^2 / 2 = p_o - \rho \omega^2 R^4 / 2r^2$$

Inside the core, we must use the simplified equations of motion. Here, the pressure-gradient force balances the centrifugal force for inviscid flow. By symmetry, p does not depend on θ. Hence,

$$\partial p / \partial r = \rho u_\theta^2 / r = \rho \omega^2 r$$

can be integrated to get

$$p = \rho \omega^2 r^2 + C_3$$

The pressures must be continuous at $r = R$, and

$$\tfrac{1}{2}\rho\omega^2 r^2 + C_3 = p_o - \tfrac{1}{2}\rho\omega^2 R^4/r^2$$

Hence,

$$C_3 = p_o - \rho\omega^2 R^2$$

We can write

$$p - p_o = \begin{cases} -\rho\omega R[1 - \tfrac{1}{2}r^2/R^2], & 0 \leq r \leq R \\ -\omega R\tfrac{1}{2}R^2/r^2, & R \leq r \leq \infty \end{cases}$$

These variables can be plotted, as in Fig. 9.21.

If we consider p_o to be the ambient atmospheric pressure away from the tornado where the velocity is small, then the pressure drops to a minimum at the center. The velocity is also zero there, however it reaches maximum not far away at the core boundary. These two effects produce the huge damage associated with a tornado.

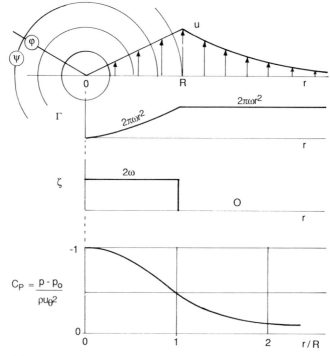

Figure 9.21   Streamlines, potential lines, velocity, circulation, vorticity, and pressure in a tornado modeled as an ideal vortex.

9.7.1 Vortex Pairs

Consider a pair of infinite *parallel rectilinear idealized vortices* of like signs with strengths Γ_1 and Γ_2, separated by distance d, with rectangular coordinates centered at 1. The corresponding velocity fields are shown in Fig. 9.22 as solid for Γ_1 and dashed for Γ_2. The sum of the implied velocities is shown in the lower figure.

For these vortices to be in equilibrium, the velocity at 2 must be $-\Gamma_1/(2\pi d)$, and at 1 it must be $+\Gamma_2/(2\pi d)$. The velocity at any point on the line joining them is (excluding the vortex centers)

$$U(x) = -\Gamma_1/(2\pi x) + \Gamma_2/[2\pi(d - x)] \qquad (9.50)$$

From Fig. 9.22, we see that for $x \le 0$, $U > 0$ and for $x \ge d$, $U < 0$. This, together with the velocity directions at 0 and d, suggests that a clockwise rotation is produced by (required for the existence of) this vortex configuration. The "center of rotation," which must lie on d, can be found from $U(x) = 0$, which implies

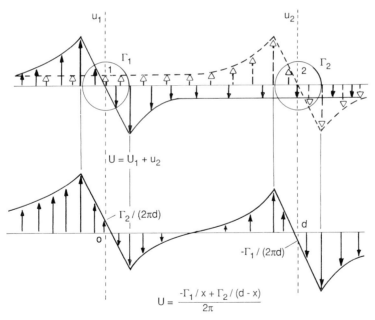

Figure 9.22 Vortex pair with same sign. Associated velocities from Γ_1 (———) and Γ_2 (– – –) and their sum (lower figure).

$$\Gamma_1/(2\pi x) = \Gamma_2/[2\pi(d - x)]$$

$$x = [\Gamma_1/(\Gamma_1 + \Gamma_2)]d$$

If $\Gamma_1 = \Gamma_2 = \Gamma$, then for x not 0 or d,

$$U(x) = -\Gamma/(2\pi x) + \Gamma/[2\pi(x - d)]$$
$$= \Gamma/\{2\pi[x(d - x)/(d - 2x)]\} \qquad (9.51)$$

In this case, at $x = d/2$, $U = 0$. At the singular points, $x = 0$ or d,

$$U(0) = \Gamma/(2\pi d); U(d) = -\Gamma/(2\pi d) \qquad \text{(not infinite or zero)}$$

The rotation about the center of the line (from ϵ to $d - \epsilon$) could be approximated by a hypothetical solid body (rotational),

$$d\theta/dt = U(0)/(d/2) = \Gamma/(\pi d^2) \qquad (9.52)$$

However, this would not be a potential flow, and would indicate incorrect velocities outside the vortices. There exists no solid-body rotation or vorticity outside the two vortex cores where potential flow is valid. Since we are dealing with potential flow, we could superimpose a uniform velocity field on this flow. The pair of vortices would then move along with this velocity and rotate about the center of rotation.

When Γ_1 and Γ_2 have opposite signs, the rotation center will lie outside of d. In fact, when they are equal and opposite, the center is at infinity and they do not rotate around each other. The total velocity field compatible with the two vortices is shown in Fig. 9.23. It suggests that the pair want to move together (upward in the figure).

9.7.2   Initial Motion

If the vortices were somehow suddenly created in a static fluid, the implied velocity inside (nearest the other vortex) would always be higher than that outside. This is due to the addition of associated velocity fields for the two vortices. Bernoulli's relation would then state that lower pressure existed inside, and they would move toward each other—toward the low pressure. This then is not a stable configuration. However, if a uniform flow field V were imposed such that the velocities in the neighborhood of the two vortices were equal, the pressure gradient would not exist. This velocity would be determined by

$$V + \Gamma/(2\pi\epsilon) - \Gamma/(2\pi d) = \Gamma/(2\pi\epsilon) + \Gamma/(2\pi d) - V \qquad (9.53)$$

where

$$V = \Gamma/(2\pi d), \qquad (d + \epsilon \approx d)$$

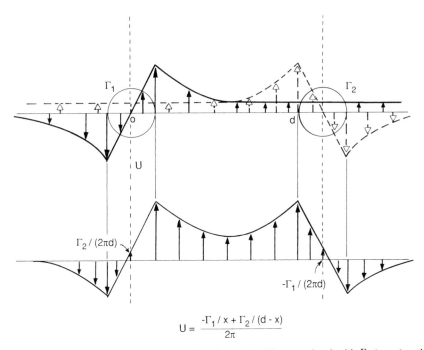

Figure 9.23 Vortex pairs with opposite signs, velocities associated with Γ_1 (———) and Γ_2 (– – –) and their sum (lower figure).

Thus, if the vortices move with velocity $\Gamma/(2\pi d)$ (or are stationary in a uniform flow of this velocity), there is no attraction. This is precisely the velocity associated with the centers of the coexisting vortices. The result merely states that to be in an equilibrium flow situation, the velocities at all points of the flow must conform to the superposed associated velocities of the vortices present.

SUMMARY

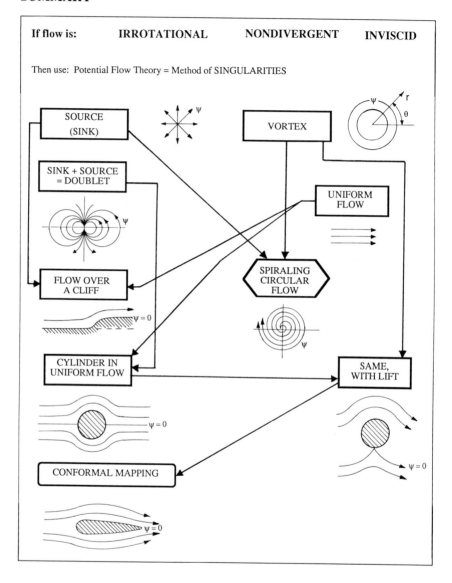

Problems

1. Given the potential

$$\varphi = ax^2y + bxy^2$$

Is it a valid potential flow?

2. Given the stream function for incompressible flow,

$$\psi = Cxy$$

Is the flow irrotational? Sketch the flow pattern. What flow might it be used to model?

3. Given the potential function,

$$\varphi = (U/r) \cos \theta$$

Find the stream function. Assume U is constant.

4. Add the Coriolis forces to the development of Eq. (9.12) to arrive at a sixth-order equation in ψ.

5. Add a simple stratification effect $(\partial T/\partial x_3 \neq 0)$ to the result of problem (4) to get an eighth-order equation in ψ.

6. Given the potential function $\varphi = C \ln r$.
Find the pressure, $p(r,\theta)$ with respect to stagnation pressure.

7. A two-dimensional flow field is described by $u = x$ and $v = -y$. Investigate the flow with respect to continuity, rotationality, and realisticity.

8. Write the potential and streamfunction for a combination of a sink and vortex. Find the velocity field. Plot the streamlines. What atmospheric phenomena might this simulate?

9. Consider a tornado as a two-dimensional, idealized vortex with core radius 30 m and a maximum wind velocity of 60 m/sec. Calculate and plot the velocity, pressure, circulation, stream function, and velocity potential.

10. Suppose you're given a potential $\varphi = f\{x, y\}$ for flow about a two-dimensional configuration. Explain the steps you take to get the flow field, the stream lines, and the forces on the configuration.

11. Given that a tornado has velocity components in polar coordinates.

$$(V_r, V_\theta) = (-A/r, -B/r)$$

Get the equation for the streamlines of the flow.

12. Consider the wind blowing perpendicular to a Quonset hut (a half-cylinder, as in the sketch below). Use ideal potential flow to determine the location of minimum pressure and p_∞. Calculate the net force on the hut. $P_\infty = 1000$ mb, $d = 10$ m, length of hut $= 30$ m.

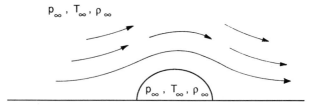

P_∞ , T_∞ , ρ_∞

P_∞ , T_∞ , ρ_∞

13. Calculate the lift and drag forces on the cylinder with circulation in uniform flow. In Example 9.6, add circulation Γ to the cylinder. This is the *Magnus effect* and the result for the lift,

$$F_{\mathrm{L}} = \rho U \Gamma$$

is known as the Kutta–Joukowski formula.

14. Consider the potential and stream functions,

$$\varphi = C r^n \cos(n\theta); \qquad \psi = C r^n \sin(n\theta)$$

Show the flow solutions depicted when $n = 2$, 3, $\frac{2}{3}$, and $\frac{3}{2}$.

Chapter 10   Perturbation Equations

10.1   The Mean Flow

Observations in the atmosphere and the ocean suggest that the flow is often irregular, either wavy or turbulent. The mathematical description for these motions is assumed to be contained in the continuum equations of motion. Whether or not it is, depends on the validity of the closure assumption. This depends on the characteristics and scale of the turbulence. A complete solution may only be obtainable in numerical integrations with sufficiently large computers. Still, this is difficult. From a practical standpoint, the entire volume of the atmosphere cannot be represented numerically on small enough grid scales to account for small-scale turbulence and dissipation Thus a choice of a finite volume must be made for the domain of calculation.

In general, we do not have the ability to recognize the myriad of forces, instabilities, and boundary conditions that determine the flow, let alone place them in a computational program. Therefore the boundary conditions, both spatial and temporal, are frequently unknown. On the scales at which we most often deal, the flow may appear random. There are unpredictable variations on many scales. Sometimes the flow under observation may not be truly random, yet it can be effectively random due to the finite time and space scales of the calculation and the limited information available to the observer. One practical result of this indefinite character of the flow condition is the lack of a precise definition of turbulence. With the goal of defining turbulence, we begin by defining the mean flow. When this is done, random turbulence and coherent waves will be what is left over.

There is evidence of both turbulence and a steady mean flow existing on almost every scale. To visualize this, consider flying a kite. It responds mainly to a mean wind. But there is usually evidence in its motion of the random turbulence in the air. If the kite has a long tail, the turbulence is often exaggerated in the motion of the tail. Or consider the wind surfer who relies on an average component of the wind for balance. She may fall when turbulent gusts hit the sail. An anemometer records the surface layer winds, and its pointer may oscillate randomly. Still, there is generally a clear average velocity. When the mean is sufficient information, the turbulence is often logged as noise in an experiment. However, some measure of the turbulence might be given. An example is the "gust factor" of the wind.

The mean value is clearly a function of the averaging period. The average may be taken over an arbitrary interval of time, Δt. This is shown by two possible cases in Fig. 10.1. It is clear that the value of the mean depends on the chosen Δt interval.

As the forces acting on a fluid increase, the flow may pass through several stages. These may be lumped into general categories:

1. Static fluid
2. Laminar flow
3. Unsteady and wavy flow
4. Chaotic and turbulent flow.

Each of the flow groups includes several flow regimes. Within the laminar flow category, the slow and orderly regimes called creeping flows yield particular solutions of forms of the Navier–Stokes equations. They are ob-

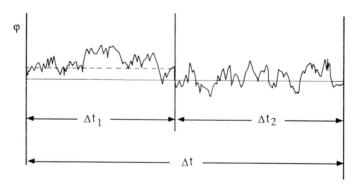

Figure 10.1 Typical data record versus time. Dashed line shows mean values for various Δt (zero for second segment).

tained by taking the approximation of the equations for a very small Reynolds number. These solutions provide accurate descriptions of the flows shown in Figs. 1.1–1.4. The flow regimes of potential flow theory are also laminar flows. An eddy viscosity concept is employed to extend this category to include eddy-laminar. This is a steady layer-like flow with respect to small-scale turbulent eddies. Then we may include the geostrophic, gradient, and ideal vortex flows in the laminar category.

The last two of the four flow categories are best described by the perturbation equations. Here, the wave component, coherent or turbulent, is separated from the average flow by considering it as a perturbation on the mean.

Atmospheric flow is seldom steady or linearly varying for great distances or periods. Still, there is often an *average* of the flow, as in the latitudinal bands of easterlies or westerlies around the globe. However, as more sensitive measurements are taken of virtually any phenomenon, a variation about the mean flow is inevitably found. In the case of the westerlies, one variation is the mid-latitude storm systems. They may appear regularly on an average of once every few days in certain locations, such as Seattle, Washington. Hurricanes are dramatic variations on the tropical easterlies. They do not have a well-defined frequency.

On the other hand, the *local* small-scale flow may also appear to be steady for hours as a large-scale weather system passes by. Yet when a larger space is considered or a longer time, there will be significant variations. For instance, the moth that lives its life cycle in the summer months will have no care or knowledge of the cold of winter. Similarly, Homo sapiens' current culture has bloomed in a warm interglacial period. Thus it gives little care to the long-term periodic return to an ice age. It is even difficult to generate regard for the possibility of anthropomorphic change of the climate, as in the greenhouse effect.

There are many examples of distinctively separate regimes of flow occurring as the flow responds to small variations in the boundary conditions. They may occur with either regular or random frequencies. Each regime may exist long enough to exhibit a quasi-steady character, and then quickly convert to another quasi-steady field. These examples also suggest caution toward applying a long-term average to a short-term phenomenon. For instance, the mean temperature of an interval that spans both glacial and interglacial periods is of little significance in describing either period.

The determination of a mean atmospheric flow is often difficult due to the sparcity of the data. The limited data prevents one from obtaining a repeatable average as defined in Chapter 1. To get some idea of the scope of this problem in the atmosphere we can look at several common categories of atmospheric flow:

Table 10.1

Categories of Atmospheric Flow and Associated Phenomena.

	Surface layer	PBL	Synoptic weather	Global weather	Global climate
Space	100 m vertical	5 km vertical	10,000 km horizontal	25,000 km horizontal	25,000 km horizontal
Time	1 hr	24 hr	7 days	1 yr	10–10^9 yr
Phenomena	Ocean waves	Air pollution	Storms	El Niño	Greenhouse effect; ozone; interglacial ice ages
Eddy scale	10 m 30 min	100 m 30 min	400 km 6 hr	500 km 1 month	1000 km 10^n yr

The estimates for eddy scale in the various categories is determined from the maximum turbulence scale that allows a mean derivative to be defined. This will be the scale of the eddy continuum. For instance, in the PBL, the vertical derivative of mean velocity, dU/dz, can only be defined when Δz is sufficiently smaller than the domain height that a limit $\Delta z \to 0$ can be approximated. This Δz must contain sufficient eddies to yield a uniform average as discussed in Chapter 1.

Since there are frequently not enough observations to adequately determine the averages or the eddy domain, the next step is to determine the assumptions needed to remedy the shortages in observations. This analysis is greatly helped by examining the governing equations. With a systematic scaling analysis, the possibilities of assuming horizontal homogeneity, periodicities, and appropriate mean values may be revealed.

Example 10.1

Consider the problem of instrumenting a 100-meter tower in the atmospheric surface layer (see Fig. 10.2). Discuss the intervals for placing sensors and measuring times for typical winds of 10m/sec. What assumptions on eddy size must be made?

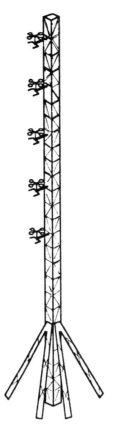

Figure 10.2   Sketch of a 100-meter tower in the atmospheric PBL with wind and temperature sensors.

Solution

We would like to have a sufficient number of sensors to define the mean variation of the wind through the 100-m layer. Sensors placed at 10-m intervals would allow the roughest approximation of dU/dz assuming that $\Delta z/H = 10/100$ and $\Delta U/U \ll 1$.

To average over typical 10-m diameter eddies in a 10-m/sec wind requires 1 sec for an eddy to pass. In 2 min one gets 120 eddies to average. Thus a 2-min average at 10-m intervals over the 100-m depth should allow a mean velocity profile with respect to an eddy continuum with 10-m maximum size eddies. Larger eddies will not allow an eddy continuum to be defined. A smaller eddy spectrum would allow shorter averaging times.

We can evaluate the significance of an average value only with respect to the measurement interval and the number of points. However, there are also restrictions placed on the existence of average flows by the governing equations. When all of the terms in an equation are placed under the same average, certain terms will dominate over the others; that is, the relative value of the terms will be determined by the choice of characteristic values (often average values) of the primary variables, as discussed in Section 3.3. For instance, a turbulence phenomenon with a particular scale (say 1 m) can be considered part of the mean flow to a much smaller scale domain (1 cm). It is dominant at its own scale, and negligible on very large scales (kilometers).

When the scales are chosen as characteristic of the height above a surface in the atmosphere, different equations and solutions emerge for different heights. The thin molecular layer (millimeters) adjacent to the surface may be described by equations with laminar solutions. A little farther out, the boundary layer (meters to kilometers) equations can ignore molecular viscous effects, but they will require direct inclusion of the turbulence. However, on the slightly larger scales of the PBL, the eddy-laminar equations may be a good approximation. Finally, the equations for the troposphere (10 km) will ignore this scale (meters to kilometers) of turbulence. Even the eddy-viscous terms can often be dropped for the mean equations of motion.

The solutions to each set of approximate equations will be valid only when the assumed characteristic values hold. In this chapter we will use perturbations on a mean flow to investigate limitations on the range of application of a given mean flow solution.

10.2   Waves

Even in the smoothest of irrotational flows there are seeds of irregular flow. Small disturbances are omnipresent. These disturbances, or perturbations, are "constantly testing the flow field to see if it will allow them to grow." If conditions are conducive to growth, the small disturbance may grow to a wave, a wave train, or a set of many different waves. Often the growth continues until the waves break and small-scale turbulence results. However, sometimes the wave will come to equilibrium at a finite amplitude, and thereby become part of the larger-scale mean flow. For instance, the earliest satellite photographs showed the frequent presence of orderly lines of clouds known as cloud streets, shown in Fig. 10.3. These pictures suggested that the flows contain waves that are regular and persistent. They are definitely not random turbulence. The observations provoked a theory and solution for

Figure 10.3 Atmospheric flow marked by sharply defined clouds. The clouds sit at the top of the PBL at about 2 km height. The flow is parallel to the "cloud streets," a "southerly" from bottom to top—from over the ocean to over the east coast of the United States. The wavelength (cloudband width plus separation) is about 2 km.

these flows based on the scaling and perturbation techniques discussed in this chapter.

However, as parameters change, the equilibrium conditions are altered and the waves may simply die out, or they may change to another quasi-steady equilibrium state. Or they may develop more growth and break. An example of such a transition to the "cloud streets" is shown in Fig. 10.4. Here, the stratification conditions change as the flow continues over a warm surface. A new flow regime results—dictated in the equations by the increased importance of terms involving density and temperature changes.

Figure 10.4 Atmospheric flow marked by cloud waves. In this case the flow is northerly, top to bottom, from over pack ice to over ocean. The wavelengths vary from about 4 km near the ice to 40 km downstream, where the flow has slowed and entered a new flow regime. Note the Karman vortices superimposed on the PBL flow in the wake of Bear Island.

A familiar example of waves in geophysical flows is found on the interface between the ocean and the atmosphere. The flow of wind over water evidently has a solution that contains waves on many scales and regimes. These waves, and the relation to the driving force, the wind, are very important in many contemporary problems of atmospheric and oceanic dynamics.

The sailor looks for the freshening (rising) wind by looking for the "cat's paws" on the water surface. What he or she sees is a darker area, marking the reduced reflection off the surface of the water in regions where the wind has generated capillary and short gravity waves. These are wavelets with wavelength in the few-centimeter range. They are generated on the water surface in direct correlation with the wind velocity. The theory behind the generation is lengthy and laborious, and not yet definitive. (In fact, the generation is possibly proportional to the acceleration/deceleration of the winds rather than to the steady wind.) However, the simple but sound correlation between these waves and the wind is the basis for the sailor's inference. It has also been extended to the interpretation of satellite microwave radar measurements as marine winds.

We can expect the solution for the flow of the atmosphere and the ocean to emerge from the Navier–Stokes equations for the continuum fluid domain, which includes the air and water. Although there is a sharp discontinuity in the density profile at the transition from air to water, the velocity and stress can be assumed to be continuous at the surface.

The flow solution must include the development of the sea-surface waves. When the mean flow is a strong wind at the surface, the velocity associated with the waves can be considered as a *small perturbation* on the mean. The flow description then becomes a *stability* problem. In this analysis, the equations are examined to see whether various perturbations will grow or decay. When the waves grow exponentially, they may break and create turbulence.

10.3 Turbulence

The wind blowing across a corn field, or blowing snow, dust, or candy bar wrappers, provides familiar visual observations of the variability of wind. Similar observations are made on larger scales in the cloud motions by sensors on satellites or airplanes. In addition to random plumes, organized cells and/or linear features appear on many scales in these natural "flow visualization experiments."

When the mean is calculated by taking the average over an interval of time or distance, the turbulent and periodic motions have been averaged out. Some of these perturbations can be brought back into the equations by separating the velocity at any time or point into the mean plus variations about the mean. The perturbations include the organized periodic motions and the apparently disorganized motions of turbulence. It will not always be evident whether or not the variations are organized or not. If we cannot identify the variations with predictable characteristics, we must call them turbulence. Although this is not a very satisfactory definition, it is adequate for practical purposes. Turbulent flow can also be recognized by the following characteristics:

1. Strongly nonlinear behavior
2. Rotational
3. Apparently random in time and space
4. Intermittent, three-dimensional, and chaotic
5. Associated with vortices, a continuous spectrum of eddies, and a mean flow shear
6. Strong mixing, the primary instrument in atmospheric diffusion

From this behavior, it is evident that the process of turbulence analysis is that of seeking universal characteristics of the turbulence. If we can then describe this behavior, it may be possible to remove this aspect of the flow from "turbulence." From the last two characteristics, it is evident that turbulence analysis and description is crucial to any regions involving flow regime boundaries. Turbulence is a dominant consideration in boundary-layer dynamics. It is essential to the descriptions of pollution dispersion, agricultural microclimates, and forces on buildings, ships, and aircraft.

There are many levels of turbulence analysis. They range from the modeler's desire for a quick and simple parametrization of its diffusive action to the mathematician's description of the statistics of the phenomena. We will concentrate on the former. This depiction follows closely the earlier derivation of the molecular basis of diffusion. It can be assumed as an *ad hoc* analogy of molecular diffusion. Considering the relative tenuous state of the eddy-continuum compared to that of the molecular continuum, this is an audacious assumption. However, Stokes' viscosity assumptions were no less bold analogs of the parametrizations of solid-state mechanics. The big problem with the eddy-diffusion concept involves scales, and the basic requirements of the continuum, to have the Navier–Stokes equations be valid. In this concept, we think of turbulence as a small-scale momentum transport mechanism and interpret it as exerting a viscous force.

10.3.1   The Eddy Continuum

The concept of a laminar flow, tied to its origins in molecular fluid dynamics, can sometimes satisfactorily be extended to an eddy-laminar definition. The averaging period chosen to establish the mean flow must sample a sufficient number of turbulent elements to be in the "continuum" domain of these eddies. Yet it cannot be so long (in time or space) as to encounter larger-scale trends. The goal is to parametrize the mixing effect of the small-scale turbulent elements in a fashion similar to the molecular flux parametrization. There, we used molecular viscosity as a parameter to relate the aggregate momentum flux by the molecules (the internal stress force) to the mean flow velocity gradient. The fact that observations show that there exist laminar-like flows on large scales lends support to the hypothesis of eddy-laminar flow.

Just as in the molecular domain, to use the Navier–Stokes equations there must exist an eddy-continuum such that a meaningful average of the relatively small-scale eddies can be obtained. There are many observations of orderly linear flow on large scales that also must include strong small-scale turbulence. These observations occur in a wide diversity of domains throughout the atmosphere. Thus, it is likely that an eddy-laminar approximation is an adequate first approximation for many flow situations. An approximate solution for geophysical flows is obtained by solving the Navier–Stokes equations with an eddy-viscosity representation for the turbulent fluxes.

More often than not, in large-scale atmospheric flow problems the viscous term can be completely neglected. This is because the nondimensional coefficient of the viscous term, $1/\text{Re} = K/(UL)$, is small in large-scale flows. The smallness of this ratio between the viscous and inertial forces may be due to the absence of small-scale turbulence (e.g., due to the damping effects of stable stratification). It could also be due simply to the large characteristic scale L of the flow. The wave and turbulent features are then considered as perturbations embedded in these mean flows, with a zero average.

10.4 Reynolds Averaging—Flux

We will start by separating the flow parameters into a mean part plus another component that oscillates about the mean. We are immediately confronted with the problem of what exactly is the mean flow. It will be defined with a scale that is large with respect to that of the perturbation phenomena we are investigating.

The mean of a function[1] Φ may be written

$$\Phi = (1/\Delta T) \int_t^{t+\Delta T} \Phi \, dt \qquad \text{or} \quad [1/(n\lambda)] \int_x^{x+n\lambda} \Phi \, dx \qquad (10.1)$$

---

[1] Note that φ and Φ in this chapter have no relation to the potential function of Chapter 9.

where λ is the wavelength associated with the oscillating part and n is a number ≈ 10. The period, ΔT, or the length, $n\lambda$, must be long with respect to the values associated with any perturbations. An overhead bar will be used to denote the average.

We then have the following rules:

$$\bar{\bar{f}} = \bar{f}; \qquad \overline{f+g} = \bar{f} + \bar{g}; \qquad \overline{\bar{f}g} = \bar{f}\,\bar{g}$$

$$\overline{\partial f/\partial s} = \partial \bar{f}/\partial s; \qquad \overline{\int f\,ds} = \int \bar{f}\,ds \qquad (10.2)$$

When the flow parameter is separated into a mean plus a perturbation, $f + f'$,

$$\bar{f} \equiv (1/\Delta t) \int_t^{t+\Delta t} f\,dt = (1/\Delta t) \int_t^{t+\Delta t} (\bar{f} + f')\,dt$$

$$= (1/\Delta t)\left[\int \bar{f}\,dt + \int f'dt\right]$$

$$= \bar{f} + (1/\Delta t)\int f'\,dt = \bar{f}$$

since

$$(1/\Delta t)\int f'\,dt = \bar{f'} \equiv 0$$

In general, the dependent variables will be separated into a mean plus an oscillating part, (φ represents any variable)

$$u_i = \bar{u}_i + u'_i; \qquad p = \bar{p} + p', \qquad \text{etc.}; \qquad \text{where } \bar{\varphi'} = 0$$

$$u^2 = (\bar{u} + u')^2 = \bar{u}^2 + 2\bar{u}u' + u'^2$$

and when an average is taken,

$$\overline{u^2} = \bar{u}^2 + \overline{u'^2} \qquad (10.3)$$

Similarly,

$$\overline{u_iu_j} = \bar{u}_i\,\bar{u}_j + \overline{u'_i u'_j} + \overset{0}{\overline{u'_i}\,\bar{u}_j} + \overset{0}{\bar{u}_i\,\overline{u'_j}} \qquad (10.4)$$

The primed quantities can represent waves or small-scale random turbulence. If both are present, we could separate the flow into three parts, using φ for the mean, a φ' for the waves, and φ'' for the small-scale turbulence. Such a development involves a proliferation of terms. For simplification in learning the technique, we will restrict analysis to a single perturbation for either waves or turbulence. In practice, when the circumstances are clear, often capitals are used for the mean flow, lower case for the perturbation.

Example 10.2

Consider the triple correlation $\rho u w$, and calculate the averaged perturbation.

Solution

Substitute $\rho = \bar{\rho} + \rho'$, $u = U + u'$, and $w = W + w'$,

$$\rho u w = (\bar{\rho} + \rho')(U + u')(W + w')$$

$$= \bar{\rho}UW + \bar{\rho}Uw' + \bar{\rho}u'W + \bar{\rho}u'w' + \rho'UW + \rho'Uw' + \rho'u'W + \rho'u'w'$$

Taking the average,

$$\overline{\rho u w} = \bar{\rho}UW + \bar{\rho}\overline{u'w'} + U\overline{\rho'w'} + W\overline{\rho'u'} + \overline{\rho'u'w'}$$

When a substitution is made in the equations and the mean terms are subtracted out, this leaves four perturbation terms,

$$\overline{\rho u'w'} + \overline{U\rho'w'} + \overline{W\rho'u'} + \overline{\rho'u'w'}$$

When one of these variables is constant, the perturbation terms are reduced to one only. It is clear that there is a quick proliferation in terms when higher-order correlations are considered.

Consider the flow through a surface in the fluid with arbitrary orientation (Fig. 10.5).

From Section 6.1 we recall that, similar to the mass flux/unit-time $= \rho u\, dA$, the u-momentum flux/unit-time is

$$dJ_x = \rho u u\, dA$$

$$dJ_y = \rho u v\, dA$$

$$dJ_z = \rho u w\, dA$$

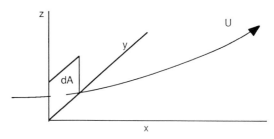

Figure 10.5   Flow through an arbitrary surface dA.

The time average is

$$dJ_x = \rho(\overline{U^2} + \overline{u'^2})\, dA$$

$$dJ_y = \rho(\overline{UV} + \overline{u'v'})\, dA$$

$$dJ_z = \rho(\overline{UW} + \overline{u'w'})\, dA$$

To get flux per unit area, divide by dA to arrive at

$$dJ/dA\,|_x = \rho(\overline{U^2} + \overline{u'^2})$$

$$dJ/dA\,|_y = \rho(\overline{UV} + \overline{u'v'})$$

$$dJ/dA\,|_z = \rho(\overline{UW} + \overline{u'w'})$$

Now the momentum flux/unit area is associated with an equal and op-posite force on the surroundings $[d(mv)/dt = F = \text{stress}]$. This force/unit area is defined as the stress. Thus the flux of momentum/unit time through an area produces a force on the area that is equal and opposite to the stress force on an element surface. There is a contribution to the force from the mean flow, u^2, uv, and uw, and from the perturbation correlations, $\overline{u'^2}$, $\overline{u'v'}$, $\overline{u'w'}$. For the surface shown above this may be written

$$\{\tau_{11},\ \tau_{12},\ \tau_{13}\} = \{-\rho(U^2 + \overline{u'^2}),\ -\rho(UV + \overline{u'v'}),\ -\rho(UW + \overline{u'w'})\} \quad (10.5)$$

We now have the internal stress terms expressed as partly due to mean flow and partly due to the perturbation flow. The mean forces are the ones that we have parametrized with Stokes's law of friction and viscosities. The new fluxes, or eddy-stresses, will have to be related to the mean flow pa-rameters to obtain a closed set of equations.

Example 10.3

Verify that the momentum flux due to the correlation between u' and w' can be nonzero in the simple shear flow of Fig. 10.6 by considering the small motions of the parcel.

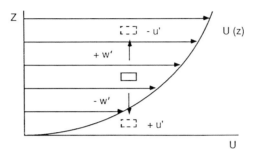

Figure 10.6   Parcel with random motion in a mean shear flow. Moving up or down with $\pm w'$, it exhibits a $\mp u'$ velocity in its new environment.

Solution

Consider the signs of the perturbation velocities of the parcel as it moves up and down, and the consequent sign of the correlation $u'w'$.

1. Movement up: $w' \geq 0$, $u' \leq 0$; $\rightarrow u'w' \leq 0$.
2. Movement down: $w' \leq 0$, $u' \geq 0$; $\rightarrow u'w' \leq 0$.

There is a correlation between the u' and the w' such that if there is a positive $U(z)$ gradient, then the average product contributes a negative momentum flux. The stress is the reactive force on the parcel,

$$\tau'_{xz} = -\rho u'w' \geq 0 \tag{10.6}$$

There are many uses of the perturbation procedure in fluid dynamics. Initially, the perturbation quantity was used by 0. Reynolds to evaluate the effects of small-scale turbulence. In this way he was able to show that when turbulence is considered as a random small perturbation it could be included in the equations of motion as a stress force contribution. Later, to get the PBL equations, we will investigate whether a flow is stable or unstable to a small disturbance, and we would like to use this procedure with the perturbation as a time-dependent simple waveform. Finally, we would like to let the perturbation grow to the point where it is interacting with the mean flow parameters sufficiently to change them. In this case, we will be considering a nonlinear problem. However, it is possible that not all of the nonlinear terms will be important. This can be investigated using scaling principles.

When we place the perturbation format into the equations of motion, we get a lengthy equation with mean terms, perturbation terms and mixed terms results. When the equations are averaged, the mixed terms frequently vanish, since averages of perturbations alone are zero. However, the mean flow equations may now contain *cross* and *autocorrelation* terms, since the products of perturbations do not necessarily average to zero. If the mean flow equation is subtracted out of the complete perturbed equation, equations in the *perturbation* result. The terms with combined mean and perturbation variables appear in these equations. A *linearized* perturbation equation can be obtained by neglecting all products of perturbations.

Example 10.4

Find the various perturbation equations for the representative equation

$$\partial u/\partial t + u \, \partial u/\partial x + bu^2 = 0 \qquad (10.7)$$

Solution

The perturbed equation is obtained by substituting perturbation expressions for u in Eq. 10.7.

$$\partial(U + u')/\partial t + (U + u') \, \partial(U + u')/\partial x + b(U + u')^2 = 0$$

or

$$\partial u/\partial t + \partial u'/\partial t + U \, \partial U/\partial x + U \, \partial u'/\partial x + u' \, \partial U/\partial x + u' \, \partial u'/\partial x$$

$$+ \, b[U^2 + 2U^2 u' + u'^2] = 0 \qquad (10.8)$$

The Mean Flow equation is obtained by averaging Eq. (10.8),

$$\partial U/\partial t + U \, \partial U/\partial \dot{x} + \overline{u' \, \partial u'}/\partial x + b(U^2 + \overline{u'^2}) = 0 \qquad (10.9)$$

The Perturbation equation is Eq. (10.8) − (10.9),

$$\partial u'/\partial t + U \, \partial u'/\partial x + u' \, \partial U/\partial x + u' \, \partial u'/\partial x - \overline{u' \, \partial u'}/\partial x \qquad (10.10)$$

$$+ \, b[2Uu' + u'^2 - \overline{u'^2}] = 0$$

The linearized perturbation equation is obtained from (10.10) by neglecting perturbation products. [Note that it is not equal to the linearized perturbed equation, (10.8). The latter will have several mean flow terms corresponding to terms in the linearized mean flow equation.]

$$\partial u'/\partial t + u' \, \partial U/\partial x + U \, \partial u'/\partial x + b2Uu' = 0 \qquad (10.11)$$

Note that the new mean flow equation (10.9) has picked up new terms from each of the nonlinear terms. The perturbation equation (10.10) involves differences between the mean values of the correlations and the point values of the same products. Finally, the linearized equation (10.11) picks up three new terms. These represent interactions between the mean and perturbation velocities.

We are now prepared to write the equations in general forms appropriate to investigations of perturbation phenomena. We will start with the *Basic (N–S) equations,* which have been derived for laminar flow. We need only to add the perturbations to get the *Perturbed equations.* The *Mean Flow equations* are derived by taking the time average.

Then, subtract the Mean Flow equations from the Perturbed equations to get the *Perturbation equations.* These are equations describing the dynamics of the perturbation quantities. Finally, consider the perturbation to be small with respect to mean values, neglect perturbation products and get the *Linearized Perturbation equations.* They will describe the initial dynamics of an infinitesimal perturbation.

10.5   The Set of Perturbed Navier–Stokes Equations

In this section we will obtain the four versions of the basic equations when perturbations of each of the dependent variables are permitted. These are the momentum, continuity, energy, and state equations. This is a straightforward but cumbersome procedure, resulting in a large number of terms. These equations are quite general. However, in particular situations many of the terms may be negligible compared to the rest.

To obtain the perturbed set of equations governing the flow we will substitute

$$u_i = U_i + u'_i; \qquad p = P + p'; \qquad \rho = \bar{\rho} + \rho'; \qquad T = \bar{T} + T'$$

Here,

$$\bar{u}' = \bar{p}' = \bar{\rho}' = \bar{T}' = 0$$

so that

$$u^2 = (U + u')^2 = U^2 + 2Uu' + u'^2 \tag{10.12}$$

and

$$\overline{u^2} = U^2 + \overline{u'^2} \tag{10.13}$$

$$\overline{u_i v_j} = U_i V_j + \overline{u_i' v_j'} \tag{10.14}$$

The procedure we will follow for all of the equations is

1. First substitute the perturbations.
2. Then average.
3. Then subtract out the mean flow equations to obtain the perturbation equations.

This process is relatively simple in the application to the basic equation of state. We will use this equation as an example.

The Basic equation

$$p = \rho R T \tag{10.15}$$

The Perturbed equations

$$P + p' = (\bar{\rho} + \rho')\, R(\bar{T} + T') = \bar{\rho} R \bar{T} + \bar{\rho} R T' + \rho' R \bar{T} + \rho' R T'$$

Or, dividing by $\bar{\rho} R \bar{T}$,

$$\frac{P}{\bar{\rho} R \bar{T}} + \frac{p'}{\bar{\rho} R \bar{T}} = \frac{\bar{\rho} R \bar{T}}{\bar{\rho} R \bar{T}} + \frac{\bar{\rho} R T'}{\bar{\rho} R \bar{T}} + \frac{\rho' R \bar{T}}{\bar{\rho} R \bar{T}} + \frac{\rho' R T'}{\bar{\rho} R \bar{T}}$$

we get

$$-P/(\bar{\rho} R \bar{T}) + p'/(\bar{\rho} R \bar{T}) = 1 + T'/\bar{T} + \rho'/\bar{\rho} + \rho' T'/(\bar{\rho}\bar{T}) \tag{10.16}$$

Then take the average of all terms to derive the *mean* equations.

$$P/(\bar{\rho} R \bar{T}) + \overline{p'}/(\bar{\rho} R \bar{T}) = 1 + \overline{T'}/\bar{T} + \overline{\rho'}/\bar{\rho} + \overline{\rho' T'}/(\bar{\rho}\bar{T})$$

or

$$P/(\bar{\rho} R \bar{T}) = 1 + \overline{\rho' T'}/(\bar{\rho}\mathbf{T})$$

or

$$P = \bar{\rho} R \bar{T} + R\overline{\rho' T'} \tag{10.17}$$

The perturbation equations are obtained by taking (10.16) − (10.17).

$$p'/(\bar{\rho} R \bar{T}) = T'/\bar{T} + \rho'/\bar{\rho} + (\rho' T' - \overline{\rho' T'})/(\bar{\rho}\bar{T}) \tag{10.18}$$

The equations can be linearized by neglecting all perturbation products. In Eq. (10.18), this means dropping the last term and retaining the first three. The linearized equation is often obtained directly by considering only infinitesimal perturbations from the beginning (no perturbation products). We then obtain from the basic equation,

$$P + p' = (\bar{\rho} + \rho') R(\bar{T} + T') = \bar{\rho}R\bar{T} + \bar{\rho}RT' + \rho'R\bar{T} \tag{10.19}$$

and the averaged equation is

$$P = \bar{\rho}R\bar{T} \tag{10.20}$$

Subtracting this averaged equation from Eq. (10.19), leaves the linearized perturbation equation,

$$p'/P = \rho'/\bar{\rho} + T'/\bar{T} \tag{10.21}$$

We can now apply the same procedures to the momentum, continuity, and energy equations.

The Basic Equations

$$\partial u_i/\partial t + u_j\,\partial u_i/\partial x_j + f\epsilon_{ijk}\mathrm{n}_j\mathrm{u}_k + 1/\rho\,\partial p/\partial x_i$$

$$+ g\delta_{i3} - \nu\,\partial^2 u_i/\partial x_j\,\partial x_j = 0 \tag{10.22}$$

$$\partial\rho/\partial t + u_j\,\partial\rho/\partial x_j + \rho\,\partial u_j/\partial x_j = 0$$

$$P = \rho RT; \qquad f = 2\Omega\sin\theta; \qquad n = (0,\,0,\,1)$$

$$\partial T/\partial t + u_j\,\partial T/\partial x_j - K_h\,\partial^2 T/\partial x_j\,\partial x_j - (RT/c_p)\,\partial u_j/\partial x_j = 0$$

The Perturbed Equations

$$\partial U_i/\partial t + \partial u_i'/\partial t + (U_j + u_j')\,\partial(U_i + u_i')/\partial x_j + f\epsilon_{ijk}\mathrm{n}_j(u_k' + U_k)$$

$$+ 1/(\bar{\rho} + \rho')\,\partial(P + p')/\partial x_i + g\delta_{i3} - \nu\,\partial^2(U_i + u_i')/\partial x_j\,\partial x_j = 0$$

$$\partial(\bar{\rho} + \rho')/\partial t + (U_j + u_j')\,\partial(\bar{\rho} + \rho')/\partial x_j + (\bar{\rho} + \rho')\,\partial(U_j + u_j')/\partial x_j = 0$$

$$\partial(\bar{T} + T')/\partial t + (U_j + u_j')\,\partial(\bar{T} + T')/\partial x_j$$

$$- (P + p')/[c_p(\bar{\rho} + \rho')]\,\partial(U_j + u_j')/\partial x_j - K_h\,\partial^2(\bar{T} + T')/\partial x_j\,\partial x_j = 0$$

$$(P + p')/(\bar{\rho}R\bar{T}) = \rho'/\bar{\rho} + T'/\bar{T} + \overline{\rho'T'}/(\overline{\rho T}) \tag{10.23}$$

These equations are general so far, with no restrictions on the "perturbation" terms. However, there is considerable simplification if we make an exception in the case of ρ to retaining the possibility of finite magnitudes for the perturbations. This can be done since $\rho'/\rho \ll 1$ is a very good approximation for nearly all atmospheric problems. Thus, we will substitute from the approximation:

$$1/(\bar{\rho} + \rho') = (1/\bar{\rho})[1 - \rho'/\bar{\rho} + (\rho'/\bar{\rho})^2 + \ldots] \approx (1/\bar{\rho})(1 - \rho'/\bar{\rho}) \quad (10.24)$$

The Mean Flow Equations

$$\partial U_i/\partial t + U_j\,\partial U_i/\partial x_j + \overline{u'_j\,\partial u'_i}/\partial x_j + 1/\bar{\rho}\,\partial P/\partial x_i$$
$$- \rho'/\bar{\rho}^2\,\partial p'/\partial x_i + f\epsilon_{ijk}n_j U_k + g\delta_{i3} - \nu\,\partial^2 U_i/\partial x_j\,\partial x_j = 0 \quad (10.25)$$
$$\partial\bar{\rho}/\partial t + \partial(\rho U_j)/\partial x_j + \partial(\overline{\rho'u'_j})/\partial x_j = 0$$
$$\partial\bar{T}/\partial t + U_j\,\partial\bar{T}/\partial x_j + \overline{u'_j\,\partial T'/\partial x_j} - [P\,\partial U_j/\partial x_j + \overline{p'\,\partial u'_j/\partial x_j}] \times$$
$$[1 + \rho'/\bar{\rho}]/c_p - K_h\,\partial^2\bar{T}/\partial x_j\,\partial x_j = 0$$
$$R\,\overline{\rho'T'} = P$$

The Perturbation Equations (with the primes dropped, and the bars kept for the mean when capitals are not available)

$$\partial u_i/\partial t + u_j\,\partial u_i/\partial x_j - \overline{u_j\,\partial u_i/\partial x_j} + u_j\,\partial U_i/\partial x_j$$
$$+ U_j\,\partial u_i/\partial x_j + f\epsilon_{ijk}n_j u_k + 1/\bar{\rho}\,\partial p/\partial x_i - \rho/\bar{\rho}^2\,\partial P/\partial x_i \quad (10.26)$$
$$+ 1/\bar{\rho}^2(\overline{\rho\,\partial p/\partial x_i} - \rho\,\partial p/\partial x_i) - \nu\,\partial^2 u_i/\partial x_j\,\partial x_j = 0$$
$$\partial\rho/\partial t + \partial(\rho U_j)/\partial x_j + \partial(\bar{\rho}u_j)/\partial x_j + \partial(\overline{\rho u_j})/\partial x_j - \partial(\rho u_j)/\partial x_j = 0$$
$$\partial T/\partial t + U_j\,\partial T/\partial x_j + u_j\,\partial\bar{T}/\partial x_j - \overline{u_j\,\partial T/\partial x_j} + u_j\,\partial T/\partial x_j$$
$$- R/c_p[\bar{T}\,\partial u/\partial x_j + T\,\partial U_j/\partial x_j + T\,\partial u_j/\partial x_j - \overline{T\,\partial u_j/\partial x_j}]$$
$$- K_h\,\partial^2 T/\partial x_j\,\partial x_j = 0$$
$$\frac{p}{\bar{\rho}RT} = \frac{\rho}{\bar{\rho}} + \frac{T}{\bar{T}} + \frac{\rho T - \overline{\rho T}}{\overline{\rho T}}$$

The Linearized Equations

$$\partial u_i/\partial t + u_j\,\partial U_i/\partial x_j + U_j\,\partial u_i/\partial x_j + f\epsilon_{ijk}n_j u_k$$
$$+ 1/\bar{\rho}\,\partial p/\partial x_i - \rho/\bar{\rho}^2\,\partial P/\partial x_i - \nu\,\partial^2 u_i/\partial x_j\,\partial x_j = 0 \quad (10.27)$$
$$\frac{\partial\rho}{\partial t} + \frac{\partial\rho U_j}{\partial x_j} + \frac{\partial\bar{\rho}u_j}{\partial x_j} = 0$$
$$\partial T/\partial t + U_j\,\partial T/\partial x_j + u_j\,\partial\bar{T}/\partial x_j$$
$$- R/c_p[\bar{T}\,\partial u_j/\partial x_j - T\,\partial U_j/\partial x_j] - K_h\,\partial^2 T/\partial x_j\,\partial x_j = 0$$
$$\frac{p}{\bar{P}} = \frac{\rho}{\bar{\rho}} + \frac{T}{\bar{T}}$$

Example 10.5

Discuss the vertical transport of (a) momentum; (b) moisture, through an area dA shown in Fig. 10.7. Relate the transport by perturbation velocity w' to the averaged flux. Consider the average U and W to be zero, density to be constant.

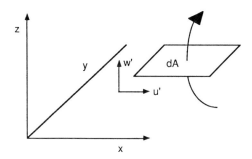

Figure 10.7   Flux through an arbitrary horizontal area dA.

Solution

(a) The mass of fluid flowing vertically through dA in time Δt is

$$\rho\, dV = \rho w'\, dA\, \Delta t$$

This mass of fluid is carrying momentum,

$$u'\, dM = u'\rho w'\, dA\, \Delta t$$

The average of this expression for the momentum transport is

$$\overline{\rho u'w'}\, dA\, \Delta t$$

Thus, there is

$$\overline{\rho u'w'}$$

momentum transported vertically per unit time and area. We noted in Chapter 6 that this is the stress component, τ_{zx}.

(b) For the transport of moisture, simply substitute humidity for momentum. Let specific humidity be

$$q = \bar{q} + q'$$

Then the volume of fluid passing through dA in time Δt is

$$dV = w' \, dA \, \Delta t$$

The moisture of this fluid is $\rho q \, dV$. Hence, the vertical flux of moisture through dA in the time Δt is

$$\rho q w' \, dA \, \Delta t$$

We get the mean moisture flux Q by averaging this expression,

$$Q = \overline{\rho w' q'}$$

It is clear that one can substitute any passive conservative quantity for $\rho u'$ or $\rho q'$ in these expressions and arrive at the vertical flux of heat, pollutant, insects, etc.

Example 10.6

Consider the perturbation equations for two-dimensional, incompressible, inviscid flow. Let the perturbation be a wave propagating in the x-direction with a magnitude that decays in the z-direction. Thus, perturbations are of the general form

$$\varphi = \Phi_\varphi(z) \, e^{ik(x-ct)}$$

where $\Phi_\varphi(z) = C_\varphi \, e^{pz}$ and $c = c_r + i c_i$. Here, Φ represents u, w, or p. The real part of c is the wave velocity; the imaginary part indicates the wave growth or decay with time.

At a fixed point in the flow, the perturbation oscillates with a frequency ωc_r as waves with wave number k pass by. This format assumes that a disturbance can be decomposed into normal modes of various wavelengths. In this example c is real.

Obtain the equation for w in the layer of fluid adjacent to a solid boundary, and the wave amplitude. For simplicity, assume that $u(\text{total})_i = U_i + u_i$ and $U_i = 0$. Assume that the scale is small enough that Coriolis force is negligible. (See Fig. 10.8.)

Solution

In this case, Eqs. (10.27) reduce to

$$\partial u/\partial t + (1/\rho) \, \partial p/\partial x = 0$$

$$\partial w/\partial t + (1/\rho) \, \partial p/\partial z = 0$$

$$\partial u/\partial x + \partial w/\partial z = 0$$

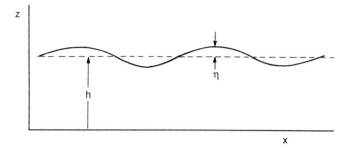

Figure 10.8   Horizontal wave motion in a fluid of depth h.

These equations are first order in the unknowns u, w, and p. To close them, we need boundary conditions on u, w, and p plus initial conditions.

When we substitute the perturbations and cancel the exponential terms that occur in each term,

$$-ickC_u + ikC_p = 0$$

$$-ickC_w + pC_p = 0$$

$$ikC_u + pC_w = 0$$

Thus,

$$C_u = C_p/c; \qquad C_w = (p/ick)C_p; \qquad \text{and} \quad p = \pm k$$

Since we are dealing with the linearized equations, we can add the two solutions for $\pm k$, to get

$$w = (C_{w+}e^{kz} + C_{w-}e^{-kz})\, e^{ik(x-ct)}$$

At the surface, $w(0) = 0$ is satisfied if $C_{w+} = -C_{w-} \equiv C$ and

$$w = C(e^{kz} + e^{-kz})\, e^{ik(x-ct)} = C2 \cosh kz e^{ik(x-ct)}$$

The boundary condition on the pressure is obtained by considering the pressure at $z = h + \eta$, where η is the wave height. For instance, the pressure on the top of ocean waves is the constant atmospheric pressure p_0. Then the pressure along the wave at the z is $p(z) = p_0$ plus the hydrostatic contribution, $-\rho g(z - h) = -\rho g \eta$. Thus,

$$P(z) = \bar{p} + p = p_0 - \rho g \eta + \rho C_{p+}\, 2 \cosh kz e^{k(x-ct)}$$

For small perturbations, $\eta \ll h$, and

$$\eta = 2C_{p+}/g \cosh kh e^{ik(x-ct)} \equiv a e^{ik(x-ct)}$$

where a is the height of the wave at the top.

Finally, the second boundary condition on the velocity appears as

$$\partial\eta/\partial t = w$$

Substitution of η yields

$$-ikcae^{ik(x-ct)} - ga/(ic)\ \tanh kh e^{ik(x-ct)} = 0$$

or

$$c^2 = (g/k)\ \tanh kh$$

In terms of maximum wave height,

$$w = W(z)\ e^{ik(x-ct)}$$

where $W(z) \equiv -(iag/c)\ (\sinh kz/\cosh kh)$.

The speed of propagation of the wave depends on g, h, and wavelength $\lambda = 2\pi/k$. The wave travels with the same speed in either direction.

===

10.6   The Eddy Viscosity Assumption

The Mean Flow equations (10.25) are the same as the basic momentum equations with the addition of the mean perturbation product terms. For the incompressible case, there is only the additional $\overline{u_j'\ \partial u_i'/\partial x_j}$ term. Since the perturbation continuity equation becomes $\partial u_j'/\partial x_j = 0$, the additional term in the momentum equation can be written

$$\overline{\partial u_j' u_i'/\partial x_j} = \overline{u_j'\ \partial u_i'/\partial x_j} + \overline{u_i'\ \partial u_j'/\partial x_j} \qquad (10.28)$$

This term represents the gradient of the momentum flux per unit mass by the perturbation velocities. This is of similar form to the molecular viscous term. We can move this term to the right side of the equation and consider it in combination with the molecular viscous term.

$$(\mu/\rho)\ \partial^2 U_i/\partial x_j\ \partial x_j - \overline{\partial u_i' u_j'/\partial x_j} = (1/\rho)\ \partial/\partial x_j[\mu\ \partial U_i/\partial x_j - \overline{\rho u_i' u_j'}] \qquad (10.29)$$

The perturbation terms represent momentum fluxes by the small turbulent eddies. Thus it is natural to consider representing them as the stresses on the parcel produced as a reaction to the eddy momentum flux through the surfaces of the parcel. In this case, to maintain the continuum concept, the turbulent eddies must be much smaller than the parcel dimensions.

The eddy stress tensor is,

$$\sigma' \equiv \begin{bmatrix} \overline{u'^2} & \overline{u'v'} & \overline{u'w'} \\ \overline{u'v'} & \overline{v'^2} & \overline{v'w'} \\ \overline{u'w'} & \overline{v'w'} & \overline{w'^2} \end{bmatrix} \tag{10.30}$$

The assumption that the eddy stresses were related to the mean shear in a manner similar to the molecular formula was first proposed by Boussinesq in 1877. With this assumption one can then write the viscous terms as

$$\overline{\rho u_i' u_j'} = -\bar{\rho} K \, \partial U_i / \partial x_j = \tau' \tag{10.31}$$

and

$$\partial/\partial x_j [(\nu + K) \, \partial U_i / \partial x_j] \approx \partial/\partial x_j [K \, \partial U_i / \partial x_j]$$

The added stresses are called Reynolds stresses, or sometimes apparent stresses. The equations with this modification for the viscous term are called the Reynolds equations.

The analogy was pursued further by Prandtl when he introduced a length scale—the mixing length—which is analogous to the mean free path of the molecules. The mean free path is the distance that the molecule travels on the average before it hits another molecule and exchanges momentum. However, an eddy of air is not a solid entity. Momentum transfer must be more complicated. The mixing length is a scale representing the average distance an eddy travels before giving up its momentum to the surroundings. This concept has shed some light on the turbulent flow analysis and is employed in boundary layer analysis. However, it gains little practical value over the *ad hoc* employment of the eddy-viscosity in place of molecular viscosity.

Example 10.7

Discuss the exchange coefficient for the moisture flux in the previous example. Use Prandtl's mixing length ℓ_q, as the distance the mass of fluid moves before giving up its moisture excess or deficit to the surroundings. Write a general diffusion relation for moisture transport Q_i in the i-direction.

The gradient of humidity at any point is shown in Fig. 10.9.

Solution

The excess moisture of a mass of fluid which has risen the distance ℓ_q is

$$q' = \ell_q \, \overline{\partial q} / \partial z$$

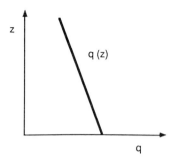

Figure 10.9 A linearly decreasing humidity in the vertical direction.

Thus, the vertical flux of humidity may be written

$$Q = -\rho w' \ell_q \, \partial \bar{q}/\partial z \, \Delta t$$

The negative sign is necessary because a negative moisture gradient implies a positive moisture flux. The average value is

$$Q = -\overline{\rho w' \ell_q} \, \partial \bar{q}/\partial z$$

As with the momentum transport, we can relate the moisture flux to the mean gradient with a diffusion coefficient,

$$Q \equiv -K_q \, \partial \bar{q}/\partial z$$

From these two relations,

$$K_q = \overline{w' \ell_q}$$

We can see that for a scalar like moisture there are three possible K_q's associated with the three possible directions of turbulent flux. The mixing length must also be allowed to vary in different directions.

Since the coefficient can also vary with the gradient, we must write

$$Q_i = -K_{ij(q)} \, \partial \bar{q}/\partial x_i$$

The exchange coefficient for a scalar quantity is a second order tensor. In isotropic turbulence, the moisture exchange coefficient is simply $K_{ij} = K_q = $ constant.

___

10.7 Summary

The equations in this chapter have been developed to allow calculation of flows that consist of a mean plus a perturbation component. The perturbation

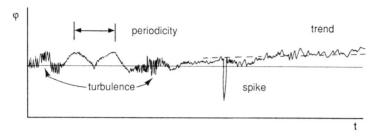

Figure 10.10 Typical data record of $\varphi(t)$ in the atmosphere. In this record are evidence of periodicity, bursts of turbulence, an aberrant point (spike), and an increasing trend.

can be random turbulent eddies or some organized periodic motion. The separation of the basic flow into parts is summarized by looking at a typical data record in Fig. 10.10. This might be a recording of velocity or temperature taken at a point several times/minute for several hours. Or it might be an airplane measurement, taken at very short time intervals over a distance of 100 km.

In the data analysis, obvious errors such as the spike shown in Fig. 10.10 are often removed as spurious. For instance, spikes may arise due to high-frequency radio transmissions interfering with the data transmission. The process is called de-spiking, and must be done with care and proper justification. There are examples where routine de-spiking has eliminated an important flow phenomenon. Then, this aspect of the flow remained to be found by an experimenter with more stringent requirements before eliminating such data.

Averaging can be done over intervals of a few points to the entire data record. The interval must be chosen with respect to the phenomena being investigated. If this is unknown, then a survey using a sequence of intervals can be done. In the typical data record shown in Fig. 10.10, we can assume that there is an overall mean for the entire record. As we choose shorter averaging intervals, periodic motion will be revealed when the period is significantly greater than the averaging interval. To be resolved, the record must contain several complete periods. The averaging process will average out the small-scale turbulence, producing "smoothing" of the record. This will also show any long-term trends in the mean (which might be due to a small portion of a much longer periodicity being measured). The waves are revealed in the averaging interval shown in Fig. 10.11, where the interval is 1/10 the entire data record interval.

Frequently the trend is removed, and the turbulence spectrum is examined by taking intervals as short as possible. But the interval still must contain

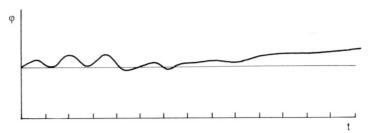

Figure 10.11   The de-spiked and simply smoothed data record.

sufficient data points to constitute a reasonable average, eliminating the small-scale turbulent fluctuations. The periodic phenomena would then be best resolved and the data record would look like Fig. 10.12.

Many other special circumstances arise in examining such a data record. The perturbations may have periodicities that approach the record length and are therefore poorly defined. The random turbulent perturbations may have wavelengths of the same order as the organized perturbations. In this case, special averaging techniques must be designed to resolve the random from the periodic signals. This is much easier to accomplish if the period of an organized perturbation is known, either from observations or theory. The latter can frequently be derived from the perturbation equations using the techniques of stability theory. The simplified PBL equations offer a fertile field for these investigations.

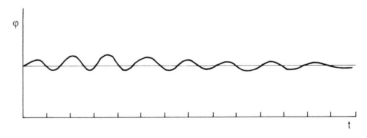

Figure 10.12   The data record with random turbulence averaged out completely and the trend removed.

DEFINITION OF THE MEAN (AVERAGED) FLOW
- → Waves
- → Turbulence
- → Eddy Continuum

REYNOLDS AVERAGING
- → Flux
- → Perturbation Equations

PROCEDURE FOR OBTAINING PERTURBED FLOW EQUA-
TIONS FROM THE NAVIER–STOKES EQUATIONS

Add a perturbation   → *Perturbed Equations*

$$\Phi = \phi + \phi'$$

Take Reynold's average   → *Mean Equations*

$$\overline{\phi'} = 0, \qquad \overline{\phi'p'} \neq 0$$

Neglect all perturbation products   → *Linearized Equations*

$$\phi'p' = 0$$

Subtract mean from perturbed   →*Perturbation Equations*

Problems

1. Perturb the three quantities a, b, and c. Perform Reynolds averaging.

   (a) $ab + c^2$
   (b) a^2bc
   (c) $(ab)^2$
   (d) $(ab)^2$

2. What is the Reynolds averaged version of ρu^2?

3. Consider the thermal energy equation written in terms of potential temperature, θ.

$$\partial\theta/\partial t + \partial\theta u_i/\partial x_i = K\, \partial^2\theta/\partial x_i\, \partial x_i$$

where K is thermal diffusivity and radiation is neglected. Obtain the perturbed equation (in u and θ). What is the equation for steady state and horizontal homogeneity?

4. What is a singular perturbation of an equation? What problems does it cause?

5. Show that the perturbation continuity equation for incompressible turbulent flow in polar coordinates is

$$u_\theta/r + \partial u_r/\partial r + (1/r)\,\partial u_\theta/\partial\theta + \partial u_z/\partial z = 0$$

6. The average value of turbulence is zero and the rms is a measure of intensity. Calculate these for the turbulent velocity data given in the sketch. Here, $u = u_{mean} + u'$. Measurements are taken once per second.

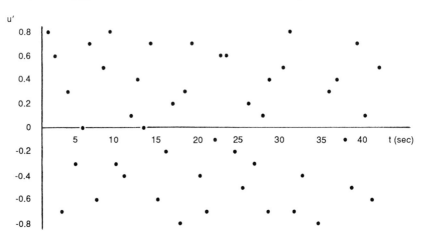

7. Consider the steady, two-dimensional turbulent flow between parallel plates. Derive the Reynolds stress equations for this flow. Obtain an approximate analytic solution for the flow.

8. Derive the perturbed energy equation.

9. For large scales, the vertical component of pressure gradient approximately balances the gravitational acceleration. Perturb the pressure and density from a base state, $p = p_0(z)$ and $\rho = \rho_0(z)$, such that $dp_0/dz = -\rho_0 g$, to obtain a perturbed momentum equation. You should obtain a buoyancy force balancing a perturbation pressure gradient.

10. Consider the u-momentum equation (10.22) only. Separate the perturbation into an organized part, $\mathbf{u}'$, plus a random part, $\mathbf{u}''$. Hence

$$\mathbf{u} = \bar{\mathbf{u}} + \mathbf{u}' + \mathbf{u}''$$

Derive the perturbed momentum equation.

11. The equation for the velocity covariances can be obtained from the momentum equation in u_i multiplied by u_k plus the equation in u_k times u_i. Thus, one gets an equation in $\partial u_i u_k/\partial t$. Obtain the equation for the averaged covariance. List the seven terms and their functions. (e.g., buoyancy production terms).

Chapter 11 | Boundary Layers

11.1 Introduction

We found that when the flow is inviscid and irrotational, the potential flow solutions use elegant mathematics to obtain solutions to simplified equations. These results have important uses in many applied problems. They provide flow solutions that fit well with the observations as long as the observations are done on the proper scale and in the proper domain. However, the potential solutions are often severely limited in both time and spatial domains. For instance, the simple potential flow solutions apply to large-scale (synoptic) flows. But these equations develop instabilities. This results in pockets of irrotationality and regions of strong divergence.

When large-scale numerical models are integrated over long periods, they usually must include viscous effects. This helps prevent spurious wave growth (viscosity has a damping effect) and accounts for dissipation in the system.

Finally, the conditions for potential flow are not valid when the flow velocity is forced to undergo large changes in a relatively short distance. In this case, the velocity shear must become large. Hence viscous terms like $v \, du/dx$ become large, and rotation is usually present.

Thus, one condition for potential flow to be a viable solution is that the domain not contain a boundary between fluid regimes that have significantly different flow parameters. In the atmosphere and ocean, stratification effects

411

can result in adjacent horizontal layers with large differences in dynamic and thermodynamic states. When stratification is stable in a layer, the instabilities that lead to small-scale turbulence are suppressed. The flow becomes eddy-laminar, with little diffusive mixing. Two layers with very different properties can then flow side by side. The relatively thin layer between them must have flow that depends on viscous forces to balance the large shear in the layer. The dynamics of each of the adjacent layers can often be treated as inviscid to obtain their large-scale flow. The boundary is then viewed as a discontinuity where the two solutions are patched. However, the boundary layer between the two flows is often the source of large-scale waves, which may grow to produce the dominant flow character of the boundary region. In this way, boundary-layer dynamics is often behind the basic wave systems of the atmosphere.

One case of freestream boundary flow occurs at the edge between two large-scale fluid masses. A vertical edge near the surface is called a frontal region. Many important weather phenomena are mainly associated with the properties of such boundaries. Since gradients are largest normal to such fronts, we can expect the equations that govern the flow in these regions to be susceptible to simplifications that emphasize the importance of the one-dimensional gradients.

One place where the shear is inevitably large is in the region adjacent to a solid surface. Here, the flow must feel the effects of the surface as frictional forces must act to bring the flow to a halt. Such boundary layers characterized by large velocity gradients are extremely important in many disciplines. However, boundary layers can be associated with large gradients in temperature, pollution, or any other parameter.

We live in a high-velocity shear layer that is hundreds of meters to kilometers thick. The fluxes that drive the freestream flow must pass through this layer and will depend on its characteristics. This atmospheric boundary layer is often called the planetary boundary layer, or PBL. Most of the discussion in this chapter will involve the PBL. Still, the concepts of boundary layer theory apply equally well in all relatively thin layers that contain large gradients in fluid properties.

Since boundaries were always present in practical flows, there was a feeling between the 1750s and 1900 that the potential flow solutions were simply curiosities, with no practical importance. How valuable could a solution for the flow over a wing be if it predicted no drag? Many felt that this serious flaw indicated that the entire solution had no validity. The arrival of the boundary layer concept in 1904 was a masterstroke that rescued the potential flow solutions for use in calculating the general mean flow. This idea opened up a new method of flow analysis for flows that include boundaries.

The idea that results in a *boundary layer* can be seen as a logical extension of concepts already developed in the previous chapters. Much basic work was done by Ludwig Prandtl. He was concerned with laboratory flows of water and air. At the same time, V. Walfrid Ekman found a derivation for the ocean PBL. Ekman had to include the virtual force in his equations to accommodate a rotating frame of reference. In both cases, the continuity equation was assumed to be simply the condition of nondivergence. However, the assumption that the flow was irrotational, which yielded the inviscid Euler equations, was relaxed and the viscous term now appears in these equations.

The Reynolds number (Re $= \rho UL/\mu$) is the ratio of inertial to viscous forces. Inertial terms are of order ρU^2. Viscous terms are of order $\mu U/L$). Re is used as an inverse measure of the importance of the viscous term in the momentum equations. It is very large in atmospheric problems. This is often because L is very large, but it is also because μ is small. The molecular viscosity is very small, to the point that a Reynolds number based on μ has little or no meaning in atmospheric problems.

At the high Re of atmospheric flow, transition to turbulence is usually assured. The Reynolds stress terms of Chapter 10 must be used. Even with the *ad hoc* eddy-viscosity parametrizations, where the eddy viscosity is of order $10^6\mu$, the Reynolds numbers are generally huge. This suggests that the viscous forces should be negligible with respect to inertial terms. Indeed the inviscid approximate equations have met with great success for applications in the atmosphere and ocean. These results include the geostrophic flow solutions, which are a balance between pressure gradient and Coriolis force. There are many inviscid modifications to these equations to include various stratification and inertial terms. Thus the potential flow solutions might be expected to be reasonable approximations if one were interested in the flow around small objects (say a mountain) within the large-scale flows.

However, the viscous force due to small-scale turbulence is related to the vertical derivative, as discussed in Sections 1.11.5 and 10.6. When the flow is near a surface, there is a source of mechanically generated turbulence, hence eddy viscosity, at the surface. The large magnitude of the eddy-viscosity coefficient K reflects the large transport capability of the eddy motion. The no-slip boundary condition, $u \to 0$, is a good approximation for the surface. The velocity shear is confined to a thin layer of depth H. Hence, the vertical stress term, $K\,du/dz$, gets large near the surface. In this case, the Reynolds number UH/K can get arbitrarily small as the boundary is approached and H decreases. The result is that the viscous terms become important in the equations. In fact, as one approaches *very* close to the surface (within a centimeter for the atmosphere), there is no room for turbulence and a laminar *sublayer* exists where only molecular viscosity is significant.

The potential flow solution is quite successful in describing the flow patterns around and very close to objects in the laboratory, where molecular viscosity sets the magnitude of the viscous term in a laminar flow. This suggests that the domain where the viscous terms become important and must be included in the equations is very thin. This is the essential hypothesis of Prandtl in his boundary layer concept for the molecular boundary layer.

The question of how thick is the geophysical turbulent boundary layer would seem to involve the distribution of turbulence, hence the value of the eddy viscosity. The inviscid equations may be assumed to apply in the geophysical freestream. We assume that $K \approx 0$ far away from where the turbulence is being generated at the boundary. In a remarkable exact solution of the N–S equations, Ekman showed that the departure of the PBL velocity from the freestream velocity decayed exponentially with distance from the surface. Hence the PBL is thin.

The implication that the PBL is thin is true independent of the details of the assumptions for the eddy-viscosity distribution. Ekman's solution applies for horizontally homogeneous, steady-state, two-dimensional flows in a geophysical (rotating) frame of reference. The main viscous forces are those associated with the vertical shear, $\approx \rho K \, dU/dz$. Since the vertical shear becomes small in the freestream, the value of K becomes moot at the top of the PBL. For this reason, a constant K assumption yields an adequate first approximation for the PBL equations. The only requirement is that there be an eddy continuum.

The two contemporary analyses by Prandtl and Ekman were, and still are, essential to the development of fluid dynamic solutions in most practical applications. Although many solutions may proceed without direct note of the boundary effects, they all depend on the concept for justification of the inviscid solutions. When the boundary-layer effects are ignored, the solutions ultimately fail in some respect. Frequently the success or failure of a large-scale numerical integration of the complete Navier–Stokes equations will depend on how well the boundary layers are parametrized.

The importance of the boundary-layer effect can be seen in the observation that the fluid must ultimately come to rest at the surface. This is nature's no-slip condition, and it is also a not-so-evident idea. But if it is true, it strongly implies that there is a thin layer with very large shear right next to the surface. The boundary conditions at the surface must be changed from the potential flow requirement that the normal component of velocity is zero (the surface is a streamline). The new condition is that the velocity *vector* must be zero at the boundary. The added boundary condition on the velocity can only be satisfied by the higher-order viscous equations. This is related to the fact that a differential equation is solved for the dependent variable by integrating. It must be integrated as many times as the highest-order term in the equation. Each integration of the equation yields a constant to be evaluated by a boundary condition. Thus, if we have approximated a fourth-order equation with a third-order equation, we have lost the ability to satisfy a boundary condition. This may, or may not, be important to the solution.

In the case of the potential flow equations, the highest-order terms, the viscous terms, have been dropped from the momentum equations. The no-slip boundary condition cannot be satisfied, and no surface shear stress results. The mathematical term for this approximation to the full equations is a *singular perturbation* (of the equations).

Once more, it should be stressed that this chapter deals with concepts that apply to thin layers where rapid transition between different boundary conditions takes place. Such boundaries abound in geophysical flows. The important ideas are those that lead to simplification of the basic equations.

There is an added reason for including this chapter in a text on fundamentals. Most of the concepts introduced in this text are employed in getting the PBL equations and the solutions to these equations. Thus this chapter provides a long exercise in the use of the material in this text.

Example 11.1

Viscosity is a measure of the fluid ability to transmit forces laterally to the mean flow. In a boundary-layer flow, one generally assumes horizontal flow. Sketch the expected velocity profile for a laminar and a turbulent boundary layer. First, consider molecular viscosity. Then, discuss PBL turbulence with different magnitudes. Comment on the stratification effect on K and the surface stress variation. Assume that the freestream velocity is fixed at 10 m/sec.

Solution

The upper boundary condition is the constant freestream flow. Thus, there is a fixed source of momentum. A laminar layer transmits momentum by diffusion, layer to layer, following

$$\tau = \rho v \, dU/dz$$

When small-scale turbulence is parametrized as diffusion,

$$\tau = \rho K \, dU/dz$$

We have seen that $K \gg \nu$, so the turbulent boundary layer is much more efficient in transmitting momentum toward the surface. Figure 11.1. shows the effect on the velocity profile. Note that the slope at the surface must be much larger for the turbulent layer. Hence,

$$\tau_0 = \rho K \, dU/dz(0) \gg \nu \, dU/dz(0)$$

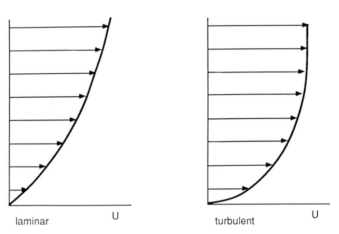

laminar   U   turbulent   U

Figure 11.1 Velocity profiles for laminar and turbulent boundary layers.

When we are dealing with PBLs, turbulence is inevitable and an eddy viscosity is used.

$$\tau = \rho K \, dU/dz$$

The eddy viscosity can depend on time, as stratification conditions change. Thus, when the layer is stably stratified, the small-scale turbulence is suppressed, K is smaller, and profiles are more like the laminar in Fig. 11.1. The expected profiles for the unstably stratified case will be more like the turbulent profile in Fig. 11.1, and surface stress will be greater.

There is another factor affecting the value of K. The small-scale turbulence is generated near the surface, either mechanically or by dynamic instabilities. Thus, $K \propto z$. In stable stratification, K may decrease rapidly outward, so that the PBL is thin. The retarding friction of the surface is not felt as deep into the freestream. This effect tends to allow high-momentum flow nearer to the surface and would seem to increase the velocity gradients there. However, the net effect is reversed in the layer very near the surface—the surface layer. The result is that the PBL is thicker and the surface stress is higher in unstable stratification.

11.2 The Boundary Layer Concept

In 1904, Prandtl presented his boundary layer hypothesis: The boundary layer is thin, such that it does not significantly alter the shape of a body with regard to the flow around it. Therefore the potential flow solution is valid everywhere except within this thin region.

Although this concept may seem evident in view of the discussion in Section 11.1, Prandtl went on to explain how to use this idea to get practical solutions. Since the boundary layer is thin and the main action in the layer is to bring the velocity to a halt, one should be able to simplify the equations by keeping only vertical changes. This leaves us with a two-dimensional (horizontally homogeneous), parallel flow. However, the acceleration (deceleration) terms must be important in addition to the viscous terms. The density is generally assumed constant (and when we extend the theory to the atmospheric boundary layer, we assume density has a small change across the thin layer). The horizontal pressure distribution can be assumed to be fixed by the large-scale flow dynamics and to be impressed on the boundary layer. That is, pressure distribution doesn't vary in the vertical. The static pressure gradient of the freestream flow is equal to the static pressure gradient at the surface. The process of approximating the complete Navier–Stokes equations to obtain the boundary layer equations uses the techniques of dimensional analysis, dynamic similarity, and asymptotic approximations.

Flow visualization techniques have substantiated many of these assumptions. Figures 11.2 and 11.3 clearly show the thinness of the boundary layer and the strong velocity shear to zero slip at the surface. The sketch in Fig. 11.4 suggests that several boundary layer concepts may be needed to span the huge scales involved in the atmosphere. A very thin layer is associated with the roughness parameter Z_0. This is a characteristic height scale arising from the mathematics but which may be related to the roughness elements of the surface. Many important phenomena in the boundary layer depend significantly on the stratification of the layer. If this is negative, with colder air overlying warm air, one might expect more turbulence to develop with the help of positive buoyancy. Positive stratification would be expected to suppress turbulent growth. Many parameters and phenomena are found to change with stratification. These include the similarity-derived Obukhov characteristic height L the secondary flow (large-eddy) effects, heat flux, and convective towers. However, the concepts of the boundary layer approximation are easier to deal with in the neutrally stratified case.

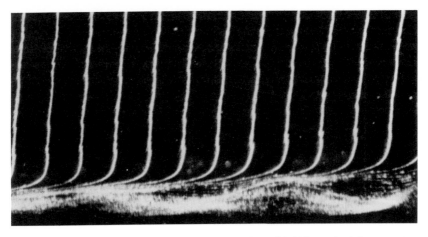

Figure 11.2 Boundary-layer flow over a wall. Lines of bubbles emitted from a cathode wire mark the velocity profiles. (Photograph by H. Bippes, NASA TM-75243, 1978.)

Figure 11.3 Cone cylinder in supersonic free flight. The laminar boundary layer before the vertex is too thin to see. The turbulent BL behind the vertex is evident (as is the wake). (Photograph by A.C. Charters; from "An Album of Fluid Motion," assembled by M. Van Dyke and published in 1982 by Parabolic Press, Stanford, California.)

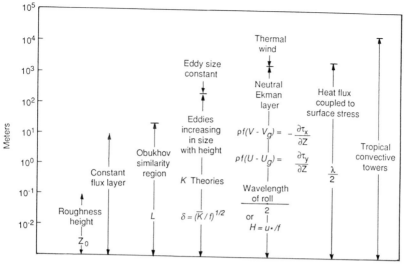

Figure 11.4.   Sketch of various vertical scales in the PBL.

11.3   Boundary Layer Equations

Now we assume that the flow outside a thin boundary layer is solvable with potential theory, and we seek a separate solution for the thin layer adjacent to a surface. We then hope to match this solution to the outer solution at a small distance from the surface, equal to the height of the boundary layer.

For the thin boundary layer region, the viscous terms must be brought back into the equations, and the inertial terms are retained. Later, we may wish to approximate the equations for the case of steady-state flows.

Since we are concerned with a thin layer near the surface, all vertical distances are relatively small, and we assume ρ is constant. The horizontal extent of the problem is also usually small enough that one can assume the Coriolis force is constant.

The most important feature from the scaling point of view is that there are different characteristic scales for vertical and horizontal directions. We name horizontal characteristic scales of length and velocity as X and U and the vertical scales as H and W. Since we are considering a thin layer with nearly horizontal flow, H/X and W/U are much less than one.

Now we can nondimensionalize the terms in the momentum equations (Chapter 6) with the characteristic values of all variables. With hats denoting dimensional variables, let

$$u = \hat{u}/U, v = \hat{v}/U, w = \hat{w}/W, x = \hat{x}/X, y = \hat{y}/X,$$
$$z = \hat{z}/H, p = \hat{p}/P, \text{ and } t = \hat{t}/T \qquad (11.1)$$

Generally in this section, all dependent variables are nondimensional. The cumbersome "hats" can be dropped except where the distinction between dimensional and nondimensional must be made.

The equations in terms of the nondimensional variables are obtained by substituting Eqs. 11.1 into the basic momentum equations. There arise coefficients for each term that are combinations of the characteristic parameters.

$$[U/T] \partial t + [U^2/X](u \, \partial u/\partial x + v \, \partial u/\partial y) + [UW/H] \, w \, \partial u/\partial z$$
$$-[Uf] \, v + [P/\rho X] \, \partial p/\partial x - [KU/X^2](\partial^2 u/\partial x^2 + \partial^2 u/\partial y^2)$$
$$-[KU/X^2 \cdot (X/H)^2] \, \partial^2 u/\partial z^2 = 0 \qquad (11.2)$$
$$[U/T] \partial v/\partial t + [U^2/X](u \, \partial v/\partial x + v \, \partial v/\partial y) + [UW/H] \, w \, \partial v/\partial z$$
$$+ [Uf] \, u + [P/\rho X] \, \partial p/\partial y - [KU/X^2](\partial^2 v/\partial x^2 + \partial^2 v/\partial y^2)$$
$$-[KU/X^2 \cdot (X/H)^2] \, \partial^2 v/\partial z^2 = 0$$
$$[U/T] \partial w/\partial t + [U^2/X](u \, \partial w/\partial x + v \, \partial w/\partial y) + [W^2/H] \, w \, \partial w/\partial z$$
$$+ [P/(\rho U^2) \cdot U^2/H] \, \partial p/\partial z - [KW/X^2](\partial^2 w/\partial x^2 + \partial^2 w/\partial y^2)$$
$$-g - [KW/X^2 \cdot (X/H)^2] \, \partial^2 w/\partial z^2 = 0$$

For simplicity, we assume that T, the characteristic time scale, is large, and assume steady-state, $T \to \infty$.

These are the complete non-dimensionalized equations. Since each term has been nondimensionalized with its characteristic value, we assume that all of the independent and dependent variables are of order unity.

Our goal is to estimate the order of each term in Eqs. 11.2. This will be made easy by dividing by the coefficient of one term. When this is done, the term with no coefficient is of order unity, and the magnitude of the coefficient of all other terms apportions their importance relative to unity.

The choice of reference term is arbitrary. For our example we pick the inertial terms in the horizontal equations and the pressure gradient in the vertical momentum equation. Thus, in the first case, multiply the equation by X/U^2, and in the second, multiply by H/U^2.

$$(u \, \partial u/\partial x + v \, \partial u/\partial y) + [X/H][W/U] \, w \, \partial u/\partial z - [fX/U]v$$
$$+ [P/(\rho U^2)] \, \partial p/\partial x - [K/(UX)](\partial^2 u/\partial x^2 + \partial^2 u/\partial y^2)$$
$$- [K/(UX) \cdot (X/H)^2] \partial^2 u/\partial z^2 = 0$$

$$(u\, \partial v/\partial x + v\, \partial v/\partial y) + [X/H][W/U]\, w\, \partial v/\partial z + [fX/U]\, u$$

$$+ [P/(\rho U^2)]\, \partial p/\partial y - [K/(UX)](\partial^2 v/\partial x^2 + \partial^2 v/\partial y^2)$$

$$-[K/(UX)\cdot(X/H)^2]\, \partial^2 v/\partial z^2 = 0$$

$$[H/X \cdot W/U](u\, \partial w/\partial x + v\, \partial w/\partial y) + [(W/U)^2]w\, \partial w/\partial z$$

$$+ [D]\, \partial p/\partial z - [K/(UX)\cdot W/U \cdot H/X](\partial^2 w/\partial x^2 + \partial^2 w/\partial y^2)$$

$$- [gH/U^2] - [K/(UX)\cdot W/U \cdot (X/H)]\, \partial^2 w/\partial z^2 = 0 \qquad (11.3)$$

We will now simplify these equations by employing the boundary layer assumptions,

$$(1) \qquad H/X \ll 1$$

$$(2) \qquad W/U \ll 1 \qquad\qquad (11.4)$$

We have already assumed that ρ is constant. Now use (1) and (2) to neglect any terms with coefficients much less than 1. The pressure terms must be retained as they can be arbitrarily large, depending on the imposed pressure gradient. Note that in the viscous terms, the last term is very much larger than the first, and only it needs to be retained. This can be seen from

$$\frac{K}{UK} \ll \frac{K}{UX}\left[\frac{X}{H}\right]^2 \approx 1 \qquad\qquad (11.5)$$

For the boundary layer, the viscous term must be the same order (unity) as the other terms,

$$\mathrm{Re} = UX/K \approx (X/H)^2 \qquad\qquad (11.6)$$

Therefore, this is known as the large Reynolds number approximation. The equations can be written in terms of the conventional dimensionless parameters,

$$D \equiv P/(\rho U^2), \qquad \mathrm{Ro} \equiv U/(Xf), \qquad \text{and} \quad \mathrm{Re} \equiv UX/K$$

The steady-state, incompressible momentum equations are

$$(u\, \partial u/\partial x + v\, \partial u/\partial y) + [X/H][W/U]\, w\, \partial u/\partial z - [1/\mathrm{Ro}]\, v$$

$$+ [D]\, \partial p/\partial x - [1/\mathrm{Re}\cdot(X/H)^2]\, \partial^2 u/\partial z^2 = 0$$

$$(u\, \partial v/\partial x + v\, \partial v/\partial y) + [X/H][W/U]\, w\, \partial v/\partial z + [1/\mathrm{Ro}]\, u$$

$$+ [D]\, \partial p/\partial y - [1/\mathrm{Re}\cdot(X/H)^2]\, \partial^2 v/\partial z^2 = 0$$

$$\partial p/\partial z - [\rho g H/P] - [1/\mathrm{Re}\cdot W/U \cdot (X/H)^2]\, \partial^2 w/\partial z^2 = 0 \qquad (11.7)$$

Since most of our flows are pressure-gradient driven, we can always include the horizontal pressure-gradient term by assuming that $D \approx 1$. Hence $P \approx \rho U^2$, called the dynamic pressure.

We have made the assumption that the pressure is impressed on the layer. In our equations, this is equivalent to assuming that $\rho g H \ll P$. However, for the PBL, we can relax this assumption and retain the hydrostatic relation without coupling the z-momentum equation to the horizontal equations. This is because the vertical momentum equation has the additional factor W/U in the viscous term. This is generally very small, hence the vertical momentum equation is simply

$$-\partial p/\partial z = \rho g \qquad \text{(or 0)} \tag{11.8}$$

Employing the same scaling procedure, the continuity equation is:

$$\partial u/\partial x + \partial v/\partial y + [X/H \cdot W/U]\, \partial w/\partial z = 0 \tag{11.9}$$

The *dimensional equations* to be solved for the boundary layer are obtained by substituting $\hat{u} = uU$, etc., in Eqs. (11.7)–(11.9). In Eqs. 11.10, all variables are dimensional (hats have been dropped).

$$u\, \partial u/\partial x + v\, \partial u/\partial y + w\, \partial u/\partial z - fv + 1/\rho\, \partial p/\partial x - K\, \partial^2 u/\partial z^2 = 0$$

$$u\, \partial v/\partial x + v\, \partial v/\partial y + w\, \partial v/\partial z + fu + 1/\rho\, \partial p/\partial y - K\, \partial^2 v/\partial z^2 = 0$$

$$\partial u/\partial x + \partial v/\partial y + \partial w/\partial z = 0 \tag{11.10}$$

$$\partial p/\partial z = 0$$

Although many terms are gone, these are nonlinear equations and thus not easily solved. The pressure gradient is usually given as a result of the freestream flow solution. The Coriolis parameter is given. And the coefficient of the viscous term is assumed to be known. For laminar laboratory flows, it is simply the molecular kinematic viscosity. For turbulent flow, it is an eddy viscosity. The boundary conditions are $\mathbf{u}(0) = 0$ and $\mathbf{u}(\infty) = U$. The set of equations is closed and amenable to numerical solutions. Several particular solutions to these equations have been found for special conditions, but no general solutions exist. If we assume horizontally homogeneous flow, $\partial/\partial x = \partial/\partial y \approx 0$, the nonlinear terms are negligible and we can obtain the planetary boundary-layer equations of Ekman.

Example 11.2

Consider a PBL with the following conditions (see Fig. 11.5):

Height = 1 km; horizontal scale = 1000 km; geostrophic flow at the top

= 10 m/sec; mean vertical flow = 1 cm/sec; eddy viscosity = 50 m^2/sec; and constant density = $1.23 \cdot 10^{-3}$ gm/cm^3. Latitude is 45°.

Calculate the parameters that determine whether the boundary layer approximation is valid. Evaluate the magnitude of the terms that can be neglected in the momentum equations.

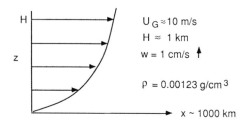

Figure 11.5   Sketch of a typical planetary boundary layer velocity profile.

Solution

The basic relations used in the boundary-layer approximation are:

$$f = 2\Omega \sin 45° = 2\pi/(24 \text{ hr}) = 1.0 \cdot 10^{-4} \text{ 1/sec}$$

$$H/X = 1 \text{ km}/1000 \text{ km} = 0.001$$

$$W/U = 0.01 \text{ m/sec}/10 \text{ m/sec} = 0.001$$

In Eqs. (11.3), the horizontal advection terms are of order unity. We can substitute to get the relative value of each of the other terms. The u-momentum equation is (v-momentum is similar)

$$(u\, \partial u/\partial x + v\, \partial u/\partial y) + [1000][0.001]\, w\, \partial u/\partial z - [1 \cdot 10^6 \cdot 1.0 \cdot 10^{-4}]v$$

$$+ [D]\, \partial p/\partial x - [50/(10 \cdot 1000 \cdot 10^3)](\partial^2 u/\partial x^2 + \partial^2 u/\partial y^2)$$

$$- [50/(10 \cdot 10^6) \cdot (1000/1)^2]\, \partial^2 u/\partial z^2 = 0$$

or

$$(u\, \partial u/\partial x + v\, \partial u/\partial y) + [1]\, w\, \partial u/\partial z - [100]v$$

$$+ [D]\, \partial p/\partial x - [3 \cdot 10^{-6}](\partial^2 u/\partial x^2 + \partial^2 u/\partial y^2)$$

$$- [5]\, \partial^2 u/\partial z^2 = 0$$

The value of D is equal to the coefficient of the Coriolis term at the top of the layer, where the geostrophic balance prevails. Hence $D \approx 100$. The other term that may become important is the vertical viscous term. Since this coefficient is $[KX/(UH^2)]$, it evidently becomes more important as the layer becomes thinner. Hence there is a balance between Coriolis, pressure gradient, and viscous forces.

The vertical momentum equation is

$$[1](u\ \partial w/\partial x + v\ \partial w/\partial y) + [10^{-6}]w\ \partial w/\partial z$$

$$+ [D]\ \partial p/\partial z - [5 \cdot 10^{-6}](\partial^2 w/\partial x^2 + \partial^2 w/\partial y^2)$$

$$- [10\ \text{m/sec}^2 \cdot 10\ \text{sec}^2/\text{m}] - [5 \cdot 10^{-3}]\ \partial^2 w/\partial z^2 = 0$$

For $D \approx 100$, the vertical pressure-gradient term must be balanced by the gravitational force term. This leaves the hydrostatic balance.

These scaling arguments tell more than the basic balance. They show that the hydrostatic balance terms are usually an order of magnitude greater than any others, and that only the horizontal advection terms might have an effect. In the horizontal advection equations, the viscous term is only slightly greater than advection terms. They quite likely may be important if horizontal gradients are a little larger than indicated by the given assumptions.

11.4 Ekman's Planetary Boundary Layer Solution

Around the time when Prandtl was presenting his boundary layer concepts in Heidelburg, Ekman was publishing his specific solution for the PBL. This particular approximation of the Navier–Stokes equations was suggested to Ekman by Fjortoft Nansen, who had returned from his two-year voyage across the Arctic with his ship frozen in the pack ice. Nansen provided the information that the drift of the ice was always at a significant angle to the right of the surface-wind direction. The deviation was evidently due to the earth's rotation. This observation was all Ekman needed to simplify the N–S equations to the point that he was able to find an analytic solution. Note that Ekman was working with the oceanic PBL; however, he recognized that the equations were equally applicable to the atmospheric PBL.

Ekman did not pursue the formalized scaling arguments, as did Prandtl. He simply postulated that the flow in the PBL was slowly varying (steady-state), horizontally homogeneous (eliminating the inertia terms), and there was negligible vertical motion relative to the horizontal. He was left with a balance between the Coriolis, pressure gradient, and viscous terms. The uniform horizontal flow left him with essentially a one-dimensionally vertically varying, two-dimensional flow problem. Ekman's equations are

$$-fv + 1/\rho \; \partial p/\partial x - K \; \partial^2 u/\partial z^2 = 0$$

$$fu + 1/\rho \; \partial p/\partial y - K \; \partial^2 v/\partial z^2 = 0 \qquad (11.11)$$

$$\partial u/\partial x + \partial v/\partial y = 0$$

Example 11.3

Obtain Eqs. (11.11) from Eqs. (11.2) by formally using the methods of nondimensionalization and scaling.

Solution

We need only to put Ekman's assumptions into corresponding assumptions on the characteristic scales. Since the pressure gradient term is going to be retained as the driving force, we choose to multiply the horizontal equations by the inverse of the coefficient of this term $[\rho X/P]$. Likewise, multiply the vertical momentum equation by H/U^2. This yields

$$[\rho XU/PT] \; \partial u/\partial t + [\rho U^2/P](u \; \partial u/\partial x + v \; \partial u/\partial y) + [\rho XUW/HP]w \; \partial u/\partial z$$

$$-[\rho XUf/P] \; v + \partial p/\partial x - [\rho KU/XP](\partial^2 u/\partial x^2 + \partial^2 u/\partial y^2)$$

$$- [\rho KU/XP \cdot (X/H)^2] \; \partial^2 u/\partial z^2 = 0 \qquad (11.12)$$

$$[\rho XU/PT] \; \partial v/\partial t + [\rho U^2/P](u \; \partial v/\partial x + v \; \partial v/\partial y) + [\rho XUW/HP]w \; \partial v/\partial z$$

$$+ [\rho XUf/P] \; u + \partial p/\partial y - [\rho KU/XP](\partial^2 v/\partial x^2 + \partial^2 v/\partial y^2)$$

$$- [\rho KU/XP \cdot (X/H)^2] \; \partial^2 v/\partial z^2 = 0$$

$$[H/X \cdot W/U](u \; \partial w/\partial x + v \; \partial w/\partial y) + [(W/U)^2]w \; \partial w/\partial z$$

$$+ [D] \; \partial p/\partial z - [K/(UX) \cdot W/U \cdot H/X](\partial^2 w/\partial x^2 + \partial^2 w/\partial y^2)$$

$$-[gH/U^2] - [K/(UX) \cdot W/U \cdot (X/H)] \; \partial^2 w/\partial z^2 = 0$$

Now examine each term as represented by the size of its coefficient:

1. The characteristic time T is assumed very long, so that steady state is obtained, or $[\rho UX/PT] = 0$.

2. Now the coefficient of the inertial terms is $\rho U^2/P \equiv 1/D$, where D is the ratio of the pressure to the dynamic pressure, ρU^2. When we wished to retain the pressure gradient term as the same order of magnitude as the

inertial terms, we set $D \leq 1$. However, in the boundary layer we are assuming that the inertial terms are small, and $D \gg 1$. Thus $[\rho U^2 / P] \approx 0$, and also $[\rho XUW / HP] = [\rho U^2 / P \cdot W / U \cdot X / H] \approx 0$.

3. The flow is assumed horizontal, so that characteristic scale $W \leq 0$, and hence the only terms left in the vertical momentum equation are

$$\partial p / \partial z = [gH / U^2 D]$$

Note that the largeness of D and the smallness of H both imply that pressure variation across the boundary layer is small. (This also leaves only horizontal gradients in the continuity equation.)

4. For a boundary layer solution, we must retain at least some of the viscous terms. Note that the vertical curvatures $(\partial^2 / \partial z^2)$ differ from the horizontal by a factor of $(X/H)^2$. For a thin PBL, $X/H \gg 1$, which keeps only the vertical stress gradient terms. (Thus Ekman's assumption that vertical turbulent fluxes are much greater than those in the horizontal was tantamount to assuming a thin boundary layer.)

5. Since the Coriolis force is assumed to be important, the coefficient $[\rho U^2 / P \cdot Xf / U] \leq 1$. Since $\rho U^2 / P = 1/D$ is very small, the quantity Xf / U must be very large. This is known as the Rossby number and serves as a criterion as to whether the Coriolis force is an important term in the approximate equations.

We are left with

$$- [\rho XUf / P] \, v + \partial p / \partial x - [(\rho KU / XP) \cdot (X/H)^2] \, \partial^2 u / \partial z^2 = 0$$

$$[\rho XUf / P] \, u + \partial p / \partial y - [(\rho KU / XP) \cdot (X/H)^2] \, \partial^2 v / \partial z^2 = 0$$

When the nondimensional parameters are replaced with the dimensional ones, Eqs. (11.11) result.

This systematic process appears to be quite cumbersome when compared to Ekman's direct assumption of the "important" terms. However, one of the advantages of this process is that the critical aspects of the assumptions are revealed. For instance, the smallness of H/X, $\rho U^2 / P$, and $\partial p / \partial z$, and the relative size of some of the other terms are defined.

We can solve Eqs. (11.11) with no-slip boundary conditions.

$$u(0) = v(0) = 0, \qquad v(\infty) = 0, \qquad u(\infty) = U_g$$

Here, U_g is defined as the freestream—the geostrophic wind (or, if centrifugal terms are included, the gradient wind). The coordinate axes can be aligned with the x-direction—along the U_g-direction—so that $V_g = 0$. There are no specifications on the PBL thickness.

In addition, the equations look simpler if we note that the upper boundary conditions will be the freestream flow. In this region, Eqs. (11.11) will hold true without the viscous terms. Hence,

$$fU_g = -(1/\rho)\, \partial p/\partial y, \qquad fV_g = (1/\rho)\, \partial p/\partial x = 0 \qquad (11.13)$$

Therefore we can substitute (U_g, V_g) for the impressed pressure-gradient terms in Eqs. (11.11).

Then, nondimensionalize u and v with $G \equiv |U_g|$ and z with some characteristic height H,

$$\hat{u} = u/G, \qquad \hat{v} = v/G, \qquad \hat{U}_g = U_g/G = 1, \qquad \text{and} \quad \hat{z} = z/H$$

and drop hats, since all variables are henceforth nondimensional. Divide by fG to make the Coriolis and pressure-gradient terms of order unity

$$[K/fH^2]\, \partial^2 u/\partial z^2 + v = 0$$
$$[K/fH^2]\, \partial^2 v/\partial z^2 - [u - 1] = 0 \qquad (11.14)$$

These equations can be made even simpler if the velocity is transformed to

$$\tilde{u} = u - U_g \qquad (\text{or } \tilde{u} = \hat{u} - 1)$$

Once again, coefficients are collected in one term, hats and tildes are dropped.

$$[K/(fH^2)\,]\, \partial^2 u/\partial z^2 + v = 0$$
$$[K/(fH^2)\,]\, \partial^2 v/\partial z^2 - u = 0 \qquad (11.15)$$

with

$$u(0) = -1 \qquad v(0) = 0 \qquad u(\infty) = v(\infty) = 0$$

There is no actual characteristic height of the PBL in this problem statement. We can obtain a characteristic height from a dimensional analysis of the important parameters (determined as those that appear in the equations and boundary conditions). This is $H = \sqrt{K/f} \equiv \delta$. Using $H = \delta$, we get

$$\partial^2 u/\partial z^2 + v = 0$$
$$\partial^2 v/\partial z^2 - u = 0$$

or

$$\partial^4 u/\partial^4 + u = 0 \qquad (11.16)$$

The equations now have no parameters. The solution will be a (nondimensional) $u(z)$, $v(z)$, which is valid for all flows governed by a balance between the three forces combined in Eq. (11.16). Since the height does not enter except as a scaling parameter, solutions to these equations are called *self-similar* (with respect to the height scale).

The solution to this fourth-order linear differential equation is readily available with standard techniques (which includes looking up in mathematical texts).

$$u(z) = \exp[-z](C_1 \cos z - C_2 \sin z) + \exp[z](C_3 \cos z - C_4 \sin z)$$

$$v(z) = \exp[-z](C_1 \sin z + C_2 \cos z) + \exp[z](C_3 \sin z + C_4 \cos z)$$

The boundary conditions yield

$$C_3 = C_4 = 0 \qquad C_1 = -1 \qquad C_2 = 0$$

and the solution may be written

$$u(z) = -\exp[-z] \cos z$$

$$v(z) = -\exp[-z] \sin z$$

or dimensionally,

$$u(z) = G(1 - \exp[-z/\delta] \cos z/\delta)$$

$$v(z) = -G \exp[-z/\delta] \sin z/\delta \qquad (11.17)$$

This solution of the boundary-layer approximation of the Navier–Stokes equations yields $u(\infty) = G$, $v(\infty) = 0$, with u rapidly (exponentially) approaching G away from the surface. A reasonable height of the PBL as proposed by Ekman is the e-folding depth of the exponential, $z/\delta = \pi$, or $H \approx \pi\delta = \pi\sqrt{K/f}$. The greater the eddy viscosity, the deeper the penetration of the surface effects. When the Coriolis effect goes to zero at the equator, the scaling depth becomes infinite, and the solution is no longer valid.

The angle of turning (problem 1) can be determined to be 45°. The two PBL solutions for the ocean and the atmosphere can be linked together in one continuous flow, as shown in Fig. 11.6. To do this smoothly, the frame of reference should be chosen to move with the surface current, as shown in Fig. 11.7. Note that the identical self-similar equations expand differently when the appropriate characteristic scales are used to make the solution dimensional. This is seen in Fig. 11.8, where both Ekman layers are shown to the same scale. The differences in density and eddy viscosity translate to atmospheric heights and velocities, which are about 30 times the oceanic values.

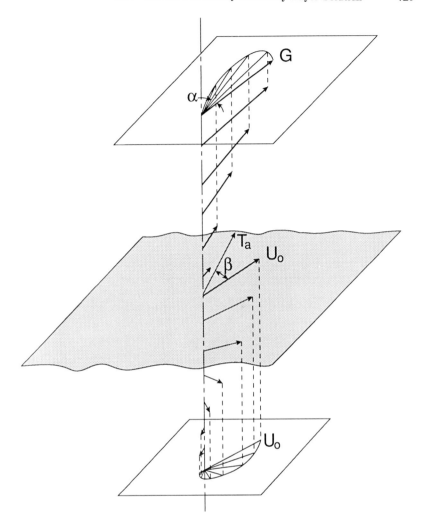

Figure 11.6   Sketch of velocity vectors in the adjacent atmospheric and oceanic PBLs. Note the change in scales.

Thus, Ekman's famous solution explained the problem posed by Nansen's observations in the Arctic. The 45° wind turning translates into a 45° angle between the surface wind and the ocean current at the surface.

Typically, turning is observed in the atmospheric PBL, in qualitative agreement with the theory. However, the height of the PBL is about 1 km, and this prevents the collection of many observations. There are several

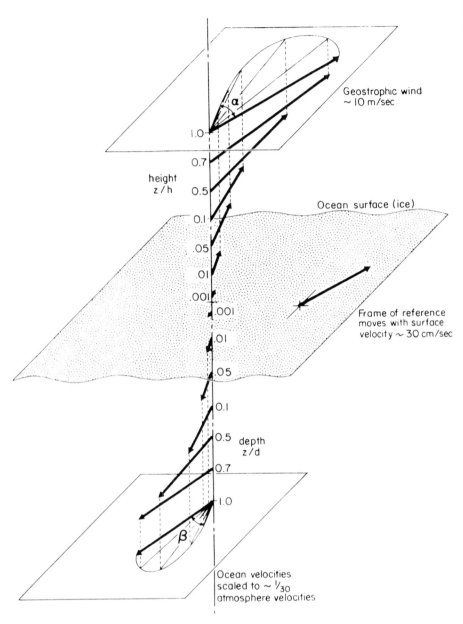

Figure 11.7 Sketch of atmospheric and oceanic PBL flow vectors when the frame of reference is moved with the surface velocity. Note the change in the oceanic hodograph.

A

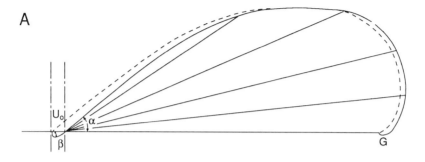

B

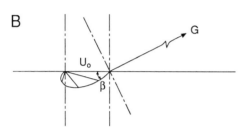

Figure 11.8 Sketch of the hodographs for the atmosphere (A) and the ocean (B) to the same scale. The dashed line is the atmospheric hodograph ignoring the surface velocity.

300-m towers that measure the lower portion of the PBL, and there have been many airplane flights at various levels through the layer. Otherwise, the main source of PBL data comes from balloon-borne sensors called radiosondes. These fairly common observations seldom indicate a steady or reproducible boundary-layer wind profile. Nevertheless, the Ekman solution remains the touchstone of many analyses of the PBL. The solution connects the surface boundary conditions to the large-scale flow. The exact analysis of the detailed character of the PBL requires an investigation into the *stability* of the Ekman solution, and into the nature of the turbulence in the PBL.

Example 11.4

Consider approximations to Eqs. (11.14) for H very large; of order $(K/f)^{1/2}$; and very small.

Solution

When H is very large, $K/(fH^2)$ is very small with respect to unity. Hence we are left with

$$v = 0; \qquad u - 1 = 0$$

Dimensionally, this is

$$(U_G, V_G) = (-P_y, P_x)/\rho f$$

Thus, for very thick layers, say 10 km (up to the tropopause), the geostrophic balance is applicable. We note that this is a singular perturbation of the equations. The ability to satisfy no-slip boundary conditions is gone. However, the approximation is still good.

When H is of order $(K/f)^{1/2}$, then $K/(fH^2)$ is of order unity. Then,

$$\partial^2 u/\partial z^2 + v = 0$$

$$\partial^2 v/\partial z^2 - [u - 1] = 0$$

The Ekman layer equations result.

When H is very small, $K/(fH^2)$ is very large, and

$$\partial^2 u/\partial z^2 = 0$$

$$\partial^2 v/\partial z^2 = 0$$

This integrates to $u = Az + B$, $v = Cz + D$. With boundary conditions $u(0) = v(0) = 0$, and $du/dz(0) = \tau_0/K$, $dv/dz(0) = 0$, we get

$$u = (\tau_0/K)z$$

We will see in Section 11.6 that this solution is inadequate.

═══════════════════════════════════

Example 11.5

Obtain an expression for the volume transport in the Ekman layer by integrating the Ekman layer equations with the viscous term written in terms of stress. First separate out the geostrophic flow so that the volume flow is relative to the pressure-driven flow. Assume constant density. Thus, Eqs. (11.11) may be written

$$\partial u/\partial t - fv + (1/\rho)\,\partial p/\partial x - (1/\rho)\,\partial \tau^x/\partial z = 0$$

$$\partial v/\partial t + fu + (1/\rho)\,\partial p/\partial y - (1/\rho)\,\partial \tau^y/\partial z = 0 \qquad (11.18)$$

Solution

Remove the geostrophic component by substituting

$$u = u^G + u^E \qquad v = v^G + u^E$$

Since

$$\partial u^G / \partial t - f v^G = (1/\rho) \, \partial p / \partial x$$

$$\partial v^G / \partial t + f u^G = (1/\rho) \, \partial p / \partial y$$

we have for the Ekman component,

$$\partial u^E / \partial t - f v^E - (1/\rho) \, \partial \tau^x / \partial z = 0$$

$$\partial v^E / \partial t + f u^E - (1/\rho) \, \partial \tau^y / \partial z = 0 \qquad (11.19)$$

$$\partial u^E / \partial x + \partial v^E / \partial y = 0$$

Integrating these equations yields

$$\frac{\partial}{\partial t} \int u^E \, dz - f \int v^E \, dz = -(1/\rho) \int d\tau^x = -\tau_0^x$$

$$\frac{\partial}{\partial t} \int v^E \, dz + f \int u^E \, dz = -(1/\rho) \int d\tau^y = -\tau_0^y$$

The integrals over the Ekman height are the *Ekman volume transport,* or, when multiplied by ρ, the mass transport.

$$(U^E, V^E) \equiv \int (u^E, v^E) \, dz = \int (u - u^G, v - v^G) \, dz \qquad (11.20)$$

The mass transport is extensively used in the oceanic Ekman layer. A sketch of the flow in both the atmospheric and oceanic layers is shown in Fig. 11.9.

11.5 The Modified Ekman Solution

Ekman's solution is a unique analytic solution to an approximate version of the Navier–Stokes equations that retains three terms. In general, boundary layer problems must be addressed by numerical means. However, an analytic solution is much simpler and provides great insight into the behavior of the flow. Unfortunately, Ekman's solution doesn't appear in the data for atmospheric or oceanic flows. Observations, usually indicate turning and

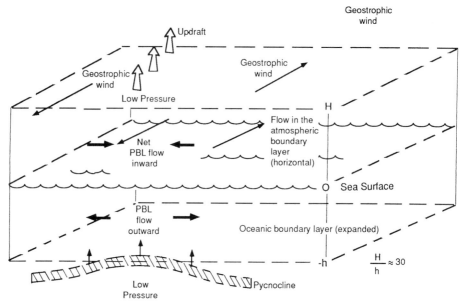

Figure 11.9 Sketch of the regions of Ekman transport in the atmospheric and oceanic PBLs in the case of a cyclone.

slowing of the wind, but seldom, if ever, show winds exactly as predicted by Ekman's solution. When the solution fails to be supported by observations, then the basic assumptions made in obtaining the solution must be checked.

The first suspect for the invalid assumption would naturally be the eddy-viscosity hypothesis. Ekman realized this, and found the solution for a simply variable eddy viscosity, dependent on the shear. Since then, there have been many assumed K distributions, some resulting in Bessel function solutions, some in simple numerical solutions. However, these empirical methods predict velocity profiles that agree only with mean flows averaged over hours or tens of kilometers. The difficulty seems to be that there are very large eddies in the PBL, with wavelengths as large or much larger than the PBL height. From our discussions on the continuum, it is clear that these eddies cannot be modeled by an eddy viscosity. The solution can be obtained only if the large eddies are directly accounted for in the equations. Fortunately, it appears that in many cases the large eddies are predictable, organized, and can be incorporated into the mean flow. The procedure includes application of most of the topics and techniques discussed in this text. While it is a specialized procedure for boundary layer flow, the basic concepts have a potential for many applications. These include the idea of isolating the large eddies for explicit description, the process of including the nonlinear interaction between these perturbations and the basic mean flow, and the parametrization of the small-scale turbulence with an eddy viscosity. Thus, we will summarize the procedure while assigning most of the details to the problems at the end of the chapter.

11.5.1   Organized Eddies (Rolls) in the PBL

1. Based on observations of persistent large-scale (2–5 km) wavelength features associated with the PBL, such as those in Fig. 11.10, it is assumed

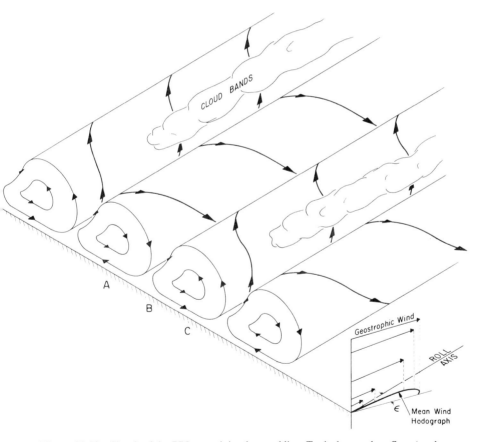

Figure 11.10   Sketch of the PBL containing large eddies. Typical secondary flow (modified Ekman layer).

that there exists a mean flow solution for the PBL that contains organized large eddies.

2. Based on the success of the surface-layer log-layer solutions obtained with an eddy-viscosity assumption (Section 11.6), and considering a continuum scale of centimeters to tens of meters (Sections 1.9 and 1.11), assume a constant eddy viscosity to obtain the mean flow solution. Thus, a parametrization of the small-scale turbulence with an eddy viscosity will yield an Ekman solution that is a good approximation of the PBL when there are no large eddies [Eq. (11.17)].

3. Since the Ekman velocity profile U_E is not observed, assume that it is unstable to infinitesimal perturbations. One can then carry out an instability analysis on the Ekman velocity profile using the linearized Eqs. (10.27), which have been modified by the boundary layer approximations (Section 11.3 and problem 3).

This reveals an inherent instability in the Ekman profile at very low windspeeds. The instability can be either entirely dynamically driven or due to convection in the presence of shear. Either of these instabilities produce large eddies that take the form of rapidly growing horizontal roll vortices with wavelengths comparable to the PBL height.

4. Based on the observations of persistence in the longitudinal cloud bands, assume that the linearized instabilities come to equilibrium at some finite amplitude. To determine the amplitude requires two more steps.

First calculate the modification to the mean profile by the now significantly large perturbations. Since observations indicate waves growing lateral to the geostrophic, there is only one Reynolds stress that contributes to the force balance. The modified Ekman equations can be written

$$-fv + (1/\rho)\, \partial p/\partial x - K\, \partial^2 u/\partial z^2 = 0$$

$$fu + (1/\rho)\, \partial p/\partial y - K\, \partial^2 v/\partial z^2 = \overline{v'w'} \qquad (11.21)$$

$$\partial u/\partial x = \partial v/\partial y = 0$$

Second, calculate the energetics of the nonlinear interaction by examining the perturbed energy equation for the system (Section 7.2 and problem 10). Assume that equilibrium is reached when the mean velocity profile has been modified to a point where the energy flux from the mean to the perturbation component of the flow becomes zero. This results in an equation of the form

$$\frac{DE}{Dt} = \int \overline{v'w'}\, \frac{Dv}{Dz}\, dz = 0 \qquad (11.22)$$

By solving these equations simultaneously, a steady-state quasi-nonlinear solution can be found. This solution has a modified Ekman mean flow with finite perturbation helical-vortex eddies that occupy the entire PBL.

This solution showed that the perturbation rolls come to equilibrium at a magnitude of about 7–10% of the mean flow velocity. The rolls are approximately aligned with the mean flow, and result in modified Ekman-type mean wind profiles, shown in Fig. 11.11, and highly variable local profiles, shown in Fig. 11.12. They also provide advective fluxes across the PBL, leading to a much more rapidly mixed PBL layer than could be accounted for by diffusion alone. The equilibrium structure is sensitive to the wind-speed, the surface roughness, and the stratification of the layer.

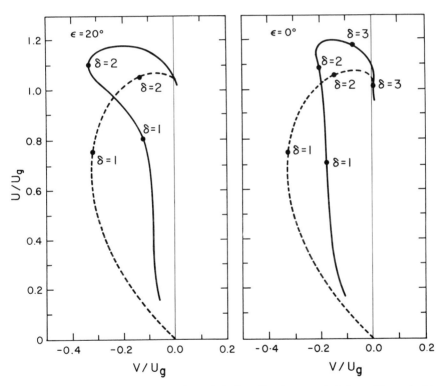

Figure 11.11   Wind profiles (hodographs) of predicted mean winds in an Ekman layer modified with organized large eddies. The δ values are nondimensional heights. The ϵ are turning angles between geostrophic and surface winds corresponding to neutral (20°) and unstably stratified (0°) PBLs.

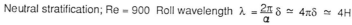

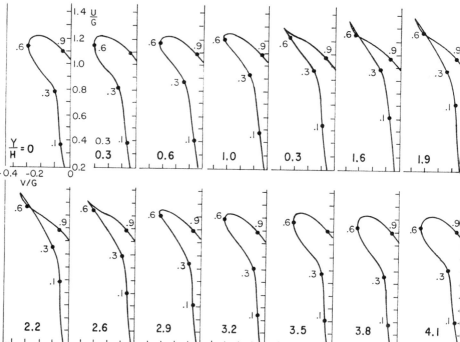

Figure 11.12 Hodographs showing the wind profile predicted at various stations in a modified Ekman PBL containing organized roll eddies. The stations are at various y/H distances with a roll wavelength of about $y/H = 4.1$. Heights are given at various z/H. Velocities are nondimensionalized with $|U_g| = G$.

Example 11.6

The Ekman layer instability is sometimes called an inflection point or vorticity maximum instability. Consider each level of the flow to be occupied by parcels (which are sections of vorticity tubes) with vorticity produced by the shear at that level. Then the net force on a displaced element can be found to be

$$F \approx 1/(\zeta_0 - \zeta) \int w^2 \, (\partial\zeta/\partial z) \, dA \qquad (11.23)$$

where w is the vertical velocity, F is the vertical force, and A is the area of vorticity perturbation. Sketch the velocity and vorticity, consider a displaced vortex parcel and the distorted vorticity elements around it (see Section 9.7), and discuss its motion based on Eq. (11.23).

Solution

A sketch of velocity and vorticity profiles with and without an inflection point are shown in Fig. 11.13.

First consider a displaced vortex ζ_0 among the constantly decreasing vorticity at level z_1. According to the forces calculated in Section 9.7, the surplus in ζ_0 will distort the vorticity nearby in a counterclockwise sense. This results in more negative vorticity filling in to the left, and more positive vorticity filling in to the right. Both of these changes lead to forces downward, returning the perturbed parcel to its original level. Similar results are obtained when the parcel is moved downward. When values are put in Eq. (11.23), the same result is obtained.

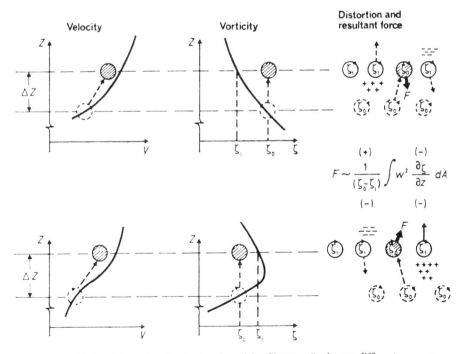

Figure 11.13   Schematic of a displaced vorticity filament, ζ_0, in two different mean velocity, and corresponding vorticity, fields.

In the case of a maximum in the vorticity field, the displaced vortex element has a deficit of vorticity. Since the inflection point has reversed the vorticity gradient, the distortion allows positive vorticity to move up to the left, negative to move downward to the right. This produces an upward force on the parcel. The resulting force is destabilizing and the parcel continues to rise until it reaches the equilibrium vorticity level.

This explanation is qualitative only. The perturbation must be finite, to move the parcel past the inflection point. It explains only limited growth. The reason for large waves in the case of inflection points must lie in the nonlinear equations, as the perturbation modifies the mean flow and the vorticity gradients.

11.6 The Surface Layer

Most boundary layer study has been done within the lowest 10–20 m of the atmosphere. Here is where the "surface" measurements of pressure, wind, humidity are taken (the standard meteorological set of data). "Upper air" data is routinely obtained from radiosonde releases. However, there is seldom more than one point recorded in the PBL. Thus most of our modeling of the atmospheric flow, numerical or otherwise, is done using conditions furnished by the surface observations. The study of the linkage between the abundant surface layer data and the geostrophic flow is an important aspect of all large-scale modeling.

We can begin by examining the "classical" derivation of the surface layer velocity profile. In fact, there are many different derivations of the surface layer equations. Nearly all result in a log-layer profile for $u(z)$ (see problem 16).

The earliest derivations simply assumed that as the surface is closely approached, there is a constant stress layer. Thus,

$$d\tau/dz = 0, \quad \text{with} \quad \tau(0) = \tau_0 = \text{surface stress} \quad (11.24)$$

From the definition of eddy viscosity, $\tau \equiv \rho K \, dU/dz$, Eq. (11.24) may be written, with constant density,

$$d(\tau/\rho)/dz = d(K \, dU/dz)/dz = K \, d^2U/dz^2 + dK/dz \cdot dU/dz = 0 \quad (11.25)$$

This equation is of the general form,

$$d^2u/dz^2 + A(z) \, du/dz = 0 \quad (11.26)$$

There are no general solutions to this equation for arbitrary $A(z)$ [or $K(z)$]. Fortunately, specific cases of $K(z)$ yield simple solutions. In particular, we expect the eddies, and hence K, to grow in size with distance from the boundary. For the simplest case, where we set $K = K_0 z$, Eq. (11.26) becomes

$$d^2u/dz^2 + 1/z \cdot du/dz = 0 \qquad (11.27)$$

This is a common form of the equation called Euler's equation. The substitutions of $du/dz \equiv V$ and $\ln z \equiv Z$ lead to the solution,

$$u = B \ln z + C \qquad (11.28)$$

However, this solution cannot satisfy the boundary condition at $z = 0$ (or $z/H = 0$ in the dimensional equation). This inability is a consequence of assuming that $K(z)$ becomes 0 at the surface, implying zero stress at the surface. If we modify $K(z)$ so that $K = K_0(z + z_0)$, we have

$$u = B \ln (z + z_0) + C$$

and $\qquad\qquad\qquad\qquad\qquad\qquad\qquad\qquad (11.29)$

$$u(0) = 0 \rightarrow B \ln (z_0) + C = 0$$

We can assume that the characteristic scale for the surface layer is $H = z_0$. As a result, the nondimensional $z_0 = \hat{z}_0/H$ in Eq. (11.29) is 1 and hence $C = 0$.

Our approximate Eq. (11.27) is second order, and the second boundary condition can be obtained from the definition of surface stress (in dimensional units). This may be written

$$[K_0 G/z_0] \, du/dz(0) = [1/G^2]\tau_0/\rho$$

or

$$(K_0 G/z_0) \, (B)/1 = \tau_0/\rho G^2$$

and thus,

$$B = (\tau_0/\rho) \, Gz_0/(K_0 G) = u^{*2} \, Gz_0/(K_0 G)$$

The definition of the friction velocity, $u^{*2} \equiv \tau_0/\rho$, has been used in this expression. Also, if we use u^* to replace G as the characteristic velocity in the surface layer,

$$B = u^* z_0/K_0 \equiv \mathrm{Re}_0, \qquad \text{a surface layer Reynolds number.}$$

This dimensionless parameter is generally written as $B = 1/k$, where k is called von Karman's constant. The surface layer velocity is

$$u = (1/k) \ln (z/z_0) \qquad (11.30)$$

Example 11.7

(a) Derive the general form of the surface layer solution for $u(z)$ using dimensional analysis and dynamic similarity. Obtain specific equations from the definition of stress in a constant stress layer,

$$K \, du/dz = \tau_0/\rho$$

and

$$du(0)/dz = \tau_0/\rho K_0$$

Assume $K(z) = (ku*z_0) \, z$.

(b) Discuss the lower boundary condition on u, $u(z_0) \equiv 0$. Do the same when the boundary condition is $u(0) = 0$. Discuss the different $K(z)$ that will satisfy these boundary conditions. [Hint: Let $K = (ku*z_0)(z + 1)$ in $u(0) = 0$ case].

Solution

(a) The only characteristic parameters for the nondimensionalization are

$$V = (\tau_0/\rho)^{1/2} \equiv u* \qquad H = (\rho K_0^2/\tau_0)^{1/2} \equiv z_0 \qquad \text{and} \quad K_0$$

The π-theorem applied to variables τ_0/ρ, K_0, u, z leads to four variables less two characteristic dimensions (L and t) or two parameters in the nondimensional relation. They are

$$\frac{u}{\sqrt{\tau_0/\rho}} \qquad \frac{z}{\sqrt{K_0^2/(\tau_0/\rho)}}$$

These may be written

$$u/u* = f(z/z_0) \qquad z_0 \equiv K_0/u*$$

Looking at dynamic similarity, the nondimensional equation is

$$(Ku*/z_0) \, du/dz = u*^2 \qquad \text{or} \quad du/dz = u*z_0/K$$

Substitute for $K = ku*zz_0$,

$$du/dz = 1/(kz)$$

There are no parameters in this equation, and a self-similar solution is

$$u = (1/k) \ln z + C$$

or dimensionally

$$u/u* = (1/k) \ln (z/z_0) + C$$

(b) The boundary condition, $u(z_0) = 0$, is satisfied with $C = 0$,

$$u/u* = (1/k) \ln (z/z_0) \tag{11.31}$$

In this case, $K = ku*z_0 z = ku*z_0^2$ when $u = 0$ and $K(0) = 0$. K is finite where u is zero and goes to zero at the surface.

To satisfy the lower boundary condition $u(0) = 0$, we note from Eq. (11.29) that $du/dz \rightarrow \infty$ at $z = 0$, an unrealistic result. However, if we assume

$$K = ku*(z + 1) z_0 \tag{11.32}$$

then

$$du/dz = 1/[k(z + 1)] \text{ and } u = (1/k) \ln(z + 1)$$

or dimensionally

$$u/u* = (1/k) \ln[(z + z_0)/z_0] \qquad u(0) = 0 \tag{11.33}$$

In this case, $K(0) = ku*z_0$. This is contrary to the expectation that eddy viscosity must go to zero at the surface. Typical values of $u* = 30$ cm/sec, $z_0 = 1$ cm, $k = 0.4$ yield $K(0) \approx 0.1$ cm^2/sec. This value of eddy viscosity has decreased to a similar order, to $\nu = 0.16$ cm^2/sec. Perhaps the log layer based on the eddy viscosity patches to the molecular layer at z_0.

Since z_0 is the height where $u = 0$, it is very small, and in general $z \gg z_0$, so that there is little difference between the expressions for $u(z)$.

When considering the steady-state, horizontally homogeneous layer, we might try to obtain the log layer as a quick result from Eq. (11.15). For $H \gg \delta$, we obtained the geostrophic balance. When we assumed $H \approx \delta$, the Ekman layer characteristic height produced the equations appropriate to the layer where Coriolis, pressure gradient, and viscous forces were all of the same order. We can enquire as to what would a choice of different H produce? Consider H equals a very small height. This would be the characteristic height of a very thin layer adjacent to the surface. In Eq. (11.15), the coefficient of the viscous term would then be very large with respect to that of the Coriolis (and the pressure gradient) terms. Alternatively, we could say that the coefficient of the Coriolis term, $[H^2/KG]$ now is very small.

Consequently only the viscous terms in the equations

$$\partial^2 u/\partial z^2 = 0$$

$$\partial^2 v/\partial z^2 = 0$$

are left. As found in Example 11.3, the same equations result from the complete steady-state equations (11.7) when we consider $H \ll 1$. However, there is a problem here. The nondimensionalization with characteristic parameters must result in individual dimensionless terms with order unity. Otherwise the basic process of getting approximate equations by using magnitude arguments based only on the nondimensional coefficients in the equations would not be valid. Yet, if these terms are of order unity, how can they be set to zero? Does this suggest the paradox that in a thin layer with shear, the viscous terms are much larger than the other forces, so large that they are zero? Evidently this is not a valid approximate equation, a fact that is apparent from Newton's second law—a single dominant force must cause acceleration. One of our assumptions is breaking down in a very thin layer.

The basic equation for the surface layer is obtained only after considering the scaling parameters much more carefully. In particular, the variation of K must be considered. We should be able to obtain the log-layer solution from the Navier–Stokes Eq. (11.7). We must first examine the characteristic values used in the nondimensionalization.

At the surface, the velocity approaches zero. The magnitude of the velocity deficit $(u - G)$ approaches G, so the velocity scaling with G seems all right. However, the eddy viscosity coefficient depends on the turbulent eddies, which have less room to exist as the surface is closely approached. Very near the surface, there must be a laminar sublayer where there are no turbulent eddies. Thus, $K \rightarrow v$ here. We can continue to pursue the surface layer solution by including these arguments.

The expression for the stress must be the Reynolds stress equation (10.31),

$$\rho \overline{u_i' u_j'} = -\rho K \, \partial u_i/\partial x_j = \tau'$$

In the boundary layer this reduces to only the component involving the vertical shear,

$$\rho K \, du/dz = \tau_{xz} \equiv \tau \tag{11.34}$$

Allowing for a variable K, the viscous force/unit mass term is obtained as the vertical gradient of the stress,

$$\partial \tau/\partial z = \partial[\rho K \, \partial u/\partial z]/\partial z \tag{11.35}$$

The variable K can be nondimensionalized with a mean value to obtain the eddy viscosity version of Eq. (11.7)

$$(u\, \partial u/\partial x + v\, \partial u/\partial y) + \frac{[XW]}{[HU]}\, w\, \partial u/\partial z - \frac{v}{[Ro]}$$

$$+ [D]\, \partial p/\partial x - \frac{[X^2]}{[ReH^2]}\, \partial(K\, \partial u/\partial z)/\partial z = 0 \qquad (11.36)$$

where

$$Re = UX/K \qquad K = \hat{K}/\bar{K}$$

As above, when $H \to 0$, this formulation implies that the viscous force term is very large. To have an equilibrium solution, another force must balance the viscous force. This could be the $\partial u/\partial t$ term, but we would like a steady-state solution. The advection terms are candidates, but let us first search for a horizontally homogeneous solution.

Very near the surface, the Coriolis force approaches zero along with the velocity. Thus, to have an equilibrium solution very near the surface, the pressure-gradient force must balance the viscous force. We assume that $K = (ku^*)z$, where the constant of proportionality ku^* is based on hindsight. This assumption, together with the assumptions that the horizontal pressure gradient is impressed on the boundary layer and that ρ is constant, will allow us to integrate the stress force term.

$$\partial(K\, \partial u/\partial z)/\partial z = -(1/\rho)\, \partial p/\partial x \equiv D \qquad (11.37)$$

(the stress force). Integrating this over z yields

$$K\, \partial u/\partial z = Dz + C$$

The constant C can be evaluated from the definition of surface stress,

$$K\, \partial u/\partial z(z = 0) \equiv \tau_0/\rho \equiv u^{*2} = C$$

Hence,

$$K\, \partial u/\partial z = Dz + u^{*2} \qquad (11.38)$$

(the stress variation).

If we now make the surface-layer assumption that $K = ku^*z$, and note that velocity is a function of z only,

$$du/dz = (D/ku^*) + u^*/(kz) \qquad (11.39)$$

Integrating Eq. (11.39), we have

$$u = D'z + (u^*/k) \ln z + C'$$

$$D' = D/ku^*$$

For the lower boundary condition, let $u(z_0) = 0$, then

$$C' = -(u*/k) \ln z_0 - D'z_0$$

and the surface-layer speed is given by

$$u = (u*/k) \ln z/z_0 + D'(z - z_0) \tag{11.40}$$

This derivation yields (once again) the log layer, plus an additional linear term. We can examine relative magnitudes of these terms by considering typical values for a 10-m/sec wind at 10 m. They are $u* = 30$ cm/sec, $k = 0.4$, $z = 10^3$ cm. The pressure gradient term D is the same order as the freestream Coriolis term,

$$fV = 10^{-4} \, [1/\text{sec}] \, 10^3 \, [\text{cm/sec}] = 10^{-1} \, [\text{cm/sec}^2]$$

Thus,

$$D' = 10^{-1} \, \text{cm/sec}^2/(0.4 \cdot 30 \, \text{cm/sec}) = 8 \cdot 10^{-3} \, 1/\text{sec}$$

$$D'z = 0.008 \cdot 10^3 \, \text{cm} = 8 \, \text{cm/sec}$$

This is compared to the first term,

$$u*/k \ln z/z_0 = 30 \, \text{cm/sec}/0.4 \ln 10^3/1 = 518 \, \text{cm/sec}$$

Thus, the linear term is often omitted, especially in unstably stratified or neutrally stratified conditions. However, it is significant in stably stratified conditions.

In summary, we have found that in the boundary layer very near the surface, a constant pressure-gradient force is balanced by a constant viscous force. The latter is the gradient of a very large internal stress. The predicted logarithmic velocity profile has been well verified by observations.

11.6.1   Summary of Force Balance through the PBL

The change in the force balance as the parcel descends through the PBL can be seen in the sketch of Fig. 11.14.

11.7   The Mixing Length

The concept of the mixing length was introduced to provide a physical analog to the molecular mean free path. The latter is the distance a molecule travels before hitting another molecule and exchanging momentum. It is assumed that an eddy will travel an average distance called the mixing length before it exchanges momentum with the surrounding eddies. For the fluid parcel, the momentum will be exchanged in a complicated mixing and dif-

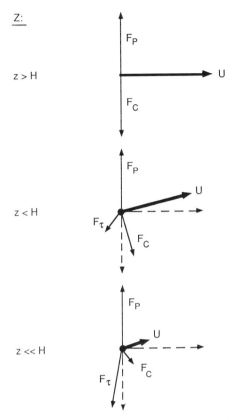

Z:

z > H

z < H

z << H

Figure 11.14 The force balance in the entire PBL of height H. F_p is the pressure gradient force, F_τ is the viscous force $d\tau/dz$, and F_C is the Coriolis force.

fusion fashion. Consequently, this process is ill defined compared to the relative simplicity of the molecular interaction. But both definitions rely on statistical averaging.

Consider the velocity shear in the two-dimensional parallel flow in Fig. 11.15.

Prandtl visualized a thought experiment where the fluid parcel is an entity that moves a distance ℓ keeping its original momentum. The velocities of the parcels arriving at height z in Fig. 11.15 will be those of the neighboring layers a distance ℓ away. These velocities can be expressed in a Taylor series about $z \pm \ell$ (the points $\pm \ell$ away from z) (note: $u(z)$ means u at z here).

$$u(z)_+ \approx [u(z + \ell) - (du/dz)(z + \ell) \, \ell]_+ = u(z + \ell) + u'_+ \quad (11.41)$$

$$u(z)_- \approx [u(z - \ell) + (du/dz)(z - \ell) \, \ell]_- = u(z - \ell) + u'_-$$

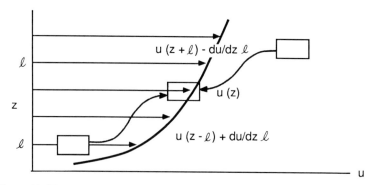

Figure 11.15   A parcel in a shear flow with movement from $\pm\ell$, a mixing length scale. The velocities of parcels arriving from above and below height z are given.

Thus the neighboring parcels arriving at height z contribute the turbulent velocity components to level z. Assume that the distance ℓ is the mean distance that the parcel travels before it transfers its momentum to the surroundings. Then the velocity at z may be written as the average of the nearby mean velocities plus a perturbation. The latter is an average of the perturbations contributed from parcels arriving from above and below. Hence, from Eq. (11.41),

$$u(z) = \frac{u(z)_+ + u(z)_-}{2}$$

or

$$u(z) = \frac{u(z + \ell) + u(z - \ell)}{2} + \frac{[|u'_+| + |u'_-|]}{2}$$

or

$$u(z) = \bar{u}(z) + u' \tag{11.42}$$

We can also write $u(z)$ in a Taylor expansion about z.

$$u(z) \approx \bar{u}(z) + \ell\, du/dz = \bar{u} + u' \tag{11.43}$$

where

$$u' = \ell\, du/dz \tag{11.44}$$

For isotropic turbulence, $u' \approx v' \approx w'$ and the Reynolds stress term is

$$\tau/\rho = \overline{u'w'} \approx \ell^2(du/dz)^2 = K\, du/dz \tag{11.45}$$

This gives a relation between eddy viscosity and mixing length,

$$K = \ell^2\, du/dz \tag{11.46}$$

There is no advantage to using ℓ over K unless there exists a fundamental relation for ℓ as a more "basic" parameter. Von Karman attempted to obtain such a relation by dimensional arguments. In a free shear layer, the only parameters available to influence the mixing length are the shear and derivatives thereof. Thus, DA using the lowest order derivatives to form a scale length yields

$$\ell = k\,\frac{du/dz}{d^2u/dz^2} \tag{11.47}$$

as a functional equation for the mixing length. The constant of proportionality is called the von Karman constant.

Using the definition of $u^* \equiv \sqrt{\tau/\rho}$ in Eq. (11.47),

$$u^* = \ell\,du/dz \tag{11.48}$$

this can be differentiated to obtain

$$d\ell/dz \cdot du/dz + \ell\,d^2u/dz^2 = du^*/dz = 0$$

or

$$\ell = \frac{-d\ell}{dz}\,\frac{du/dz}{d^2u/d^2z} \tag{11.49}$$

Thus, comparing Eqs. (11.47) and (11.49), the von Karman constant,

$$k = d\ell/dz \tag{11.50}$$

is a measure of the change in mixing length with height.

The mixing length could be expected to be zero at the surface and grow linearly with distance away from the surface.

$$\ell = kz \tag{11.51}$$

In this case the solution for $u(z)$ is the same logarithmic relation as obtained in Section 11.5. In general, there is no better understanding or hypothesis for the distribution of ℓ in a fluid then there is for K. Its only claim toward being a "fundamental" parameter is its analogy to the mean free path for molecular interaction.

However the more severe restriction on the use of mixing lengths is that they apply only to isotropic, small-scale turbulence. We have noted that when large eddies are present, modeling with the diffusion equation is incorrect. K-theory cannot correctly account for the advective transport by the large eddies. The eddy continuum does not exist. Similarly, the mixing length must depend on the eddy size. Large eddies will move fluid large distances before there is mixing with the local fluid. The large eddies must be handled explicitly, as can be done with the organized eddies in Section 11.5. However, when the large eddies are random, there is no analytic solution for them, and numerical parametrizations that include the eddy size-dependent mixing length must be employed—or else the eddies must be explicitly resolved.

11.8 Summary

The material in this chapter deals with flow in a thin layer with large gradients. Although the concepts are very general, we have concentrated on the thin layer next to a surface. We have obtained two different velocity-shear boundary layers for the neutrally stratified, horizontally homogeneous, steady-state atmospheric flow above a surface.

One is the modified Ekman solution, which contains Coriolis, pressure gradient, and viscous forces. It includes the entire layer from the surface to the freestream flow. In the atmosphere, this is typically a distance of 1–2 km. It is known as the planetary boundary layer (PBL), the atmospheric boundary layer, the Ekman layer, or the mixed layer.

The second boundary layer we obtained was the surface layer solution, or the log-layer. It is a venerable, often-measured, wind profile, existing in the lower 10–100 m of the PBL. The assumptions used in the derivation indicate that a logarithmic profile is a likely solution for any thin layer. In relatively thin regimes the viscous forces are important by virtue of high-velocity gradients. Thus a log layer applies to the boundary between two separate flow domains. Since the atmosphere and ocean are prone to flow in layers with different stratification regimes, this solution probably applies in the thin regions seaming two adjacent flows.

The principal shortcoming for both of these solutions is the lack of stratification effects. The stratification effect is added to the surface-layer solution with an empirical modification of the log-layer profile. The effect in the modified Ekman layer appears in the explicit consideration of the dynamic and convective eddy structure. The large eddies called rolls are more vigorous and alined with the geostrophic flow when convective energy is available. They become completely damped out in moderately stable stratification. In strongly convective conditions, large eddies occur randomly, and chaotic flow is likely.

We have seen that the solution for the flow near a boundary evoked many of the principles and methodologies discussed in this text. The same needs will appear in the derivation of complex interactions on many scales in the atmosphere. Although the details of each application may become complex, the concepts are the same. They provide the framework for the method of solution and evaluation of the basic validity of the governing equations for a wide range of fluid dynamics problems.

Example 11.8

Consider the u-momentum equation. Cases are marked with x where the terms must be considered. What are the special cases?

	$\dfrac{\partial u}{\partial t}$	$u\dfrac{\partial u}{\partial x}$	$v\dfrac{\partial u}{\partial y}$	$w\dfrac{\partial u}{\partial z}$	$-fv$	$=$	$-\dfrac{1}{\rho}\dfrac{\partial p}{\partial x}$	$K\Bigl(\dfrac{\partial^2 u}{\partial x^2}$	$+\dfrac{\partial^2 u}{\partial y^2}$	$+\dfrac{\partial^2 u}{\partial z^2}\Bigr)$
1	x	x	x	x	x		x			
2		x	x	x	x		x			
3		x	x	x	x					
4					x		x			
5	x	x	x	x	x		x	x	x	x
6					x		x			x
7							x			x
8	x	x	x	x			x			x

Solution

The terms apply for special cases that have been mentioned throughout Part II. Cases 1 and 2 constitute the upper boundary conditions for the various boundary layer equations of cases 3–6. Applicable flow situations (not exclusive) and their names are

1. Atmospheric or oceanic flow outside of the PBL.
2. Same as 1 for steady flow (gradient).
3. Same as 2 with small pressure gradient (cyclostrophic).
4. Same as 1 for steady, horizontal homogeneous flow (geostrophic).
5. Flow in the PBL.
6. Ekman layer equations; steady, horizontal PBL.
7. Surface layer equations.
8. Boundary layer in laboratory, Prandtl's equations.

SUMMARY OF PBL THEORY

Working Equations	Assumptions Decreasing Validity $\rightarrow$
$\rho\, D\mathbf{u}/dt = \rho\mathbf{F} + \nabla\boldsymbol{\sigma}$	
$D\rho/Dt + \rho\,\nabla\cdot\mathbf{u} = 0$	*Conservation laws*
$\rho\, DE/DT = DQ/Dt - \boldsymbol{\sigma}\cdot\nabla\mathbf{u}$	
Navier–Stokes equations	*Stokes' assumptions:*
$\rho\, D\mathbf{u}/Dt = -\rho\mathbf{g} - \nabla p + \mu\,\nabla^2\mathbf{u}$	continuum
$D\rho/Dt + \rho\,\nabla\cdot\mathbf{u} = 0$	linearly viscous fluid
$\rho c_v\, DT/Dt = K_h\,\nabla^2 T - p\,\nabla\cdot\mathbf{u}$ $\quad + Q_r + Q_p + \Phi$	static fluid has pressure only
where	stress $\propto$ strain rate
$\boldsymbol{\sigma} = -p\mathbf{I} + \boldsymbol{\tau}$	
$\boldsymbol{\tau} = 2\mu\,\text{def}\,\mathbf{u} - \dfrac{2}{3}\,\mu\,\text{div}\,\mathbf{uI}$	
$\Phi = \boldsymbol{\tau}\,\nabla\cdot\mathbf{u}$	
Atmospheric Equations	*Atmospheric assumptions:*
$\rho\, D\mathbf{u}/Dt + \rho 2\boldsymbol{\Omega}\times\mathbf{u} =$ $\quad -\rho\mathbf{g} - \nabla p + \nabla(K\,\text{def}\,\mathbf{u})$ $\quad + \nabla(K'\,\text{div}\,\mathbf{u})\,\mathbf{I}$	turbulence
$D\rho/Dt + \rho\,\nabla\cdot\mathbf{u} = 0$	noninertial frame
$\rho c_v\, DT/Dt = \nabla(K_h\,\nabla T) - p\,\nabla\cdot\mathbf{u}$ $\quad + Q_r + Q_p + \Phi$	eddy-continuum eddy-viscous (variable K replaces ν)
where	
$\boldsymbol{\tau} = K\,\text{def}\,\mathbf{u} + K'\,\text{div}\,\mathbf{uI}$	
Eddy viscous Navier–Stokes equations	*Closure assumptions:*
$D\mathbf{u}/Dt + 2\boldsymbol{\Omega}\times\mathbf{u} =$ $\quad -\mathbf{g} - \nabla p/\rho + K\,\nabla^2\mathbf{u}$	eddy-continuum K = constant

SUMMARY OF PBL THEORY—(Continued)

Working Equations	Assumptions Decreasing Validity $\rightarrow$
$D\rho/Dt + \rho\, \nabla \cdot \mathbf{u} = 0$	$K'\,\mathrm{div}\,\mathbf{u} = 0$
$\rho c_v\, DT/Dt = K_h\, \nabla^2 T$ $\quad -(RT/c_v)\, \nabla \cdot \mathbf{u}$	neglect radiation and dissipation
Boundary layer equations	*Boundary layer assumptions:*
$U_t + UU_x + VU_y - fV =$ $\quad -(1/\rho)p_x + KU_{zz}$ $V_t + UV_x + VV_y + fU =$ $\quad -(1/\rho)p_y + KV_{zz}$	$W \ll U, V \qquad z \ll x, y$ Boussinesq assumption (ρ changes are small)
$U_x + V_y = 0$	neutral stratification
$\quad -\rho g = p_z$	
where	
$\mathbf{u} = \mathbf{U}$	
Ekman equations	*Ekman assumptions:*
$fV + KU_{zz} = 0$ $fU - KV_{zz} = 0$	steady state small eddies
where	
$\mathbf{u} = \mathbf{U} - \mathbf{U_G}$	horizontal homo- geneity
$\mathbf{U_G} = (-p_y, p_x)(\rho f)$	
Modified Ekman equations	*Modified Ekman assumptions:*
$fV + KU_{zz} = 0$	explicit large eddies
$fU - KV_{zz} = A(z)$	steady state
where $A(z)$ from stability analysis and $\quad$ energy equations	

Problems

1. Determine the angle of turning of the velocity through an Ekman layer.

2. What are the forces in the Ekman equations? Discuss the problems for these equations and the solution as the equator is approached.

3. Ekman's equations, and solution, are self-similar. What is the factor that allowed this solution? (It is not available in the Prandtl or classic flat-plate boundary layer equations or solutions.)

4. The method of "normal modes" is used to determine whether a given mean flow is stable or unstable to infinitesimal perturbations. It involves placing in the equations for the flow a perturbation of the form

$$u' = \phi \exp[i\varphi t]$$

Do this for the Ekman layer equations to obtain an equation for the perturbation.

5. Consider the laminar sublayer immediately adjacent to a surface. Assume the stress is constant in this thin layer and determined by $\tau = \mu \, du/dz$. Show that the velocity distribution follows

$$u/u^* = z/(v/u^*)$$

6. Show that when one assumes that $K = K_0$ or z, where K_0 is a constant, Ekman's equations become Bessel's equation. Show that Bessel's equation results when $K = Cz^n$ is assumed. Consider complex $Q = u+iv$, and $\mathbf{u} = \mathbf{u} - \mathbf{u_G}$.

7. According to Ekman's solution, if you are standing with your back to the wind, looking up at the clouds moving by with the wind at the top of the PBL, what direction are they moving with respect to the direction that you're facing?

8. Consider an inversion in the freestream (e.g., the top of the PBL often has a 10–50-m thick stably stratified layer; or a pycnocline in the ocean; or the tropopause). Assume it is thin with respect to the vertical and horizontal coordinates of your basic problem. Write a simplified version of the steady-state u-momentum equation for this layer.

9. Integrate the boundary layer equations to obtain the momentum integral equation. Boundary conditions are $U = V = 0$, $\tau = \tau_0$ at $z = 0$; $U = G$, $V = 0$, $\tau = 0$ at $z = H$. The u-momentum equation is

$$\frac{d}{dx}\int_0^H u(U_G - u)\, dz + \frac{dU_G}{dx}\int_0^H (U - u)\, dz - \int_0^H fV\, dz = \frac{\tau_0}{\rho}\bigg|_{x=0}$$

where H is the height of the PBL, U_G is the geostrophic flow and τ_0 is the surface stress.

10. Derive the boundary-layer form of the energy equation.

11. In Ekman's boundary-layer solution a constant geostrophic flow is assumed. When the pressure gradient is not constant with height, the geostrophic flow will be a function of z. The variable pressure gradient is associated (via the equation of state) with a horizontal temperature gradient. Hence the *variation* in the geostrophic wind is called the *thermal wind*. Show that the vertical change in geostrophic flow is

$$dU_G/dz = (U_G/T)\, dT/dz - g/(fT)\, dT/dy$$

$$dV_G/dz = (V_G/T)\, dT/dz + g/(fT)\, dT/dx$$

12. Use the results of problem 11 and the fact that the second terms on the right side of these equations are much greater than the first to show that the thermal wind can be linearly added to Ekman's solution.

13. Show that the Ekman mass transport of the atmospheric plus the oceanic PBLs is zero. Is this true of the volume transport? Why?

14. Show that in steady conditions, the Ekman mass transport is directed at right angles to the surface stress. Discuss the directions with respect to geostrophic for atmospheric and oceanic flow.

15. A measure of the vertical velocity needed to replace fluid transported from or to regions with different stress can be found by integrating the continuity equation and using the results of Example 11.5. Show that the Ekman pumping velocity, ρwE, is approximately $1/(\rho f)$ times the curl of the surface stress.

16. Show that the asymptotic limit of the Ekman solution for $z \rightarrow 0$ yields a logarithmic velocity profile.

Suggested Reading

Chapter 1

The following texts are recommended as supplements to the material in this book.

Gerhart, P.M. and Gross, R.J. *Fundamentals of Fluid Mechanics*, Addison–Wesley, 1985. This is an excellent introduction to engineering fluid dynamics. An undergraduate focus for engineers, it lacks tensors, turbulence and rotation.

Fox, R.W. and McDonald, A.T. *Introduction to Fluid Mechanics*, Wiley & Sons, 1978. A very good introduction to engineering fluid dynamics at the undergraduate level.

Batchelor, G.K. *An Introduction to Fluid Dynamics*, Cambridge University Press, 1967. A classic mathematical treatment of fluid dynamics at the graduate level. The best source for more details for the advanced student. Extensive treatment of vorticity.

Wallace, J.M. and Hobbs, P.V. *Atmospheric Science, an introductory survey* Academic Press, 1977. An introduction to the dynamics and thermodynamics of the atmosphere largely from a phenomenological viewpoint. Large-scale emphasis, with good clouds, radiation and thermodynamics. Can be studied simultaneously with this text.

Fleagle, R.G. and Businger, J.A. *An Introduction to Atmospheric Physics*, Academic Press, 1980. This text provides details of the atmospheric phenomena which compliment the development of the equations in this text.

Haltiner, G.J. and Martin, F.L. *Dynamical and Physical Meteorology*, McGraw Hill, 1957, 470 pp. Good thermodynamics.

Hess, S. *Introduction to Theoretical Meteorology*, Holt, 1959. A classic basic text which provides an emphasis on atmospheric thermodynamics.

The following texts are recommended for additional study in specific areas. This book is meant as a front-end preparation for these texts.

Holton, J. *An Introduction to Dynamic Meteorology*, Academic Press, 1972. This text provides a natural following course for atmospheric scientists interested in large-scale dynamic meteorology.

Pedlosky, J. *Geophysical Fluid Dynamics*, Springer–Verlag, 1979. An emphasis on large-scale dynamics. Specific aspects of the equations for large-scale dynamics, Rossby waves, inertial boundary currents, the beta plane, wave interactions. A sequel to this text for those going into large-scale dynamics.

Gill, A. *Atmosphere-Ocean Dynamics*, International Geophysics Series, **30**, Academic Press, 1982. A good advanced treatment of ocean (70%) and atmospheric (30%) applied physics.

Palmén, E. and Newton, C.W. *Atmospheric Circulation Systems*, International Geophysics Series, **13**, Academic Press, 1969. A classic presentation of large-scale atmospheric phenomena.

Haltiner, G.J. *Numerical Weather Prediction*, Wiley, 1971. An excellent graduate-level text showing the equations applied to numerical weather analysis.

Thompson, P.D. *Numerical Weather Analysis and Prediction*, Macmillan, 1961. A comprehensive graduate level introduction to the application of finite-differencing to the equations of motion.

Stull, R.B. *An Introduction to Boundary Layer Meteorology*, Kluwer, 1988, 666 pp. This is a good follow-on text for students interested in graduate-level analytic treatment of small to meso-scale meteorology.

Chapter 2

Feynman, R.P., R.B. Leighton and M. Sands, *The Feynman Lectures on Physics*, **1**, Calif. Inst. of Tech., Addison–Wesley, 1963. Deals with gravitational forces and non-inertial frame of references.

Kittel, C., W.D. Knight and M.A. Ruderman, *Berkeley Physics Course*, Mechanics Vol. **1** UC Berkeley, McGraw–Hill, 1965. Explanations of physics, gravity and non-inertial frame of reference.

Chapter 3

Buckingham, E. in *Transactions ASME*, **37**, 263–296, 1915.

Chapter 4

Panton, R.L. *Incompressible Flow*, Wiley, 1984. An excellent treatment of classic flow topics; graduate level engineering fluid dynamics. A rigorous tensor approach, with comprehensive coverage of classical laboratory experiments.

Chapter 9

Lamb, H. *Hydrodynamics*, Dover, 1945. Classical inviscid flow.

Milne–Thompson, L.M. *Theoretical Hydrodynamics*, Macmillan, 1938. Another classic of potential flow theory. Uses vector notation to discuss vortex motion in an ideal fluid.

Greenspan, H.P. *The Theory of Rotating Fluids*, Cambridge University Press, 1968. Advanced text on flow in rotating fluids.

Lumley, J.L. and Panofsky, H.A. *The Structure of Atmospheric Turbulence*, Wiley, 1964. Particularly good observational material of the atmospheric turbulence.

Chapter 10

Meyer, R.E. ed., *Transition and Turbulence*, Academic Press, 1981. The latest concepts on transition to turbulence, including vortex interactions, rotation effects, and large-scale eddies.

Swinney, H.L. and Gollub, J.P., eds., *Hydrodynamic Instabilities and the Transition to Turbulence*, Springer–Verlag, 1981. Modern approaches to instabilities and turbulence, including strange attractors, bifurcation techniques and chaos.

Chapter 11

Panofsky, H.A. and Dutton, J.A. *Atmospheric Turbulence; Models and Methods for Engineering Applications*, Wiley, 1984. The application of statistics to atmospheric turbulence modeling. Great observational details, design criteria, state-of-the-art models.

Atmospheric Turbulence and Air Pollution Modeling, Nieuwstadt, F.T.M. and van Dop, H., eds., Reidel, 1982. Applications in diffusion modeling.

Schlicting, H., *Boundary Layer Theory*, McGraw-Hill, 1959. A classic treatment of laboratory boundary layers.

Sorbjan, Z. *Structure of the Atmospheric Boundary Layer*, Prentice–Hall, 1989. Details on the modeling of turbulence, diffusion; similarity, and measurements in the surface layer and the PBL.

Arya, P. *Introduction to Micrometeorology*, Academic Press, 1989. A summary of the classical approaches (primarily empirical modeling) to treating turbulence in the boundary layer. Good practical modeling.

Brown, R.A. *Analytical Methods in Planetary Boundary Layer Modeling*, Adam Hilger, 1973. Application of the equations of motion to model the coherent large-eddies of the PBL. The mathematical solution for PBL flow follows directly from the equations developed in this text.

Answers to Selected Problems

Chapter 1

1. $\rho_{air} = p/RT = 103{,}000/[287(10 + 273)] = 1.268 \text{ kg/m}^3$
   $\rho_{water} = 1000 \text{ kg/m}^3$. The $\rho_w/\rho_a = 1000/1.268 = 788$
   μ_{air}, $20°C = 1.81 \cdot 10^{-5} \text{ Ns/m}^2$; $\nu = 1.51 \cdot 10^{-5} \text{ m}^2/s$
   μ_{water}, $20°C = 1.0 \cdot 10^{-3} \text{ Ns/m}^2$; $\nu = 1.0 \cdot 10^{-6} \text{ m}^2/s$
   $\mu_a/\mu_w = 1.81 \cdot 10^{-2}$; $\nu_w/\nu_a = 0.066$

2. The design of a boat's keel is made to affect the following functions:

   (a) To balance the air force on the sail and keep the boat from traveling sideways.

   (b) To provide a torque between the keel center of force and the sail center of force such that the board can be turned.

   (c) To lower the center of gravity.

3. The irregular motion of the ball is due to transition to turbulent boundary layer flow taking place nonuniformly around the ball because of irregular distribution of seams, which trip the laminar flow in the boundary layer. Transition depends on Re. The lateral forces change in response to varying drag and pressure distribution.

   Action takes place when Re is critical.

$$\text{Re} = \rho U L/\mu = U L/\nu$$

$$1.8 \cdot 10^5 = U \cdot 3/12[\text{ft}]/1.6 \cdot 10^{-4} \text{ ft}^2/s$$

$$U = 1.8 \cdot 10^5 \cdot 4 \cdot 1.6 \cdot 10^{-4} = 115 \text{ ft/s}$$

or

$$= 78 \text{ mph}$$

4. The major difficulty is to obtain sufficient points. Since the cycle of the mountain-valley wind is 24 hrs, an hourly average would provide 24 points. This is enough to define the cycle. However, to get an ensemble of points, measurements must be made on successive days. In this case, the large-scale weather must be nearly the same. This will require a criterion for large-scale steady state and mean state. Then daily averages at each hourly interval can be accumulated.

5. Transition takes place as the Reynolds number, $U L/\nu$, reaches a

critical value. (a) Increasing the scale factor L increases Re—more likely to have transition. (b) Helium has less density so Re decreases—transition is less likely. (c) Doubling the wind doubles Re—transition is more likely. (d) Increasing temperature at constant pressure must decrease density, from the perfect gas law—transition is less likely.

6. The floating ice displaces its equivalent weight of water. Thus when it melts, it simply replaces this amount of water. There is no change in water level. Thus, the melting pack ice will not raise the ocean level. However, melting of the vast amounts of ice above sea level in Greenland and Antarctica would raise the ocean level.

7. Pool level will be less because the boat rises by a volume of water equal to the weight of the anchor, but the anchor displaces only its volume. Since water is less dense than the anchor, the water volume lost is greater than the volume of the anchor. The pool level will drop.

8. Ideal gas: $p = \rho RT$.
Hence,

$$dp/dz = -\rho g = \gamma = -pg/(RT)$$

$$\int_{p_0}^{p} dp/p = \ln(p/p_0) = -\int_{0}^{z} g/(RT)\, dz$$

$$T = T_0 - \gamma z$$

leads to

$$p/p_0 = (1 - \gamma z/T_0)^{[g/(\gamma R)]}$$

If $T = T_c$ (constant)

$$\text{and} \quad z_0 \leq z \leq z$$

$$p/p_0 = e^{[-g(z-z_0)/(RT_c)]}$$

9. (a) The momentum exchange with the surface.
   (b) The momentum flux across an imaginary plane at the point.
   (c) The kinetic energy of the molecules in the volume considered.

10. What is viscosity? A measure of the internal stickiness of a fluid; the proportionality factor between stress and rate-of-strain; a constant coefficient in a Taylor expansion; a fudge factor to accomplish closure . . . .. It is measured by using the definition,

$$\tau = \mu\, du/dz$$

and measuring τ (force/unit area) and velocity shear.

11. This experiment is a way of producing a thin layer of uniform shear across a layer. Use the formula

$$\tau = \mu \, \Delta U/\Delta z, \qquad \Delta U = \tau \times \Delta z \times 1/\mu$$

$$\Delta y = (6 - 5.995)/2 = 0.0025 \text{ in}$$

$$\tau = F/A = 5 \text{ lb}/(2 \times \pi 6) = 0.1326 \text{ lb/in}^2$$

$$\Delta U = 0.1326 \times (0.0025)/(7 \cdot 10^{-5}) \times 144 \text{ in.}^2/\text{ft}^2$$

$$= 682 \text{ in /sec}$$

This is the difference in the velocity of the oil adhering to the cylinder wall and that adhering to the weight. Since the cylinder is stationary, the ΔU is the velocity of the weight.

12. For $\tau = \mu \, du/dr$, $\mu = \tau/(du/dr)$,

$$u_{(R_i)} = R_i\Omega, \quad u_{(R_0)} = 0, \quad u = u_\Theta, + \text{ clockwise}$$

For $R_i < r < R_0$, $u = R_i[(r - R_i)/(R_0 - R_i)]\Omega - R_i\Omega$

$$du/dr = R_i/(R_0 - R_i) \, \Omega = R_i\Omega/d$$

Or, the shear across the gap is $du/dr = [u_{(R_i)} - u_{(R_0)}]/d = R_i\Omega/d$.

τ is the force on the wall at R_i. The total force required to drive the apparatus is $F = 2\pi R_i H\tau_{(R_i)}$. The torque is $T = FR_i$.

Hence, $\mu = F/(2\pi R_i H) \, d/(R_i\Omega) = Td/(2\pi R_i^3 H\Omega)$.

13. The eddies must be (a) large enough to be measured; (b) small enough to allow a sufficient number in the parcel to guarantee a uniform average (an eddy continuum); and (c) in a uniform, steady state.

14. (a) As sand starts to fall, part is in freefall and thus does not appear in the weight; $W_1 < W_0$.

(b) As sand impacts, $F = \Delta mv/\Delta t$ is added to the weight:

$$v_f = v_0 + at; \qquad s = s_0 + \frac{1}{2}at^2;$$

$$v_f = gt_f = (2gH)^{1/2}, \, t_f = (2H/g)^{1/2}$$

$$\int_0^M dm = \int_0^t F/V_f \, dt$$

Hence

$$M = (F/v_f)t_F = \text{column weight}$$

$$= Mg = F/v_f \, gt_f \equiv F$$

Thus, Δ(mom force) = weight of sand in column, $W_2 = W_0$.

(c) Near the end, a small amount of sand is left in the column, $M_3 < M_2$, but the momentum impact remains the same; therefore, briefly, $W_3 > W_0$.

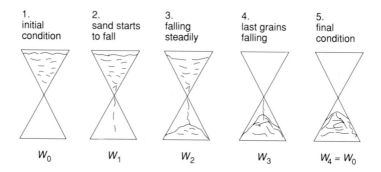

1. initial condition	2. sand starts to fall	3. falling steadily	4. last grains falling	5. final condition

W_0 W_1 W_2 W_3 $W_4 = W_0$

15. For a Newtonian fluid,

$$\tau = \mu \, du/dz = \mu U[2/H - 2z/H^2] = 2\mu U/H[1 - z/H]$$

$$\tau(0) = 2\mu U/H; \qquad \tau(H/2) = \mu U/H; \qquad \tau(H) = 0$$

16.

$$\tau = \mu \, dU/dz = \mu UA \, d/dz[2(z/h) - (z/h)^2] = 2\mu UA/h(1 - z/h)$$

$$\tau_{z=0} = 2\mu UA/h; \qquad \tau_{z=h/2} = \mu Ua/h; \qquad \tau_{z=h} = 0 = \tau_{air} = \text{the B.C.}$$

Thus, although the air stress on the water is sufficient to establish the velocity U in the water over a long time period, at equilibrium, this stress is much less than the internal water stress.

17. (1) There is a hydrostatic pressure force due to the weight of liquid above the area, ρgh, where h is the depth.

(2) There is a momentum flux normal to the surface due to the inertia of the fluid being deflected or stopped. This is

$$F_{\text{inertial}} \approx \Delta p \cdot (\text{Area}) \approx \frac{1}{2}\rho U^2 A = \rho U_L L^2$$

(3) There is a drag force along the surface due to viscous effects. These are also momentum flux of the molecules scattering in all directions off the microscopic imperfections of the surface. We parametrize these net forces with viscosity and call them viscous forces within the continuum concept.

$$F_{\text{viscous}} \approx \mu(U/L)\,L^2 = \mu UL$$

Chapter 2

1. (a)

$$\partial u/\partial t + u \, \partial u/\partial x + v \, \partial u/\partial y + w \, \partial u/\partial z = F_1$$

$$\partial v/\partial t + u \, \partial v/\partial x + v \, \partial v/\partial y + w \, \partial v/\partial z = F_2$$

$$\partial w/\partial t + u \, \partial w/\partial x + v \, \partial w/\partial y + w \, \partial w/\partial z = F_3$$

(b)

$$\partial s_3/\partial x_2 + \partial s_2/\partial x_1 + \partial s_1/\partial x_3 - \partial s_1/\partial x_2$$
$$- \partial s_3/\partial x_1 - \partial s_2/\partial x_3$$

2. (a)

$$\text{div } u = \partial u/\partial x + \partial v/\partial y + \partial w/\partial z$$

$$= 6y + (-6y) + 0 = 0; \qquad \text{nondivergent}$$

(b)

$$\text{curl } u = (\partial w/\partial y - \partial v/\partial z), \quad (\partial u/\partial z - \partial w/\partial x), \quad (\partial v/\partial x - \partial u/\partial y)$$
$$= (0 - 0), \quad (2y - 0), \quad (4y - 2z); \quad \text{rotational}$$

3. (a) curl $\mathbf{u} = (0, 0, 0);$ irrotational
   (b) div $\mathbf{u} = 10 + (-10) + 0 = 0;$ nondivergent

4. (a) Velocity is given as a function of (x, y, z, t); Eulerian.
   (b) Local: $\partial u_i/\partial t = (0, 3, 0)$
   (c) Advective: $u_j \, \partial u_i/\partial x_j = 3u + 2vz, \qquad 4uy + 4vx, \qquad 0$
   (d) (b) + (c)

5. The total derivative, $\partial T/\partial t + V \, \partial T/\partial y$, is $12/24(°C/hr) + 360 \,(\text{km}/$
   hr$)(-0.06°C/km) = -21.1°/hr$ or $-0.0586°C/km$.

6. (a)

$$\mathbf{a}_{\text{local}} = \partial\mathbf{U}/\partial t = \partial u/\partial t \, \mathbf{i} + \partial v/\partial t \, \mathbf{j} + \partial w/\partial t \, \mathbf{k}$$
$$\partial u/\partial t = \partial/\partial t[x + 2y + 3z + 4t^2]$$
$$= 8t = 8 \cdot 2 = 16$$

Likewise, $\partial v/\partial t = 1, \qquad \partial w/\partial t = 2.$ Thus, $\mathbf{a}_{\text{local}} = 16\mathbf{i}$
$+ \mathbf{j} + 2\mathbf{k}. \; a_{\text{local}} = (16^2 + 1^2 + 2^2) = 16.155.$

(b)

$$\mathbf{a}_{\text{advective}} = \mathbf{u} \cdot \nabla\mathbf{u}$$

$$a_x = u \, \partial u/\partial x + v \, \partial u/\partial y + w \, \partial u/\partial z$$

$$a_y = u \, \partial v / \partial x + v \, \partial v / \partial y + w \, \partial v / \partial z$$

$$a_z = u \, \partial w / \partial x + v \, \partial w / \partial y + w \, \partial w / \partial z$$

The x-component of advective acceleration,

$$u = x + 2y + 3z + 4t^2 = 1 + 2 + 3 + 16 = 22$$

$$\partial u / \partial x = \partial / \partial x [x + 2y + 3z + 4t^2] = 1 + 0 + 0 + 0 = 1$$

$$u \, \partial u / \partial x = 22 \cdot 1 = 22$$

Likewise,

$$v = 3, \ \partial u / \partial y = 2, \ v \, \partial u / \partial y = 6, \ w = 6, \ \partial u / \partial z = 3, \ w \, \partial u / \partial z = 18$$

$$a_x = 22 + 6 + 18 = 46; \qquad a_y = 31; \qquad a_z = 49$$

$$a_{adv} = [46^2 + 31^2 + 49^2]^{1/2} = 74$$

(c) Total acceleration,

$$\mathbf{a} = \mathbf{a}_{local} + \mathbf{a}_{adv} = 62\mathbf{i} + 32\mathbf{j} + 51\mathbf{k}$$

$$a = [62^2 + 32^2 + 51^2]^{1/2} = 86$$

8. The advective acceleration, $a = u \, \partial u / \partial x$; $\qquad u = Q/A$

$$A = A_0 - \Delta A / \Delta x \, x$$

$$= 0.00636 \text{ m}^2 - \frac{0.00566 \text{ m}^2}{0.36 \text{ m}} x = (0.00636 - 0.0157)x$$

$$A_{mid} = 0.00636 - 0.0157 \cdot 0.18 = 0.00353 \text{ m}^2$$

$$V_{mid} = \frac{A}{A_{mid}} = \frac{0.02 \text{ [m}^2/\text{s]}}{0.00353 \text{ [m}^2]} = 5.66 \text{ [m/s]}$$

$$\partial u / \partial x = \partial (Q/A) / \partial x = \partial / \partial x [Q/(0.00636 - 0.0157x)]$$

$$= 0.0157 Q / [0.00636 - 0.0157x]^2$$

$$\partial u / \partial x \ (x = 0.18 \text{ m}) = 25.1 \text{ [1/s]}$$

$$a = u \, \partial u / \partial x = 142 \text{ [m/s}^2]$$

9.

$$a_\ell = \partial U / \partial t = \partial / \partial t [0.2t/(1 - 0.5x/L)^2] = 0.2/(1 - 0.5x/L)^2$$

$$= 0.2/(1 - 0.5 \cdot 0.5L/L)^2 = 0.267 \text{ [units] (constant)}$$

At 30 sec., $U = 10.67$ units. Assume this is m/s. Hence, 0.2 [m/s^2] and $a_\ell = 0.267$ [m/s^2]

$$a_v = U(\partial U/\partial x) = [0.2t/(1 - 0.5x/L)^2] \, \partial/\partial x [0.2t/(1 - 0.5x/L)^2]$$

$$= 0.04t^2/[(1 - 0.5x/L)^5 L] = 0.04 \cdot 30^2/[(1 - 0.5 \cdot 0.5L/L)^5 \, 5.]$$

$$= 0.03 \text{ [m/s}^2\text{]}$$

Chapter 3

1. a) $\rho[M/L^3]U^2[L/t]^2/2 = \rho U^2/2 \; [M/(L^2 t)]$.
   b) $\tau[ML/t^2/L^2]/\rho[M/L^3] = \tau/\rho[L^2/t^2]$, hence $[L/t]$.
   c) Velocity gradient over distance $[L/t/L]$ or $[1/t]$.
   d) μ defined as stress$/(du/dz)[M/(Lt^2)/(1/t)]$ or $[M/(Lt)]$.

2.

$$F_D = f\{ \quad W, \quad d, \quad \mu \}$$

$$[ML/t^2] = \quad [L/t], [L], [M/(Lt)]$$

$4 - 3 = 1$ ND variable; $F_D/\mu[L^2/t]$ to get rid of M; hence $F_D/4(\mu W d) = $ constant.

4.

$$\mu = f\{ \quad T, \quad R, \quad \Omega, \quad t/H \}$$

$$[M/(Lt)] = \quad [ML^2/t^2], [L], [1/t], \text{ND}$$

$5 - 3 = 2$ ND variables.
$\mu/T \; [t/L^3]$ (cancels M); hence $\mu \Omega R^3 = f\{t/H\}$.

5. (a) The Reynolds numbers must be matched: $\text{Re}_m = \text{Re}_b$;

$$U_m = \frac{\rho_b \, U_b L_b \, \mu_m}{\mu_b \, \rho_m L_m}$$

$$U_m = \rho_b/\rho_m \, L_b/L_m \, \mu_m/\mu_b \, U_b$$

$$= 1 \cdot 10 \cdot 1 \cdot 20 = 200 \text{ m/s}$$

   (b) This is a very high windspeed to obtain. One could use more dense fluid, a bigger model, or a less viscous fluid (e.g., air at a higher temperature).

6. (a) The thickness of the active layer H (m) depends on the thermal diffusivity K (m^2/s) and the period T (s). We have three parameters:

H, K, and T; and two dimensions: length and time. There is $3 - 2 = 1$ nondimensional parameter, which is thus a constant. Therefore $K/(H^2/T) = $ constant.

Absorbing this constant into the constant K,

$$K_a = H^2/T = 2000^2/(24 \cdot 60 \cdot 60) = 46 \text{ m}^2/\text{s}$$
$$K_0 = (20/2000)^2 46 = 0.0046 \text{ m}^2/\text{s}.$$
$$K_s = (5 \ 10^{-3}/2000)^2 \cdot 46 = 2.88 \cdot 10^{-10} \text{ m}^2/\text{s}$$

(b) Molecular values are very much smaller:

$$\nu_a = 16.3 \cdot 10^{-6} \text{ m}^2/\text{s}; \quad \text{and} \quad \nu_w = 1 \cdot 10^{-6} \text{ m}^2/\text{s}$$

(c) The Coriolis parameter furnishes a time scale, $1/\text{s}$, and together with K, a height scale, $(K/f)^{1/2}$.

With $f = 1 \cdot 10^{-4} \text{ s}^{-1}$ and $L = 500$ m, $K = 25 \text{ m}^2/\text{s}$ for the atmosphere, $0.25 \text{ m}^2/\text{s}$ for the ocean.

Eddy scale transport mechanisms are on the order of 10^6 more effective in transporting momentum.

8.

$$T = f\{L, \quad \rho, \quad g, R\}$$
$$[1/t] = [\ L], [M/L^3], [L/t^2], [L]$$

$5 - 3 = 2$ ND variables.

Since only ρ has M, evidently T does not depend on ρ. Hence, $4 - 2 = 2$ variables.

The ratio L/R is evidently one possibility. Since this is extremely small, we might reconsider including R. Then, $3 - 2 = 1$ and $T/(L/g)^{1/2}$ is constant, or $T = C(L/g)^{1/2}$.

9. The drag must be a function of length, speed, diameter, density of fluid, viscosity, perhaps roughness of the surface. Hence we can write

$$D = f[L, V, d, \rho, \mu, \text{(maybe } z_0)]$$

Say it is made as smooth as possible, $z_0 = 0$. Scales are

$$m/Lt^2 = f[L, L/t, L, m/L^3, m/Lt]$$

$$n = 6, \quad m = 3, \quad n - m = 3$$
$$\pi_1 = D/(\rho V^2 L^2); \quad \pi_2 = \mu/(\rho VL); \quad \pi_3 = D/L$$

Or

$$D/(\tfrac{1}{2} \rho V^2 L^2) = f\{\mu/(\rho VL), D/L\}$$

If design (D/L) is held constant, we must match Re numbers only.

10.

$$\tau = f\{\rho, \quad U\}$$

$$[M/(Lt^2)] = [M/L^3], [L/t]$$

Hence, $\tau/(\rho U^2)$ is evidently the one ND parameter.

11.

$$fu + K\, d^2u/dz^2 = 0.$$

Keep $f[1/t]$ and $K[L^2/t]$; let $U = u/U_G$, $Z = z/H$.

$$(fU_G)\, U + (KU_G/H^2)\, d^2U/dZ^2 = 0.$$

Or

$$U + [K/(fH^2)]\, d^2U/dZ^2 = 0.$$

This equation will be self-similar if there is no coefficient in the equation and B.C. To get this, let $H = (K/f)^{1/2}$.

$$U + d^2U/dZ^2 = 0 \quad \text{(where } Z = z/(K/f)^{1/2}); \quad U(\infty) = 1, u(0) = 0.$$

12. (a) There exists DS if Re $= \rho VL/\mu$ is the same for the bridge and the model. ρ and μ are the same for air at the same P and T. Hence

$$[\rho VL/\mu]_b = [\rho VL/\mu]_m \Rightarrow V_m/V_b = L_b/L_m = 10 \Rightarrow V_m = 1000 \text{ mph.}$$

(b) Since $V_m/c_0 = 1000/770 = 1.3$, flow is supersonic; shock waves may appear around the structure. This problem can be avoided by

1. Building a bigger model;
2. Using more dense fluid;
3. Using less viscous fluid;
4. Using a combination of the above.

14.

$$u = f\{g, \quad \lambda, \quad \mu\}$$

$$[L/t] = [L/t^2, [L], [M/(Lt)]$$

$4 - 3 = 1$ ND variable. However only μ has M. We can either eliminate μ to get

$$u = c\, (g \cdot \text{wavelength})^{1/2}$$

or add ρ to get $(\mu/\rho \equiv \nu)$

$$u = f\{g, \quad L, \quad v\}$$
$$[L/t] = [L/t^2], [L], [L^2/t]$$
$4 - 2 = 2$ ND variables, $u/(gL)^{1/2}$, and $v^2/(gL^3)$.

Chapter 4

3. (a) curl (grad f) $= \epsilon_{ijk} \, \partial/\partial x_j [\partial f/\partial x_k] = \epsilon_{ijk} \, \partial^2 f/\partial x_j \, \partial x_k$.
   The i-component consists of j and k each 1 through 3, or nine parts. However, all parts are zero except for 123 and 132, 231 and 213, and 312 and 321. Since each of these pairs cancel ($\epsilon_{123} = -\epsilon_{132}$), the sum is zero.
   (b) div curl $\mathbf{u} = \partial/\partial x_i(\epsilon_{ijk} \, \partial u_k/\partial x_j \, e_i)$. Writing these out and using the properties of ϵ_{ijk} shows that this expression is identically zero.

5. (a) Left side: $[A_{11}U_1 + A_{21}U_2 + A_{31}U_3]\mathbf{i} + [A_{12}U_1 + A_{22}U_2 + A_{32}U_3]\mathbf{j} + [A_{13}U_1 + A_{23}U_2 + A_{33}U_3]\mathbf{k}$.
   Right side: $[A_{11}U_1 + A_{12}U_2 + A_{13}U_3]\mathbf{i} + [A_{21}U_1 + A_{22}U_2 + A_{23}U_3]\mathbf{j} + [A_{31}U_1 + A_{32}U_2 + A_{33}U_3]\mathbf{k}$.
   They are not equal.
   (b) $u_j A_{ij} = A_{ji}u_j$ is not true.
   The physical positions of u and $\mathbf{A}$ matter in vector/matrix form, but not in indicial notation, where the indices determine the outcome.
   (c) $\mathbf{uv}$ does not equal $\mathbf{vu}$.
   (d) $u_i v_j = u_j v_i$.

6.

$$\partial u/\partial t + u\partial u/\partial x + v\partial u/\partial y + w\partial u/\partial z = (-\partial p/\partial x)/\rho + fv \quad + v(\partial^2 u/\partial x_i \, \partial x_i)$$
$$[L/t^2] + [L/t^2] \quad + [L/t^2] \quad + [L/t^2] \quad = [L/t^2] \quad + [L/t^2] + [L/t^2]$$

The terms are all accelerations and scalars (the indices are summed).

7. 1, 3, 9, 27, 81 (3^n).

Chapter 5

2. Only the flow at the entrances and exits needs to be considered.

$$V_1 A_1 + V_2 A_2 + V_3 A_3 = 0$$

$$-20 \cdot 25 \cdot H + 19 \cdot 25 \cdot H + V_3 \cdot 8 \cdot H = 0$$

$$V_3 = 25(20 - 19)/8 = 25/8 = 3.13 \text{ m/s}$$

3. Liquid displaced upward: $V_{up}(D^2 - d^2) \pi/4$; displaced downward: V_{down} $\pi d^2/4 = V_{up}(D^2 - d^2) \pi/4$. Hence $2d^2 = D^2$; $D = \sqrt{2} d$.

By symmetry, $z/D = 24d/2d$; $D = z/12$.

Eliminate D: $z = 12 \sqrt{2} d$.

4. $\mathbf{u} = -C_r/r \, \mathbf{e}_r - C_\theta/r \, \mathbf{e}_\theta$;

The mass coming in through the perimeter/meter height is equal to that going up the center.

$$2\pi r u_r h = \pi r^2 w$$

$w = 2U_r h/r = 2C_r h/r^2 = 2(300/30^2)$ or 0.67 m/s per meter height.

$u_r(30) = -30$ m/s; for a 100-m high tornado, $w = 67$ m/s.

5. $\partial u_i/\partial x_i = U(3x^2 + y^2) + U(3y^2 + x^2) + 0 = 4U(x^2 + y^2)$.

This doesn't satisfy continuity, hence it is not a legitimate flow.

Adding $\partial w/\partial z = -4Ur^2$ makes $\nabla \cdot \mathbf{u} = 0$, and it is then a legitimate incompressible flow.

6. Use continuity; entrainment around the circumference,

$$(\rho_a + \rho_{p0}) V_0 A_0 + \rho_a V_a A_a = (\rho_a + \rho_{px})(V_x A_x)$$

$$A_a = \pi \int D \, dx = \pi \int (D_0 + 0.02x) \, dx = \pi(D_0 x + 0.01x^2)$$

Divide through by ρ_a.

$$(1 + \rho_{p0}/\rho_a)(V_0 A_0) + V_a A_a = (1 + \rho_{px}/\rho_a)(V_x A_x) \tag{1}$$

We have everything except V_x; we could solve for ρ_{px}/ρ_a. However, if we assume $\rho_p/\rho_a \ll 1$, we get

$$V_0 A_0 + V_a A_a \approx V_x A_x$$

Thus, Eq. (1) becomes

$$\rho_{px}/\rho_a = (\rho_{p0}/\rho_a)V_0 A_0/(V_x A_x)$$

$$= \rho_{p0}/\rho_a \, 1/[1 + V_a A_a(V_0 A_0)]$$

$$= \rho_{p0}/\rho_a\{1/[1 + 0.00067(x + 0.005x^2)]\}$$

7. $\sum UA = 0$; density is constant here.

$4 \cdot 8 \, [\text{cm}^3/\text{s}] - 3 \cdot V_3 - 2 \cdot 4 \, [\text{cm}^3/\text{s}] = 0$. (The angle does not enter.)

$$V_3 = \{(32 - 8)/3\} = 8 \text{ m/s}$$

Chapter 6

2. Momentum equation in one-dimension (along x-axis):

$$\partial u/\partial t + u\, \partial u/\partial x + v\, \partial u/\partial y + w\, \partial u/\partial z = -1/\rho\, \partial p/\partial x + fv + K\, \partial^2 u/\partial z^2$$

$$0 + u\, \partial u/\partial x + 0 + 0 = -1/\rho\, \partial p/\partial x + 0 + 0$$

Thus, along a streamline,

$$\int_0^x u\, \partial u/\partial x\, dx = \int_0^x -1/\rho\, \partial p/\partial x\, dx;$$

or

$$u^2/2 - u_0^2/2 = -(1/\rho)[p - P_0]$$

$$p = P_0 - \rho[u^2 - u_0^2]/2; \qquad u = Q/A; \qquad A = \pi r^2$$

$$p(x) = P_0 - \rho Q^2/(2\pi^2) \cdot [1/r^4 - 1/r_0^4]$$

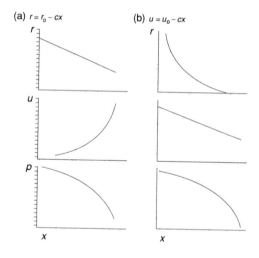

(a) $r = r_0 - cx$ (b) $u = u_0 - cx$

3. $p = \rho RT; \qquad dp/dz = -\rho g = -pg/(RT)$

$$\int_{P_0}^p dp/p = \ln\,(p/p_0) = -\int_0^z g/(RT)\, dz$$

$$= g/R \int dz/(T_0 - \gamma z) = -\ln(T_0 - \gamma z)^{g/(\gamma R)}/T_0$$

or

$$p/p_0 = [1 - \gamma z/T_0]^{g/(\gamma R)}$$

When T = constant,

$$\ln (p/p_0) = -g/(RT_c) \int_{z_0}^{z} dz = -g(z - z_0)/(RT_c)$$

$$p/p_0 = \exp[-g(z - z_0)/(RT_c)]$$

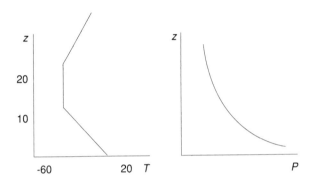

4. The problem is much simpler than a solution of the momentum equation. Given, the flow is parabolic. Get $u(z)$. Use B.C.s suggested in the problem: u_{max} at centerline by symmetry, and $u(\pm d) = 0$ for no-slip. Set up a coordinate system with the centerline as the x-axis:

$$u = Az^2 + Bz + C;$$

$u(0) = u_{max}, \quad du/dz\,(0) = 0, \quad u(\pm d) = 0$ (assuming no-slip at the plates).

$$u_{max} = C; \qquad 2A(0) + B = 0; \qquad B = 0$$

$$A\,(\pm d)^2 + u_{max} = 0 \qquad \text{yields } u = u_{max}[1 - z^2/d^2]$$

$$Q = \int u\, dA = \int u_m[1 - z^2/d^2]\, dz \text{ (flow rate } Q)$$

$$= 4/3\, du_m \qquad u_{ave} = 2/3\, u_m$$

$$\tau_{31} = \mu\, du/dz(z = \pm d) \qquad du/dz = -(2z/d^2)u_m \qquad \tau = \pm(2\mu/d)u_{max}$$

$$\text{vorticity: } \nabla \times \mathbf{u} = (2u_m/d^2)\, z\mathbf{j} = 2(u_m/d^2)z$$

8. The force (e.g., from DA): $F = \rho A V^2$

$$F = [.00123 \text{ g/cm}^3/(1000 \text{ g/kg}) \cdot 10^6 \text{ cm}^3/\text{m}^3] \cdot 5 \cdot 20 \text{ m}^2 \cdot 30^2 \text{ m}^2/\text{s}^2$$
$$= 1.11 \ 10^5 \text{ kg-m/s}^2$$

Chapter 7

1. It is important to note that there is a contribution to the radial pressure variation from hydrostatic pressure. Thus the net force across the parcel at depth z is $F_P = \partial p/\partial r + \rho g \, \partial z/\partial r$. This opposes the centrifugal force.
Or,

$$d(p + \rho gz)/dr = -a_r$$
$$= -(\rho u_\theta^2/r) = \rho r \omega^2$$

Integrate with respect to r:

$$p + \rho gz = \rho r^2 \omega^2/2 + C$$

$\omega = u_\theta/r$ and ρ is constant; $p/\rho + gz - u_\theta^2/2 = $ constant.
Note: This is similar to Bernoulli's equation, but the minus sign is crucial. This is not irrotational flow.

2. Use Bernoulli's equation with only $V = \Delta V$ and Δp given.
$$V = (2 \ \Delta p/\rho)^{1/2} = (2 \cdot 2,000/1.2)^{1/2} = 57.7 \text{ m/s}$$

3. (a) Both gauges read the same, reading hydrostatic pressure.
   (b) $\Delta C_P = (p_A - p_B)/(\rho u_0^2/2) = 1.4;$
   $$u_0^2 = 2(5,000)/[(1.5) \cdot 1.4]; \qquad u_0 = 69 \text{ m/s}$$

4. Write the *energy* equation from upstream end to downstream end. For discussing power and work, the energy equation is usually appropriate. Head loss in meters is a potential energy loss of $HL \cdot g$.

$$p_1/\rho + u_1^2/2 + gz_1 + Wk = p_2/\rho + u_2^2/2 + gz_2 + HL,$$
$$0 + 0 \quad + 0 \ + Wk = 0 \quad + u_2^2/2 + 0 \ + 0.02u_t^2/2$$

For discussing velocities through areas, the *continuity* equation is appropriate.

$$u_t A_t = u_2 A_2; \qquad u_2 = u_t A_t/A_2 = 0.5u_t = 5 \text{ m/s}$$
$$u_2^2/2 = 0.25 \ u_t^2/2.$$
$$Wk = (0.27)u_t^2/2 = (0.27)(10^2/2) = 13.5 \text{ m}^2/\text{s}^2$$

$$\text{Power} = Q\rho Wk = (10 \text{ m/s} \cdot 15 \text{ m}^2) \cdot 1.2 \text{ kg/m}^3 \cdot 13.5 \text{ m}^2/\text{s}^2$$

$$= 2430 \text{ kg/m}^2/\text{s}^3 = 2.43 \text{ kW}$$

6. The energy equation is decoupled from the momentum equations. Mechanical energy and thermal energy equations can then be written separately.

8. (a) $v^2/2 + p/\rho + gz = c$

(b) $\int v^2/2 + p/\rho + gz = c$

(c) $v^2/2 + p/\rho + gz = c$

(d) $v^2/2 + \int p/\rho \, ds + gz = c$

9. The Coriolis force is always perpendicular to the velocity. Hence, it can do no work.

10. Bernoulli's equation:

$$p_1/\rho_1 + u_1^2/2 + gz_1 = p_2/\rho_2 + u_2^2/2 + gz_2 + \Phi \text{ (dissipation)}$$

Neglect dissipation.

$$z_2 - z_1 = \{p_1/\rho_1 - p_2/\rho_2 + (u_1^2 - u_2^2)/2\}/g$$

Get $p(r)$ from problem 1.

11. The spin will cause the relative velocity difference between the free-stream and the surface (in coordinates on the ball) on one side of the ball to be greater than on the other side. The higher velocities have lower pressures according to Bernoulli's relation, giving a sidewards pressure push. This depends on the friction building boundary layers on the ball. The boundary layers depend on the relative velocity (and the seams), hence they are unsymmetric on each side of the spinning ball. If one side of the ball is laminar while the other is turbulent, the boundary layers will be quite different in thickness and velocity profile.

Chapter 8

1. Check the curl **u**.

$$\mathbf{u} = (\partial\phi/\partial x, \partial\phi/\partial y, \partial\phi/\partial z) = (2x - 2y, -2x, 0)$$

$$\text{curl } \mathbf{u} = [0, 0, -2 - (-2)] = 0$$

It is irrotational.

3. Calculate curl **u**.

4. Use a tornado approximation using line vorticity:

$$\Gamma \equiv u_\theta \, 2\pi r = 20 \cdot 2\pi \cdot 2000 = 8\pi \cdot 10^4 \text{ m}^2/\text{s}$$

$$= \text{strength}$$

$$U = \Gamma/(2\pi \, 500) = u_\theta \, 2000/500 = 80 \text{ m/s}$$

5. The total vorticity equation.

6. Vorticity can change due to divergence, $\nabla \times \mathbf{u}$, and by twisting, turning by $\nabla \mathbf{u}$ in an inviscid fluid. If it is inviscid two-dimensional, only the divergence term can change vorticity.

7. Get the velocities by differentiation. At $r = R$, use Bernoulli's equation to get the pressure–velocity relation over the surface. Integrate over θ in the x (drag) and y (lift, or lateral force) to get F_D and F_L. *Note:* Symmetry conditions eliminate some integrals (e.g., odd powers of sine). Get

$$F_L = \rho U\Gamma/\pi \int \sin^2 \theta \, d\theta = \rho U\Gamma$$

9. Think of the tube between the ground and a capping inversion at the tropopause and use Helmholz's law. $\zeta/L = $ constant. When L is increased over the plains, ζ increases, cyclone intensifies.

Chapter 9

2. Plot the streamline flow for $(u, v) = (\partial\psi/\partial y, -\partial\psi/\partial x) = (Cx, -Cy)$

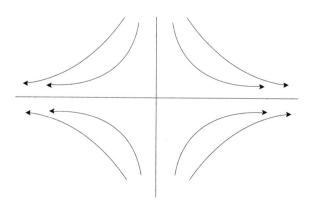

The flow is 2-D, steady-state, constant, with $\nabla \cdot \mathbf{u} = 0$. Check for incompressibility.

$$\partial u/\partial x = C \qquad \partial v/\partial y = -C \qquad \text{and} \qquad \nabla \cdot u = 0$$

Rotationality:

$$\nabla \times \mathbf{u} = \begin{bmatrix} \mathbf{i} & \mathbf{j} & \mathbf{k} \\ \partial/\partial x & \partial/\partial y & \partial/\partial z \\ u & v & 0 \end{bmatrix}$$

$$= \mathbf{i}(-\partial v/\partial z) - \mathbf{j}(-\partial u/\partial z) + \mathbf{k}(\partial v/\partial x - \partial u/\partial y) = 0$$

The flow is irrotational and satisfies continuity. It is a practical representation of several flow situations. It could represent two jets coming together (or two rivers, or air mass flows). It does well in representing a single jet impinging on a wall, since any streamline can be replaced by a solid surface. Thus, it also represents the flow in a corner. It is used in a typical prefrontal flow situation.

10. Differentiate the potential to get the velocity.

Integrate the velocity to get the streamfunction, hence the streamlines.

Use Bernoulli's relation between velocity and pressure to calculate the pressure distribution and hence the forces.

Chapter 10

1. (a)

$$\overline{(a + a')(b + b') + (c + c')^2}$$

$$= \overline{ab} + \overline{a'b} + \overline{ab'} + \overline{a'b'} + \overline{(c^2 + 2c'c + c'^2)}$$

$$= \overline{ab} + \overline{a'b} + \overline{ab'} + \overline{a'b'} + \overline{(c^2 + 2\overline{c'c} + \overline{c'^2})}$$

$$= \overline{ab} + \overline{a'b'} + \overline{c^2} + \overline{c'^2}$$

(b)

$$\overline{(a + a')^2(b + b')(c + c')}$$

$$= \overline{(a^2 + 2aa' + a'^2)(bc + bc' + b'c + b'c')}$$

$$= \overline{a^2 bc} + \overline{a^2 b'c'} + \overline{2aba'c'} + \overline{2aca'b'} + \overline{2a\,a'b'c'} +$$

$$+ \overline{bca'^2} + \overline{bc'a'^2} + \overline{cb'a'^2} + \overline{a'^2 b'c'}$$

(c)

$$\overline{(ab)^2} = \overline{[(a + a')(b + b')]^2} = \overline{(ab + ab' + a'b + a'b')^2}$$

(difficult expansion)

Or

$$\overline{(a + a')^2(b + b')^2} = \overline{(a^2 + 2aa' + a'^2)(b^2 + 2b'b + b'^2)} =$$
$$\overline{a^2b^2} + \overline{2aa'b'^2} + \overline{2ba'^2b'} + \overline{4aba'b'} + \overline{a^2b'^2} + \overline{b^2a'^2} + \overline{a'^2b'^2}$$

(d)

$$\overline{(ab)^2} = \overline{[(a + a')(b + b')]^2} = \overline{[ab + ab' + a'b + a'b']^2}$$
$$= \overline{a^2b^2} + \overline{2aba'b'} + \overline{a'^2b'^2}$$

2.

$$\overline{(\rho + \rho')(u + u')^2} = \overline{\rho u^2} + \overline{\rho'u^2} + \overline{\rho u'^2} + \overline{\rho'u'^2} + \overline{\rho uu'} + 2\,\overline{\rho'uu'};$$

averaged: $\qquad\qquad \overline{\rho u^2} + \overline{\rho u'^2} + \overline{\rho'u'^2} + 2\overline{u\rho'u'}$

3. Substitute $\Theta = \Theta + \Theta'$, $u = U + u'$ to get

$$\partial\Theta/\partial t + \partial\Theta'/\partial t + (\partial/\partial x_i)\,[(\Theta+\Theta')(U_i + u_i')]$$

$$= K[\partial^2\Theta/\partial x_i + \partial^2\Theta'/\partial x_i]\partial x_i$$

$$\partial\Theta/\partial t + \partial\Theta'/\partial t + (\partial/\partial x_i)\cdot[\Theta U_i + \Theta'U_i + \Theta u_i' + \Theta'u_i']$$

$$= K[\partial^2\Theta/\partial x_i + \partial^2\Theta'/\partial x_i]\partial x_i$$

Steady state and horizontal homogeneity $\qquad (\partial/\partial z \gg \partial/\partial x, \partial/\partial y)$

$$(\partial/\partial z)[\Theta U + \Theta'U + \Theta u' + \Theta'u'] = K\,\partial^2(\Theta + \Theta')/\partial z^2$$

4. A singular perturbation of an equation occurs when the complete governing equation appropriate to the general flow condition is approximated by an equation that does not contain the highest order terms. The solution of such a differential equation requires less integrations. It cannot satisfy as many boundary conditions as a higher order equation. The validity of the approximation will depend on the importance of the boundary conditions that no longer can be satisfied.

10. First substitute into terms, then average: u-momentum equation:

$$\partial u/\partial t + u\,\partial u/\partial x + v\,\partial u/\partial y + w\,\partial u/\partial z - fv + (1/\rho)\,\partial p/\partial x$$

$$- v(\partial^2 u/\partial x^2 + \partial^2 u/\partial y^2 + \partial^2 u/\partial^2 z^2) = 0$$

Substitute $\mathbf{u} = \mathbf{u} + \mathbf{u}' + \mathbf{u}''$ (all three velocity components), ignore turbulent (random) components of ρ and p. In this case $\mathbf{u}$, ρ, and p are the mean values without any marks on them; and primed values are due to organized perterbations.

$\partial(u + u' + u'')/\partial t + (u + u' + u'') \, \partial(u + u' + u'')/\partial x$

$\quad + (v + 'v + v'') \, \partial(u + u' + u'')/\partial y + (w + w' + w'') \, \partial(u + u' + u'')/\partial z$

$-f(v + v' + v'') + [1/(\rho + \rho')] \, \partial(p + p')/\partial x$

$-v[\partial^2(u + u' + u'')/\partial x^2 + \partial^2(u + u' + u'')/\partial y^2 + \partial^2(u + u' + u'')/\partial z^2] = 0$

or

$\partial(u + u' + u'')/\partial t + u \, \partial u/\partial x + u \, \partial u'/\partial x + u \, \partial u''/\partial x + u' \, \partial/\partial x(u + u' + u'')$

$\quad + u''\partial/\partial x(u + u' + u'') + v \, \partial u/\partial y + v \, \partial u'/\partial y + v \, \partial u''/\partial y$

$(v' + v'') \, \partial(u + u' + u'')/\partial y + w \, \partial u/\partial z + (w' + w'') \, \partial/\partial z(u + u' + u'')$

$\quad - fv - fv' - fv'' + 1/(\rho + \rho') \, \partial(p + p')/\partial x - v \, \nabla^2(u + u' + u'') = 0$

average:

$\overline{\partial u/\partial t} + u \, \partial u/\partial x + \overline{u' \, \partial u'/\partial x} + \overline{u' \, \partial u''/\partial x} + \overline{u'' \, \partial u'/\partial x} + \overline{u'' \, \partial u''/\partial x}$

$\quad + v \, \partial u/\partial y + \overline{v' \, \partial u'/\partial y} + \overline{v' \, \partial u''/\partial y} + \overline{v'' \, \partial u'/\partial y} + \overline{v'' \, \partial u''/\partial y}$

$\quad + w \, \partial u/\partial z + \overline{w' \, \partial u'/\partial z} + \overline{w' \, \partial u''/\partial z} + \overline{\omega'' \, \partial u'/\partial z} + \overline{w'' \, \partial u''/\partial z} - fv$

$\quad + 1/\rho \, \partial p/\partial x - \overline{(\rho'/\rho^2)} \, \partial p'/\partial x - v \, (\partial^2 u/\partial x^2 + \partial^2 u/\partial y^2 + \partial^2 u/\partial z^2) = 0$

Chapter 11

1. The flow turns 45° through the Ekman layer.

2. The pressure gradient force, the Coriolis force, and the viscous force form a balance. At the equator, the Coriolis force is zero and the characteristic scale height (and the Ekman solution) do not exist.

3. The self-similarity is with respect to the height when nondimensionalized by the "Ekman depth of frictional resistance," $\sigma \equiv [2K/f]^{1/2}$. The new parameter in the PBL solution is the Coriolis parameter. It provides a characteristic time scale and allows a characteristic length scale to be formed.

6. Consider writing the Ekman equations in complex form, $Q \equiv u - U_G + i(V - V_G)$ so that

$$\frac{d}{dz}\left(K\frac{dQ}{dz}\right) - i f Q = 0$$

with

$$K = K_0 z,$$

$$K_0 z(d^2Q/dz^2) + K_0(dQ/dz) - if Q = 0$$

or

$$z(d^2Q/dz^2) + dQ/dz - i(f/K_0)Q = 0$$

$$\text{Let } \eta \equiv 2(f/K_0)^{1/2}z^{1/2}$$

to get the zero-order Bessel equation

$$\eta^2 d^2Q/d\eta^2 + \eta dQ/d\eta - i\eta^2 Q = 0$$

This provides a good approximation near the surface. K grows too rapidly at higher z, but dQ/dz gets small at these heights.

7. The clouds at the top of the PBL move with the freestream wind, at an angle to the right (in the northern hemisphere). Low pressure would be to the left in a direction perpendicular to the cloud motion.

8. From Eqn (11.2),

$$(u\, \partial u/\partial x + v\, \partial u/\partial y) + \left[\frac{XW}{HU}\right] w\, \partial u/\partial z - \left[\frac{fX}{U}\right] v + \left[\frac{P}{\rho U^2}\right] \partial p/\partial x$$

$$- \left[\frac{K}{XU}\right](\partial^2 u/\partial x^2 + \partial^2 u/\partial y^2) - \left[\frac{K}{UX}\cdot\frac{X}{H}\right] \partial^2 u/\partial z^2 = 0$$

For $H/X \ll 1$ and $W/U \ll 1$,

$$(u\, \partial u/\partial x + v\, \partial u/\partial y) + \left[\frac{XW}{HU}\right] w\, \partial u/\partial z - \frac{v}{\text{Ro}}$$

$$+ D\, \partial p/\partial x - \left[\frac{1}{\text{Re}}\left(\frac{X^2}{H^2}\right)\right] \partial^2 u/\partial z^2 = 0$$

or dimensionally,

$$u\, \partial u/\partial x + v\, \partial u/\partial y + w\, \partial u/\partial z - fv + (1/\rho)\, \partial p/\partial x - K\, \partial^2 u/\partial z^2 = 0$$

16. In a coordinate system aligned with geostrophic flow, $U/G = \cos \alpha - e^\zeta (\cos \zeta - \sin \zeta) \sin \alpha$
where $\zeta \equiv z/\delta$.
Expand u in small ζ,

$$u/G = \cos \alpha - \sin \alpha\, (\cos \zeta - \sin \zeta)(1 - \zeta + \zeta^2/2 + ...)$$

$$= \cos \alpha - \sin \alpha (1 - 2\zeta + \zeta^2 + ...)$$

$$= \cos \alpha - \sin \alpha + 2 \sin \alpha(\zeta - \zeta^2/2 + ...)$$

$$= \cos \alpha - \sin \alpha + 2 \sin \alpha \ln(\zeta + 1) + O[\zeta^3]$$

Similarly, $v/G = 0 + O[\zeta^2]$

Index

International Geophysics Series

EDITED BY

RENATA DMOWSKA
Division of Applied Science
Harvard University

JAMES R. HOLTON
Department of Atmospheric Sciences
University of Washington
Seattle, Washington

*Out of print.